Science: A Second Level Course

The Open University

Biology: Brain & Behaviour

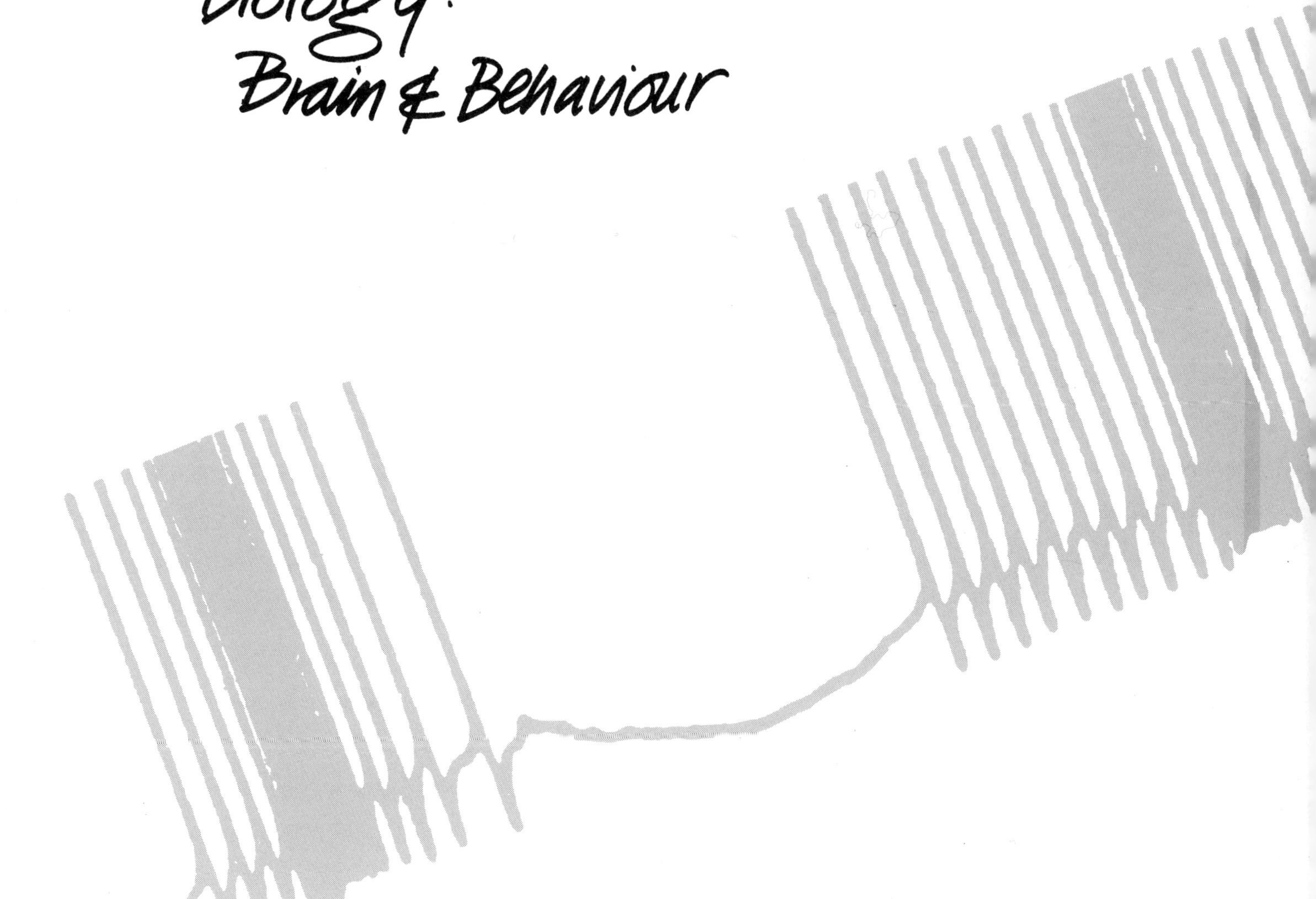

Book 6
Degeneration, Damage and Disorder

Edited by Judith Metcalfe

The SD206 Course Team

Unless otherwise stated, all contributors are members of the Open University

Chair

Tim Halliday

Course Manager

Marion Hall

Course Secretary

Allison Richens

General Editor

Frederick Toates

Academic Editors

Marion Hall (Book 1)
Tim Halliday (Books 1 and 3)
Judith Metcalfe (Book 6)
David Robinson (Book 2)
Vicky Stirling (Book 4)
Frederick Toates (Book 5)
Terry Whatson (Book 4)

Authors

Sarah Bullock (Books 2 and 6, TV)
Gillian Cohen (Book 6)
Basiro Davey (Books 2, 5 and 6)
Geoff Einon (Books 2 and 6)
Marion Hall (Book 1, TV, Video, Audiocassette)
Tim Halliday (Books 1 and 3, TV, Video)
Mike Harris, Psychology Department, University of Birmingham (Book 3)
Heather McLannahan (Books 1 and 3, TV, Audiocassette)
Judith Metcalfe (Book 6)
Robin Orchardson, Institute of Physiology, University of Glasgow (Book 3, TV)
David Robinson (Books 2 and 3, TV, Video, Audiocassette)
Steven Rose (Book 6)
Mike Stewart (Book 4)
Vicky Stirling (Books 2, 3 and 4, TV)
Alison Taylor (Book 2, Brain Video, Audiocassette)
Frederick Toates (Books 1 and 5, TV, Video, Audiocassette)
Terry Whatson (Books 1, 4 and 5, Audiocassette)
Laury Melton (Guide to Information Sources)

Consultants

Stuart Armstrong, Psychology Department, La Trobe University, Australia (Book 5)
Peter Bailey, Psychology Department, University of York (Book 3)
Lynda Birke, Continuing Education Department, University of Warwick (Audiocassette)
Alan Davison, National Hospital for Nervous Diseases, London (Book 6)
Dick Stephen, Physiology Department, University of Leicester (Book 3)
Anna Wirz-Justice, Psychiatrische Universitätsklinik Basel, Switzerland (Book 5, TV)

Editors

Perry Morley
Joanna Munnelly
Ian Nuttall
Sheila Dunleavy

BBC

Nick Watson (Series Producer)
Rissa de la Paz (Producer)
Cameron Balbirnie (Assistant Producer)
Clive Holloway (Producer)
John Stratford (Producer)
Mike Bibby (Graphic Designer)

Summer School Computer Software

Jerry Cooper
Mike Harris, Psychology Department, University of Birmingham
Jon Rosewell

Design Group

Ros Porter (Design Group Coordinator)
Sian Lewis (Designer)
Pam Owen (Graphic Artist)
John Taylor (Graphic Artist)

External Assessors

Richard Andrew, School of Biological Sciences, University of Sussex (Course Assessor)
Peter Slater, Zoology Department, University of St. Andrews (Book 1)
John O'Keefe, Anatomy Department, University College, London (Book 2)
Clive Catchpole, Zoology Department, Royal Holloway and Bedford New College, University of London (Book 3)
Lynn Bindman, Department of Physiology, University College and Middlesex School of Medicine, University College London (Book 4)
Trevor Robbins, Department of Experimental Psychology, University of Cambridge (Book 5)
Alec Jenner, Department of Psychiatry, University of Sheffield Medical School (Book 6)

The Open University
Walton Hall, Milton Keynes
MK7 6AA

First published 1992

Copyright © 1992, 1995 The Open University

All rights reserved. No part of this publication may be reproduced, stored in a retrieval system or transmitted, in any form or by any means, without written permission from the publisher or a licence from the Copyright Licensing Agency Limited. Details of such licences (for reprographic reproduction) may be obtained from the Copyright Licensing Agency Limited of 90 Tottenham Court Road, London, WC1P 9HE.

Edited, designed and typeset by the Open University

Printed in the United Kingdom by Alden, Oxford, Didcot and Northampton

ISBN 0 7492 8150 2

This text forms part of an Open University Second Level Course. If you would like a copy of *Studying with the Open University*, please write to the Central Enquiry Service, PO Box 200, The Open University, Walton Hall, Milton Keynes, MK7 6YZ. If you have not already enrolled on the Course and would like to buy this or other Open University material, please write to Open University Educational Enterprises Ltd, 12 Cofferidge Close, Stony Stratford, Milton Keynes, MK11 1BY, United Kingdom.

2.1

CONTENTS

PREFACE

Degeneration, Damage and Disorder, like any other textbook, is designed to be read on its own, but it is also the last in a series of six books that form part of *SD206 Biology: Brain and Behaviour*, a second-level course for Open University students. The first two books in the course, *Behaviour and Evolution* and *Neurobiology*, provide core material which allows the student to go on to study Books 3–6. These four books take an interdisciplinary approach to a series of subjects in the brain and behavioural sciences. *Book 3: The Senses and Communication* looks at vision, hearing, touch and pain, and the role of the senses in communication. *Book 4: Development and Flexibility* considers the development of the nervous system in the growing embryo, the development of behaviour as the individual grows and matures and the flexibility of the nervous system seen in the ability to learn. *Book 5: Control of Behaviour* looks at the role of motivation, biological clocks, aggression and stress in controlling behaviour. Finally, *Book 6: Degeneration, Damage and Disorder* considers the effects of brain disease and disorder, brain damage, and ageing on the nervous system and on behaviour.

As well as the printed text, the course involves television broadcasts, videocassettes, radio programmes and audiocassettes, plus experiments carried out at home and attendance at a one-week residential summer school. As is the case with most other Open University courses, students of SD206 are required to complete written assignments during the year and to sit an examination at the end of the course.

In this, and all other books in the course, each subject is introduced in a way that makes it readily accessible to readers without any previous knowledge of that area, other than that already presented in the previous books. Questions within the text, marked with a □, are designed to help readers understand and remember the topic under discussion. (Answers to in-text questions are marked with a ■.) The major learning objectives are listed at the end of each chapter, followed by questions (with answers given at the end of the book) which allow readers to assess how well they have achieved these objectives. Key terms are identified in bold type in the text; these are listed, with their definitions, in a glossary at the end of the book. Key references are given at the end of each chapter, where appropriate, plus, for some chapters, a 'further reading' list for those who wish to find out more about the subject of the chapter in question. A 'general further reading' list, of textbooks relevant to the whole book, is also included at the end.

The study of the brain and behaviour, as presented in this course, is an experimental science. This means that it involves the collection of observations, the formulation of specific hypotheses to explain those observations and the carrying out of experiments to test (confirm or falsify) those hypotheses, as explained in the *Introduction and Guide* to this course. Throughout the course, these different aspects of the investigative process are emphasized, often through the use of in-text questions in which students are invited to engage in the process of deductive reasoning themselves. An understanding of the scientific method, as it applies to the behavioural and brain sciences, is an important aim of this course, which is pursued not only in the text, but also in the Experiments and their accompanying Notes and in the work students do at Summer School.

This last book of the course differs from earlier books in that it is entirely concerned with *human* brain and behaviour. It discusses some of the ways in which nervous system functioning and patterns of behaviour can change or go wrong. These may be the consequence of the process of ageing, or the result of damage or disease, or precipitated by social factors or events in a person's life history, or a combination of any of these. The book focuses on explanations of such conditions and on what, in turn, can be learnt about normal nervous system functioning and behaviour patterns from studying these conditions. Some examples of disease or disorder have been selected for study in depth. The topics have been chosen to illustrate the range of degeneration, damage or disorder, and the breadth of explanations known, as well as to revise earlier material of the course.

The first chapter asks what is normal and what is abnormal in terms of nervous system functioning and behaviour and outlines the methods used to study them. Chapters 2 and 3 are concerned with the psychological and biological aspects of the process of ageing. Chapter 4 explores the genetic and environmental factors that result in disorder and disease. Chapter 5 discusses the range of damage to which the brain is vulnerable and the mechanisms of recovery from such damage. Chapter 6 looks at brain-damaged individuals and considers what such studies reveal about the phenomenon of consciousness. Chapter 7 focuses on the vulnerability of the mind. Finally, Chapter 8 departs from the rest of the book, which essentially discusses the types of damage and disease of the nervous system, and reviews the implications for treatment of the new knowledge about the brain and discusses therapy and control and the borderline between them.

The approach of Book 6 is to give general explanations that emerge from a detailed study of a great many individual cases of a particular disease or disorder. While studying this book, it is important to realize that some of the issues discussed are of a sensitive nature in terms of people's own experience of a particular disease or disorder. You may have first-hand experience of one specific case by knowing, or living with, a person who suffers from a particular condition. Clinicians, of course, can draw on their knowledge of general patterns and research findings and bring them to bear on the individual sufferer in the light of that person's unique circumstances and conditions.

Before you begin to read this book, there are some further important points that you should bear in mind.

1 Terminology and important terms

In discussions of behaviour, it is commonplace for authors to use a number of words that are used in everyday language, such as 'affect' and 'labelling'. This creates potential misunderstanding because, as used in everyday language, these words have certain meanings that may not be wholly appropriate in a scientific context. Where such words are first used in this book they are printed in bold type and are defined in terms that are specific to the scientific study of behaviour. These definitions, together with those of other, more technical, terms printed in bold type, are also given in the Glossary at the end of the book.

2 Experiments on animals

The use of living animals in research is a highly emotive, contentious and political issue. You are no doubt aware of the strong views held by animal liberationists. There is also considerable debate among scientists concerning what kinds of experiments and procedures are acceptable and what are not. Most scientists working with animals seek to minimize any suffering that animals may experience during experiments and each researcher makes his or her own judgement as to whether the suffering caused by an experiment is justified by the scientific value of the results that the experiment yields. The ethics of animal experimentation is not simply a matter of individual judgement, however, but is a matter of concern for society as a whole. In Britain and many other countries, all researchers work within strict guidelines enforced by government; for example, the Home Office licences all animal experimentation in the UK. Some academic societies, such as the Association for the Study of Animal Behaviour, and many institutions, such as medical schools, have Ethical Committees that oversee animal-based research. The OU is the first University in the UK to have its own Animal Ethical Committee, in which the experiments on animals that we conduct in our own research and which we devise for you to do at Summer School are scrutinized. In this book, a number of experiments are described; this in itself raises ethical issues because, in reporting the results of an experiment, the Course Team may be thought to be giving tacit approval to that experiment. This is not necessarily true and it should be pointed out that some of the experiments described were carried out several years ago and a number of them would not be carried out today, such has been the shift in opinion on these issues within the biological community. Paradoxically, certain experiments carried out many years ago, such as those on the effects of maternal deprivation on young monkeys, produced such strong and distressing effects on their subjects—results that were not generally anticipated—that they have had a substantial impact on the kind of experiments that are permitted today. The issue of animal experiments is a complex one and is discussed at length in the audiocassette *The Use of Animals in Research*.

3 Latin names for species

A particular individual animal belongs to various categories. If you own a pet, it may, for example, be categorized as a bitch, a spaniel, a dog, a mammal, or an animal. Each category is defined by particular features that differentiate it from other, comparable categories. The most important level of categorization in biology is at the level of the species. When a particular species of animal is referred to in this course, its Latin name is also given, e.g. robin (*Erithacus rubecula*). It is not necessary for you to learn any of the Latin names given in the text, though some, such as that of the sea slug *Aplysia*, will become familiar to you through frequent use. This particular animal is in fact rarely referred to by its English name. The use of Latin names and the categorization of animals is discussed more fully in the *Introduction and Guide* to the course.

CHAPTER 1
BRAIN DISORDER AND MIND DISORDER

1.1 Introduction

So far this course has been primarily concerned with the workings of the normal brain and its behavioural output. Book 3 has shown that pain and stress play a functional part in continued effective existence. But not all of human life experiences can be so easily dealt with. It has been said that the only certain thing about life is death. For that small proportion of the world's population fortunate enough to live to old age before dying, death itself is usually preceded by a longer or shorter period of decline, during which brain and behaviour may show characteristic forms of degeneration. Disease or accidental damage can strike long before old age, and many diseases profoundly affect both brain and behaviour. As a result of such disease or damage, or in response to what seem to be the overwhelming pressures and complexities of social living, very large numbers of people, at one time or another in their lives, find themselves overcome by mental distress or confusion, unable to function 'normally' in work, personal relations or day-to-day life. In addition, a significant number of people are born with seemingly irreversible deficits or disorders of brain and behaviour as a result of an adverse combination of genes, or of problems during pregnancy.

It is with the understanding of these conditions, of distress, disease, damage, disorder and degeneration, that this book is concerned. Although many or all of them may be as inevitable a part of the life cycle of large-brained, non-human animals as they are of humans, it is inevitably with the human experience of these conditions that this book is primarily concerned. As a result, the book differs from those earlier in the course in the extent to which it concentrates on human biology, brain and behaviour. Although there are, as you will see, some 'animal models' which can be studied to cast light on certain aspects of degeneration, disorder and disease, in most cases the understanding of these conditions begins with clinical observations among humans and the success or failure of attempts to alleviate them. Explanation of such conditions tends to follow rather than precede treatment.

The plan of this book is as follows: the remainder of this chapter sketches out the nature of the problem. Just what characterizes the borderline between 'normality' and 'abnormality'? When does a person's apparently erratic behaviour indicate a 'brain' problem? What is the relationship between a 'brain' problem and a 'mind' problem? How should such behaviour patterns be explained? What are the methods available to study them? Chapters 2 and 3 are concerned with the phenomena of ageing. What does ageing mean, socially, behaviourally, and biologically? Is growing old a period of inevitable decline and loss of powers, or is it accompanied by an increase in wisdom? Is there a relationship between the cellular and biochemical changes that occur in the brain with ageing and the changes in behaviour? How do the immune and nervous systems interact during degenerative diseases? Chapter 4 explores the variety of biological events that can result in disorders of brain and behaviour: from genetic and congenital conditions to the effects of environmental factors, such as the poison lead. Chapter 5 considers the vulnerability of the brain, the effects of conditions such as those resulting from a stroke on perception and cognition, and what these effects can tell us about the brain processes associated with such aspects of mental performance. It asks whether rehabilitation after such brain disasters is possible and discusses the significance of recent controversial studies on the effects of transplanting fetal brain tissue in the treatment of such conditions as Parkinson's disease. Chapter 6 reflects on the implications of these findings for understanding the elusive phenomenon of consciousness. Chapter 7 turns from the vulnerability of the brain to the vulnerability of the mind. What distinguishes a neurological from a psychiatric disorder? What are schizophrenia and depression, and how are they best explained? Finally, Chapter 8 reconsiders the new and increasing knowledge of the brain and behavioural sciences in terms of its implications for the treatment of human distress, the borderlines between therapy and control, and the changing understanding of people as human animals.

1.2 What are brain and mind disorders?

1.2.1 Experiencing brain and mind disorder

An elderly woman shuffles slowly and painfully down the street. Her left leg is clearly in trouble, for she drags it as she walks. Has she sprained her ankle, or pulled a muscle? A middle-aged man walks rapidly past, gesticulating wildly and talking, incomprehensibly, to no one in particular. Is he drunk? an alcoholic? disturbed? 'mad'? A growing child fails to learn to speak properly and does not seem capable of relating to or playing with other children. Does he have problems with hearing or seeing, or is he just wilful? A teenager starts to behave oddly, mooning about, lying on the bed all day, bad-tempered, slovenly and difficult, sometimes wildly aggressive. Have her parents failed her? Is it just an adolescent phase which will soon pass? A young woman sits in her chair all day, rocking quietly, sometimes crying, refusing all efforts to cheer up or to get on with her life. Has she been sacked from her job, or had an unhappy love affair? A black man with dreadlocks and a straggling beard stands in the middle of the road singing hymns, until the police try to move him on, when he responds incoherently and refuses to move. The first response of the police is to arrest him on suspicion of dealing in drugs. Their subsequent response, after apparently failing to communicate with him, is to refer him to a psychiatrist. An old man at home tries to pour a cup of tea and finds his hand shaking uncontrollably so the tea spills down the side of the cup. A pensioner in a supermarket stacks her trolley with dozens of cans of cat food and then walks out without paying; when stopped by a security guard, she claims not to remember her name or where she lives. A 40-year-old shopworker, living alone, complains to the police that his neighbours are harassing him, bugging his telephone calls and trying to break into his room at night. Is he imagining it, and if so, why?

All these are common enough events. Most of us will have directly experienced one or more of them, as participant or observer. The individual at the centre of each is clearly suffering. His or her behaviour or emotions are incompatible with effective day-to-day adult life in the midst of complex industrial societies. But what relevance do they have for a course on brain and behaviour? That question lies at the core of the concerns of neuroscientists, psychiatrists, lawyers and philosophers alike. The behaviour patterns described are in some way abnormal, in the sense that they are unusual; most people, most of the time, do not behave in these ways. For some of them there are common labels—'confused', 'autistic', 'senile', 'disturbed', 'criminal', 'paranoid', 'mad'. But what does it mean to describe behaviour patterns in this way as abnormal, and how should one respond to such abnormality? Is one dealing with inevitable aspects of the human condition, part of that sequence from birth to death in which the debilities of age eventually catch up with even the most agile, or with diseases that can be treated medically? Alternatively, are these examples of 'bad behaviour' to be dealt with by the legal system; or spiritual distress, to be offered comfort; or the inevitable consequences of the social injustices of an unfair society? These questions form the subject matter of the present book. In order to avoid prejudging them, they are described here very generally as 'brain disorder and mind disorder'. The rest of this chapter will be concerned with exploring the meaning of these terms, whether they are identical, and the several different approaches to the study of brain and mind disorders. First, however, it is necessary to explore a little further the concept of *normality* and the ways in which *abnormality* is defined.

In past times in the UK, and indeed even today in less industrialized societies than Britain, many of the behaviour patterns described in the first paragraph of this section would have passed relatively unnoticed, as part of life's rich and infinite variety and complexity. Talking to yourself in the street, forgetting your name and address, communing with your god or desperately grieving were things that old or unhappy people (or prophets) often did; they were a natural part of the human condition (in the same way that women often died young in childbirth). Modern industrial societies are less accepting of such behaviour, of deviations from the norm, and seek to find acceptable biological, psychological or social explanations for them. 'Normal' behaviour means walking steadily in the street, not talking to oneself, practising religion quietly, paying for goods at the supermarket checkout and showing decent but not excessive sadness at life's adversities. To fail to do these things is deviant or abnormal. As well as often being profoundly distressing to the individual concerned and to his or her relatives and friends, it disturbs the orderliness that modern industrialized societies demand.

1.2.2 On normality

'*Normality*' is a portmanteau term into which are packed several types of meaning. On the one hand, normal means what most people do; it can therefore be regarded as a statement about average behaviour. However, what most people do is itself not fixed; it is shaped by culture and society. Thus in pre-industrial, peasant societies it was relatively common for babies to die before they were one year old and relatively rare for people to survive beyond their forties. The death of children was thus normal and to be in one's sixties was abnormal; in the UK today, these examples of normality and abnormality are reversed. So if one is to use the term normal, one must begin by asking 'normal for whom?' This is important for example in considering the processes of ageing in Chapters 2 and 3. As ageing is a 'normal' process, perhaps the conditions that frequently go with ageing (such as memory loss) are themselves normal, the inevitable accompanying features (concomitants) of degenerative processes in a body and brain as they wear out. Or is memory loss a form of disease, requiring, or at least inviting, medical intervention? Normality is not merely a socially fluctuating concept, but it is also age-dependent. It is normal for a baby to be incontinent but a sign of disease in an adult.

However, normality also has other meanings. *Norms* are standards, for instance standards of behaviour. They describe not merely how most people *are*, but how they *should* be. Thus a person who transgresses these norms is *abnormal*, a deviant, and it is then expected that society will try to adjust the behaviour of the deviant back towards normality once more. The abnormal behaviour may first be criminal, in which case it is the concern of the courts. Second, it may be immoral, in which case it is the concern of the priesthood or its modern equivalents. Third, if it is sick behaviour it is the concern of doctors and, finally, it may simply be a nuisance to others. Which of these ways in which society might become involved in any particular case varies depending on the circumstances and current social theories. A decade ago, people defined as mentally ill were mainly confined in psychiatric hospitals, today they are more likely to be released into 'the community' and many end up wandering the streets and sleeping in cardboard boxes, while others substitute prison for the asylum. In Samuel Butler's famous book *Erewhon*, a Utopian novel written a century ago, physical illness was a criminal offence punished by imprisonment, while stealing from the bank was a sickness to be treated by a visit from the 'straightener' who was a kind of psychiatrist.

Nowhere do these problems of what constitutes normality become so acute as in the context of brain and mind disorders. Of course, some conditions are relatively clear-cut. Take, for instance, the first example described in the first paragraph of Section 1.2.1, the woman who cannot move her left leg properly. A medical examination might show that there was nothing wrong with the muscles, ligaments or bones of her leg, but that there had been damage to the motor regions of the right hemisphere of her brain, perhaps as a result of a stroke or an accident (discussed in Chapter 5). The tremor of the pensioner's hand as he pours a cup of tea could be the result of a condition called Parkinson's disease, in which there is degeneration of the neural pathways of the substantia nigra within the basal ganglia (see Book 2, Section 9.4.3 and Chapters 4 and 5 of this book). During the second half of the 19th century, a number of such relatively localized brain areas were identified, each apparently associated with specific functions. You have already met some of these, for instance Broca's and Wernicke's areas which are the left hemisphere regions associated with speech (Book 2, Sections 11.2.1 and 11.2.2, and Chapters 5 and 6 of this book). Damage to these areas results in characteristic deficits in a person's ability to speak coherently and intelligibly. No one would doubt that these are debilitating deviations from normality which require medical treatment.

1.2.3 Neurological and psychiatric disorders

However, although they were medically identified during the late 19th and early 20th century as disorders of the mind, with reasonably distinct patterns of symptoms, several of the other conditions described in the first paragraph of Section 1.2.1 did not seem to be associated with any clear-cut brain damage. Clinicians were therefore led to draw a distinction between what they saw as two distinct types of disorder, one **neurological** (sometimes called *organic*), in which a person's behaviour was affected because of damage to the brain, and the other **psychiatric** (sometimes called *psychological* or *functional*), in which a person's mental distress or disturbed behaviour was not associated with any obvious brain damage. Note that the sense in which the term functional is used here is different from that used by ethologists and discussed in Book 1. Here it implies a deficit in normal behavioural functioning without any detectable underlying physiological damage. In clinical terms the first type of disorder became the province of neurology and the second of psychiatry. Within medical science, neurology was seen as depending on neuroanatomy and neurophysiology, whereas psychology underlay psychiatry. According to this categorization, stroke, Parkinson's or Alzheimer's disease and mental handicap are neurological disorders, whereas schizophrenia, depression, 'personality disorders' and autism are psychiatric. The boundary between the two types of condition has from the start been rather flexible; thus, what was originally categorized simply as a matter for the psychiatrist and labelled 'general paralysis of the insane' was later recognized as an aspect of the brain degeneration resulting from untreated syphilis.

The distinction draws strongly on the dualist philosophical tradition which since the days of Descartes in the 17th century (see Book 2, Section 11.6.3 and Sections 6.2.1 and 8.2.1 of this book), had driven a firm line between the material and the mental: on the one side body and brain and on the other soul and mind. Today the same point is sometimes made using the language of computer technology: in neurological disorders, it is argued, the hardware of the brain is damaged; in psychiatric ones the hardware may be unaffected but the software, the program, malfunctions. Here, and in later chapters however, you will see that the neurological/psychiatric distinction cannot really withstand the integrative advances of modern neuroscience. There is ample reason to believe that 'psychiatric' disorders are associated with changes in brain chemistry, physiology and even anatomy, just as neurological disorders are associated with changes in a person's state of psychiatric health, and are not just a problem with, for example, moving a leg or enunciating words (see also the discussion of pain in Book 3, Chapter 5). The real problem becomes one of understanding just what it is that causes any specific condition: is it primarily derived from a biological disorder, or from the social and emotional experiences of a person's life?

1.2.4 The extent of psychopathology

The scale and scope of diagnosed mind and brain disorder in the UK today is sufficient to suggest that it is the normal experience of a very large proportion of the population. Fifteen per cent of all people over 85 years old eventually come to be diagnosed as suffering from Alzheimer's disease. Sixteen thousand children a year are admitted to institutions for the mentally handicapped. One in twelve men and one in eight women in the the UK will enter hospital at least once in their lives to be treated for some type of psychiatric illness. These figures are not stable, though. In the 1950s, when it was customary to hospitalize people diagnosed as suffering from mental disorders, there were some quarter of a million psychiatric patients. Over the past decades the policy has changed to one of care in the community and the psychiatric hospitals have been dramatically emptied or closed. A better measure may therefore be given by GP consultations. It is estimated that some 23% of all such visits are for advice on emotional and social problems. Many other consultations ostensibly for physical illnesses may of course be directly related to such emotional and social problems. However, it is also quite possible that doctors (still predominantly male in the the UK) tend to regard people (especially women) who consult them about what those people regard as physical ailments as being really concerned with emotional problems.

There have been many attempts over the years to estimate the total extent of 'psychopathology' in the population. In 1980 Bruce Dohrenwend and his collaborators in the USA (Dohrenwend *et al.*, 1980) collated all the estimates for Europe and North America that had been published since 1950 (a total of 27). They concluded that the median (one way of calculating the average) estimate for all psychopathology in the adult population under 65 years old was 21%. However, the range of the different estimates here was vast, from a miniscule 0.6% to a staggering 69% of the population. For the European estimates, the range was between 4% and 56%. The variations depend, among other factors, on the nature of the samples used and the diagnostic criteria employed and so should be treated with caution. Nonetheless, they give an indication both of the extent of mind and brain disorder as a social, medical and human problem, and of what the sciences of brain and behaviour are called upon to explain.

□ It has been claimed that mind disorders are much more common today than they were a century ago. Suggest three reasons why this might either be, or *appear* to be, the case.

■ 1 People are living longer and not dying of infectious diseases, and therefore have more opportunity to suffer from brain/mind disorder.

2 There is something about contemporary society which is more stressful than earlier societies and therefore there is an increased likelihood of succumbing to brain/mind disorder.

3 There is no more such disorder than there was a century ago, but many conditions which went unrecognized then are now regarded as symptoms of brain/mind disorder.

Summary of Section 1.2

This section began with some examples of the range of conditions that are embraced within the terms brain and mind disorder. It then considered the fluctuating definitions of normality against which conditions of brain and mind disorder must be viewed as abnormal. It was pointed out that normality is a socially- and culturally-bound concept. The two terms psychiatric and neurological derive from the history of the study of mind and brain disorders. However, this distinction is vanishing in the light of modern neuroscience despite its historical justification. Finally, some estimates were made of the extent of 'psychopathology' in modern Europe and the USA.

1.3 Explaining brain and mind disorder

Understanding the causes of brain and mind disorder is of course not a matter of pure scientific curiosity; the hunt for explanation is also part of a search for treatment. Knowing what causes something puts one in a better position either to prevent it happening or to alleviate its consequences when it has occurred. However, the search for such explanations has not been easy. Neurology and psychiatry have developed in rather autonomous and even mutually suspicious ways, and the problem of understanding causation and treatment for the types of 'psychopathology' exemplified at the start of Section 1.2.4 is genuinely very difficult (think back to the discussions of causation and explanation in Book 1, Section 1.1.1). Consequently, it is small wonder that numbers of competing explanations have been proposed, and that there is very little consensus among them.

In principle, however, explanations of brain and mind distress tend to involve one of three types of approach, which can be described as social, psychological or biological. Each of these can be seen as different 'levels', even though within each such level there is more than one form of explanation. (The term 'level' has appeared several times in this course so far, and some of its several meanings are discussed in the *Introduction and Guide* to the course and also in Book 2, Chapter 1.) It should also be remembered that although some conditions seem to be obviously biological today, this has not always been so in the past, nor is it always clear in all circumstances even now. The aberrant behaviour which may result from a developing brain tumour can be very similar to that seen in the psychiatric condition usually diagnosed as manic depression, and the speech defects following a stroke can resemble the consequences of acute alcohol intoxication. So each class of explanation must be considered separately.

1.3.1 Social explanations—labelling

The most uncompromising form of social explanation of some forms of mental disorder (especially though not exclusively 'madness') is that which became associated, from the 1960s on, with the name of the French social philosopher and medical historian Michel Foucault. Foucault argued, in essence, that disease categories such as 'madness' have no objective existence; one should not consider an individual to be ill, because describing someone as ill becomes a way of **labelling** them. Thus the label of madness describes not the condition of an individual but a *relationship* between that individual and the rest of the world. Consider the word *invalid*. A person who is an *invalid* has, using Foucault's interpretation, become *invalidated*—their experience and understanding of the world no longer need be taken seriously. Thus, Foucault argued, the powerful (in Western society; rich, white and male) label the powerless (poor, non-white and female) as mad, so providing an excuse to ignore them, or to incarcerate them in prison or in an asylum. Asylums and prisons serve similar purposes, because society needs to have a category of people to treat as scapegoats and to lock up. It comes as no surprise within this framework to note that as the psychiatric hospitals have emptied over the past decades, the prison population in the UK has steadily risen, standing now at around 50 000. Those labelled a few years ago as 'mad' are now labelled as 'bad' instead. This argument, backed with historical examples by

Foucault and his followers in France and later in the UK, became intertwined with certain more psychological theories favoured by a school, discussed in Section 1.3.3, which described itself as 'antipsychiatry'. Together they achieved considerable popularity in many circles during the following decades and still attract a number of followers, even though they are now well past their peak.

1.3.2 Social explanations—epidemiology

More empirically grounded forms of social explanation do not dispute the 'reality' of conditions like depression, schizophrenia or alcoholism, but seek to understand the factors that contribute to precipitating them by studying their **epidemiology**—that is, how the condition is distributed within the population and how it changes over time. Are working-class people more likely to be affected than middle-class people, women more than men, those living in Scotland more than those living in London? What is the likely age of onset of the condition and does its typical course vary from one category of individual to another? In England, women are twice as likely as men to be diagnosed as suffering from depression; working-class people are more likely than middle-class people, and black people are more likely than white people, to be diagnosed as schizophrenic; and people of Irish origin are more likely to be diagnosed as alcoholic. Do such discrepancies point to potential causes of the conditions?

1.3.3 Psychological life-history explanations

Whereas social explanations seek the origins of mental disorder within the structure of society, psychological ones search for it within an individual's own life history. There are many versions of such psychological theories and there is not enough space to consider any of them in detail here. However, the best known types of **psychological life-history explanation** tend to be variants of the psychoanalytic ones first advanced by Sigmund Freud and his successors. Although, as a young clinician at the end of the 19th century, Freud started out with a commitment to an 'organic' view of mental distress, he soon abandoned that project as too difficult, and concentrated instead on understanding a person's present in terms of their past. An adult's way of thinking, behaving and relating to others, he argued, is shaped by formative experiences during childhood. For Freud, these experiences revolved centrally around sex. During a 'normal' childhood there is, he argued, a steady developmental progression, during which children recognize their existence as a separate being from their parents, discover their own sexuality and pass through a series of stages in which sexual desires shift from self-gratification, through attempts to substitute themselves for one of their parents in the affection of the other, and finally to a 'normal' heterosexual orientation. This progression is, in Freudian theory, different for boys (who need to differentiate themselves from their mothers) and girls (who suffer from 'penis envy'). Disturbances in this pattern of progression are assumed, within Freudian theory, to lead to abnormal behaviour in adults.

Later schools of analysis differentiated themselves from Freud by isolating other developmental features (from some hypothetical birth trauma onwards) as of central importance. For instance the 'object-relations' school of psychoanalysis argues that a fundamental distinction between the psychological development of boys and girls occurs because girls can continue to identify themselves as fundamentally 'like' their mothers but boys are different and therefore need to differentiate themselves, for instance by, they claim, becoming more cognitive and less emotional. All such schools of psychoanalysis share a basic indifference either to the type of biological explanation discussed in the next section or to the epidemiological data described in the previous section.

For Freudian psychoanalysts, there is no doubt that the condition they wish to explain and treat is an abnormality to be corrected. By contrast the school of antipsychiatry that grew up in the 1960s and 1970s, drawing upon the writings of the social anthropologist Gregory Bateson in the USA and the psychiatric practice of Ronald Laing in the UK, rejected the notion that conditions like schizophrenia were disease states. Rather, they claimed, when understood correctly 'schizophrenic behaviour' could be seen as the rational response of an individual placed in irrational conditions. Laing's classic case is what he called the *double bind*, of the child who, whatever she does, is criticized by her parents, for whom she can do nothing right. If she dresses casually, she is being slovenly, if carefully, she is being sexually provocative. Schizophrenia is then seen as a rational response to the double bind, and the job of the analyst is not to 'cure' a person of their delusions, but to help them to understand and live comfortably with them. The links between this view and the Foucaultian position are clear. From the 1980s, antipsychiatry has suffered a decline in credibility and popularity, especially since Laing himself was denounced by some of his former patients and publicly renounced his earlier claims for the efficacy of his approach to the understanding and treatment of schizophrenia. However, the radical challenge that it has mounted to all forms of conventional psychiatric and biological explanations of psychic disorder retains its significance and relevance.

Not all life-history explanations, however, are so locked into the theoretical perspectives of psychoanalysis. Behaviour and cognitive therapists, for instance, regard the manifestations of mental disorder or distress as forms of learned habits which may be unlearned by appropriate patterns of treatment irrespective of whether or not they are symptomatic of more deep-seated underlying problems. Epidemiologists have endeavoured to discover what they call 'vulnerability factors' which, occurring in an individual's early life, may make them more likely to suffer psychiatric disorder in later life. Such factors may include loss of a parent through death or divorce. Events in later life ranging from social upheavals like war or disaster (fire, earthquake, surviving a crash or the sinking of a ship) to strictly personal ones like bereavement, may all be potential precipitating factors in occasioning longer or shorter periods of mental disorder.

1.3.4 Biological explanations

By contrast with the social and psychological levels, biological explanations tend to be thoroughly reductionist. The reductionist approach argues that the ultimate causes of complex phenomena such as social and individual behaviour are to be found in chemistry and physics. Thus, rejecting earlier distinctions between organic and functional disorders, biological explanations assume that if an individual shows disturbed thoughts or behaviour patterns, then this disorder must be a reflection of some underlying biological malfunction. If there is nothing 'obviously' wrong with the person's brain, in terms of gross and anatomically visible damage due to stroke or disease, then the damage must lie at some more subtle level, for instance in a biochemical disturbance or imbalance, either general or localized in particular regions of the brain.

Of course, not all behaviour originates in the brain. As you are aware from earlier books, other important bodily regulatory systems play a major part in the control and shaping of behaviour.

□ Try to name two such general mechanisms.

■ 1 *The endocrine system* Hormones can profoundly affect behaviour. A woman's behaviour is often 'explained' in terms of menstrual cycle hormones: 'it's that time of the month'; 'pre-menstrual tension'; 'post-natal depression'. Masculine aggressive behaviour is often similarly reduced to a statement about 'male hormones' like testosterone. (You will remember that the concept of 'male hormones' is erroneous because testosterone is present in both sexes.)

2 *The immune system* As you have learned (in Book 2, Chapter 5 and Book 5, Section 5.5.5) this is increasingly coming to be recognized as interacting in important ways with both neural and endocrine systems and as having potentially profound interactions with behaviour.

Biologically based views received a boost from the 1950s with the development of powerful drugs which alleviated many of the symptoms of mental distress. The first of this new generation of drugs was the phenothiazine compound *chlorpromazine*, introduced in 1952. Within a decade of its introduction chlorpromazine had been prescribed to over 50 million patients world-wide. Later it began to be realized that chlorpromazine was not quite such a wonder drug, because its prolonged use resulted in serious and seemingly irreversible brain damage, and so other drugs were substituted for it. One commonly used modern alternative is the trifluoperazine drug *stelazine*. Nonetheless, its effects were dramatic, and at the peak of enthusiasm for its use it was claimed to allay everything from depression to

> the restlessness of senile dementia, the agitation of involutional melancholia, the excitement of hypomania or mania...the impulsiveness and destructive behaviour of schizophrenics. The patients develop a kind of detachment from their delusions and hallucinations...(Curran and Partridge (1969) *Psychological Medicine*, Livingstone.)

If people suffering from mental distress do so because of a biochemical abnormality in their brains, perhaps the reason that drugs like chlorpromazine (generically known as psychotropics) work is because they interact with that same biochemical process, rectifying or neutralizing the abnormal step. Therefore, if one could find out how the drug works one would also discover the biochemical lesion or malfunction that causes the disease.

□ Outline a common mode of action of psychotropic (or mind-altering) drugs, using information you gained from Book 2, Section 4.5.8.

■ Many psychotropic drugs act as transmitter agonists or antagonists, or they interact with receptor binding or re-uptake sites (Book 2, Sections 4.2, 4.4, 4.5 and 5.4.1), especially those for dopamine, serotonin, GABA and the opioids.

Could it be therefore, that diseases such as schizophrenia or depression are, like Parkinson's disease, disorders of neurotransmission? If this is a general model of biological explanation for mental disorder, it leaves open the question of how such neurotransmission disorders are themselves caused.

□ Suggest two general types of such explanation.

■ The reasons could be either environmental or genetic.

Potential environmental factors include accident, or the consequences of high blood pressure, which can lead to stroke, or a viral infection, or poisoning by environmental pollutants, or lack of some dietary factor. Included here could even be the presence in the diet of some food giving rise to allergy, there being much concern at present about the effects of 'junk food' and dietary additives on behaviour. One well-known dietary deficiency condition with marked behavioural effects is that caused by vitamin B deficiency. This gives rise to the disease called pellagra, one of whose symptoms is apparent mental retardation, which can be reversed by adding Vitamin B to the diet.

However, there is a strong tendency within biological explanations to move further down the reductionist path and to seek to explain biochemical abnormality by genetic factors. Some disorders, such as Huntington's disease, clearly run in families; others, such as schizophrenia and manic depression, may do also. Where a condition does run in families in this way, there are good grounds for suspecting that there may be a genetic explanation for it, and many biologically oriented psychiatrists are convinced that the 'ultimate' cause for much mental disorder, from depression to alcoholism, is some type of genetic malfunction.

1.3.5 Explanations in action

To see how these seemingly incompatible types of explanation may work in action, consider some of the examples given in Section 1.2.1.

□ Describe possible social, psychological and biological explanations for the behaviour of (a) the crying young woman, and (b) the black hymn singer.

■ (a) *The crying young woman* Social explanations would include the Foucaultian interpretation that her behaviour was not abnormal at all, but that a male-dominated society could not come to terms with her needs and wishes, and therefore labelled her abnormal. What has been described as the more empirically grounded approach would point to the disproportionate number of women who become depressed and would seek to know her social situation: was she for instance afraid of unemployment, economically insecure, pregnant and alone? Psychological life-history explanations would explore her relations with her parents: were they over-domineering? had her father abused her in childhood? were there particular aspects of her life history which had predisposed her to responding excessively to problems which others would see as minor? Biological explanations might suggest she was suffering from a hormonal or neurotransmitter imbalance; was her condition transient, for instance was she just about to start menstruating? or was it more lasting as a result of a genetic predisposition?

(b) *The black hymn singer* For Foucault and his followers, there would be nothing 'wrong' with the hymn singer's behaviour. A racist society, they would argue, cannot come to terms with such forms of black religious expression (for example Rastafarianism) and therefore attempts to suppress it by legal (the police) or medical (the psychiatrist) means. From a more epidemiological perspective the man's behaviour would be regarded as an illness, perhaps schizophrenia, the predisposing cause of which could be the fact of being black in a white and frequently racist society. Psychological explanations would once again discount these possibilities in favour of an exploration of his childhood experiences and interactions with his parents. Biological psychiatrists might consider he was suffering from some form of brain damage or lesion, or, suspecting schizophrenia, might seek to discover whether his parents or siblings were similarly afflicted.

1.3.6 Integrating explanations

It would appear that such explanations cannot all be correct. Indeed in several cases they seem to be mutually exclusive. Is it possible to resolve the differences between them? Subsequent chapters will look at the evidence in favour of different types of explanation for several different common conditions, some unequivocally biological, others on the borderlines of the biological, the psychological and the social. For the moment, however, consider the general question of how to explain mind and brain disorder. There is no doubt that research and clinical opinion is divided on the matter. There are those who would argue strongly that, in the last analysis, only biochemistry or genetics counts, and equally those who are contemptuous of attempts to reduce the great variety of human experience to a matter of mere neurotransmitter imbalance. Even for unequivocally 'biological' disorders, explanations which seem to ignore the subjective experience of suffering from a particular condition (for instance the severe depression that often accompanies Parkinsonism) in favour of its mere anatomy, physiology, biochemistry or pharmacology, may seem to diminish the experience of the disorder, *explaining away* rather than explaining. The limits of merely biological explanation have been sensitively described by the neurologist Oliver Sacks in a series of books, especially his first, *Awakenings*, which has also been made into a popular film.

Some clinicians are more selective in their approach, and would accept that for certain conditions the overriding factor is biochemical and for others social. For instance, it is common practice among clinicians to divide depression into two categories, one an *endogenous* condition and the other an *exogenous* depression. The first is claimed to be caused by primarily internal biological factors, the other precipitated by external causes such as the death of a close relative, the loss of a job or even moving house. This is, however, no more than another way of restating the old neurological/psychiatric, or organic/functional dichotomy.

Neither of these solutions seems appropriate to the goal of this course, which is to provide an integrated account of brain and behaviour. There seems no reason to adopt an either/or approach. It is, for instance perfectly possible to maintain that changes in a person's experience can produce lasting and highly specific changes in their biology. There are good examples from earlier in this course.

□ Try to name some.

■ As described in Book 4, environmental enrichment and impoverishment, visual experience and deprivation, during early development, all have lasting effects at both brain and behavioural levels. Learning and memory formation (Book 4, Chapter 4) change synaptic biochemistry, morphology and physiology.

Grieving over death or loss of a job could perfectly well in principle result in biochemical changes similar to those produced as a long-term consequence of childhood experience or even from a genetic deficit, just as a biochemical imbalance can 'translate' into feelings of intense grief or depression. One of the clearest cut examples of this translation is a condition known as 'panic disorder'. In sufferers from this condition normal environmental events can precipitate a frightening sense of panic, associated with a massive increase of the glucose breakdown product lactic acid in the bloodstream. But equally, if exogenous lactic acid is injected into the bloodstream, the person experiences a similar sense of panic. Thus the cause of mental distress can be simultaneously understood at one level as biochemical and at another level as social or psychological, just as memory can be understood at one level as caused by experience and at another as stored in the form of changing patterns of synaptic connectivity. However, despite the fact that explanations for mental disorder always exist at multiple levels, for any particular person and situation, there is likely to be one level which is the most significant, thus precipitating the condition in all its biological and behavioural aspects .

Summary of Section 1.3

This section considered three general types of explanation of brain and mind disorder: social, psychological and biological. These types of explanation each involve a different level of explanation and analysis, though within each there are variants, often conflicting. Social explanations vary. Some are associated with Foucault, who denied the very existence of many of the classical categories of disease and distress, attributing them to social labelling. Others seek the epidemiological conditions which make particular conditions more probable. Psychological explanations derive from the exploration of the individual's life history, relating present-day problems primarily to childhood experience. Such explanations may be driven by theoretical preconceptions about family relationships in childhood, as in psychoanalysis or antipsychiatry, or they may be more empirical, and be based on identifying predisposing 'vulnerability factors'. Biological explanations may search for environmental or genetic causes of a condition. Often, where there is no overt brain damage, they may focus on the discovery of biochemical abnormalities, especially in the interactions of neurotransmitters and receptors. Finally, it was argued that explanations at the several levels need not automatically be contradictory, that it is possible for a condition to be manifested simultaneously in biological, psychological and social forms, and to be explicable at all levels.

1.4 Studying brain and mind disorder

1.4.1 The clinical method

Any attempt to explain a condition regarded as an 'illness' must begin by observing and categorizing it. Clinicians, that is, general practitioners, neurologists and psychiatrists, must examine a patient and come to some decision about the nature of their problem. When the condition is clearly physical such as a broken leg or acute appendicitis, this is relatively straightforward. When the manifestations of the condition are essentially behavioural it is much harder. The man found wandering unsteadily in the streets, muttering incoherently to himself and brought into hospital for examination might be drunk, might have had a stroke, or might be suffering from schizophrenia. In making such a judgement clinicians will rely on their own past experience, but also on consensus medical opinion on what constitutes a 'disease category'. Drunkenness is easy to assess, a stroke can be detected by looking for other neurological signs, or by checking for damage to particular brain areas by way of a PET (positron emission tomography) scan. But once these possibilities have been eliminated, then it has to be decided if the man is indeed really ill, especially as he might vehemently deny being so. If ill, then what is the nature of his condition? And if he denies being ill and refuses treatment, what is the responsibility of the clinician?

Over the years there has been a tendency to refine and alter the diagnostic criteria that are used within the terrain of both neurological and psychiatric disorders. In general, clinical scientists tend to be either 'lumpers' or 'splitters'. The first group tend to combine different types of signs and symptoms as being all reflective of a unitary underlying condition. The second group fragment diseases into many different subcategories. There are examples of this both in the discussions of brain damage in Book 2, Chapter 11, and later in this book (Chapters 5 and 6) in relation to aphasias and related disorders. The same problem occurs with the psychiatric disorders; thus, as will become apparent in Section 7.2.1, conditions once called schizophrenia have in recent years tended to be re-classified as manic depression, while some clinicians now doubt whether there is any such 'single' disease entity as schizophrenia at all and refer instead to 'the schizophrenias' or 'schizo-affective disorder spectrum'. Psychiatrists will try to apply behavioural criteria in order to arrive at a diagnosis. These may include trying to assess a person's general state of health; for instance, whether they have difficulty in going to sleep at night or in waking up in the morning, or whether they are suffering from a loss of libido or of interest in normal day-to-day activities. There are a number of standard psychiatric rating scales in which a patient may be asked to answer a battery of questions designed to test for anxiety, depression, and so forth, in order to provide objective (or at least standardized) diagnostic criteria. Figure 1.1 shows an example of such a rating scale. For instance, the diagnostic criteria for schizophrenia include auditory hallucinations ('hearing voices') as one of the commonest manifestations.

Sometimes diagnosis is based on a person's response to a drug: if their condition improves when they are treated with lithium, they may be defined as manic-depressive; if lithium is without effect but they respond to the new generations of phenothiazine drugs, they may be termed schizophrenic. There are, of course, hazards in allowing clinical judgement to be driven by pharmacology in this way, but what is sure is that diagnosis of psychiatric disorders is far from being a simple, unequivocal matter.

MULTIPLE AFFECT ADJECTIVE CHECKLIST

DIRECTIONS On this sheet you will find words which describe different kinds of moods and feelings. Mark an X in the boxes beside the words which describe how you feel at the present time. Some of the words may sound alike, but we want you to **mark all the words** that **describe** your feelings. Please work rapidly.

1 active
2 adventurous
3 affectionate
4 afraid
5 agitated
6 agreeable
7 aggressive
8 alive
9 alone
10 amiable
11 amused
12 angry
13 annoyed
14 awful
15 bashful
16 bitter
17 blue
18 bored
19 calm
20 cautious
21 cheerful
22 clean
23 complaining
24 contented
25 contrary
26 cool
27 cooperative
28 critical
29 cross
30 cruel
31 daring
32 desperate
33 destroyed
34 devoted
35 disagreeable
36 discontented
37 discouraged
38 disgusted
39 displeased
40 energetic
41 enraged
42 enthusiastic
43 fearful
44 fine
45 fit
46 forlorn
47 frank
48 free
49 friendly
50 frightened
51 furious
52 gay
53 gentle
54 glad
55 gloomy
56 good
57 good-natured
58 grim
59 happy
60 healthy
61 hopeless
62 hostile
63 impatient
64 incensed
65 indignant
66 inspired
67 interested
68 irritated
69 jealous
70 joyful
71 kindly
72 lonely
73 lost
74 loving
75 low
76 lucky
77 mad
78 mean
79 meek
80 merry
81 mild
82 miserable
83 nervous
84 obliging
85 offended
86 outraged
87 panicky
88 patient
89 peaceful
90 pleased
91 pleasant
92 polite
93 powerful
94 quiet
95 reckless
96 rejected
97 rough
98 sad
99 safe
100 satisfied
101 secure
102 shaky
103 shy
104 soothed
105 steady
106 stubborn
107 stormy
108 strong
109 suffering
110 sullen
111 sunk
112 sympathetic
113 tame
114 tender
115 tense
116 terrible
117 terrified
118 thoughtful
119 timid
120 tormented
121 understanding
122 unhappy
123 unsociable
124 upset
125 vexed
126 warm
127 whole
128 wild
129 wilful
130 wilted
131 worrying
132 young

Figure 1.1 Example of a questionnaire used for rating psychiatric state.

1.4.2 Epidemiology

The task of epidemiology is the study of the distribution and determinants of a disease condition within human populations, in order to find correlations between patterns of behaviour or events and the occurrence of the disease. Thus epidemiology asks *who* gets ill, with a view to answering the question of *why* they get ill.

To do this the first task is to estimate the extent of a disease within a population and then to seek to identify common features among those who succumb to it. The extent of a disease may be calculated as its **incidence**: that is, the number of new cases occurring per thousand of the population in any given period; or a person's lifetime expectancy of being diagnosed as having the condition. Such estimates need to be corrected by a variety of statistical procedures, for example for age, the confounding effect of other conditions, and so forth. Some of the complexities of interpreting epidemiological data will become apparent later in this section. Sometimes the common feature may be an aspect of a person's life-style or diet, as in the classic success stories of epidemiology: the associations between smoking and lung cancer and between dietary fat intake and coronary heart disease.

The strength of such associations is measured by statistical procedures, in particular the derivation of a *correlation*. The problem is that even if quite a strong correlation has been discovered, it may be fortuitous, rather than revealing a causal relationship. For example in the 1950s and early 1960s there was a rapid increase in the diagnosis of coronary heart disease that was almost precisely paralleled by the increase in the issuing of television licences in England and Wales. It would, however, be foolhardy to jump to the conclusion that watching television causes heart disease.

In order to derive their correlations, epidemiologists must have access to accurate medical records such as admissions to hospital or causes of death as indicated by death certificates. In addition, they must have criteria by which to divide the population: for example, in the case of cigarettes and lung cancer, into smokers versus non-smokers. Certain broad categories are of particular value in this context, as you will see in practice in later chapters.

1.4.3 Classification

□ Suggest a variety of ways in which the population of the UK might be divided up for the purposes of epidemiology.

■ You may have thought of one or more of the following: age; sex/gender; geography; class or occupation; 'race'.

Each can be considered in turn.

(a) *Age* This is the most straightforward: to ask whether a disease or disorder is, for example, more frequent in children or in the elderly is relatively unambiguous.

(b) *Sex/gender* Here you can ask whether the diagnosis is more common in women than in men, or vice versa, or whether there is no difference at all. Note that the term gender is used here. You will recall that a distinction between sex and gender was made in Book 1 (Section 9.3.1). The distinction made there was in an ethological context and so was a rather esoteric one favoured by some ethologists. The more general distinction, used in Book 4 (Section 4.5.5) and in this book, is that *sex* defines a person's primary biological characteristics (possession of XY chromosomes and male genitalia if a male and of XX chromosomes and female genitalia if a female). *Gender* relates to social ascriptions (masculinity and femininity, ways that men and women are 'normally' supposed to behave by virtue of being men and women): that is, men to be breadwinners and women homemakers, for instance. This distinction between sex and gender is important, because clearly one does not necessarily map precisely onto the other, and although sex is a more-or-less fixed category, gender characteristics are not stable, but are very much shaped by history and culture.

(c) *Geography* In the UK there are quite sharp regional variations in the incidence of particular diseases. Coronary heart disease, for example, is more frequent in Scotland than in England, while perinatal mortality is greater in the Yorkshire city of Bradford than the East Anglian city of Cambridge. There may also be differences between living in urban and rural environments.

(d) *Class and occupation* Some diseases are rather occupation-specific (lung diseases among miners and masons, for example), but there are also broader differences (as you will see, especially in Chapter 7), in the incidence of particular diagnoses in particular social classes. There is much debate about how social class should be identified and measured; the standard classification in the UK is given by the Registrar General's scheme, which allocates people to five categories on the basis of the occupation of the so-called 'Head of Household'.

Class I Professional (for example accountants, doctors, judges, even university teachers!).

Class II 'Intermediate' (for example aircraft pilots, publicans, members of parliament and school-teachers).

Class III Subdivided into Skilled Non-manual (for example cashiers, clerical workers, sales representatives and shop assistants) and Skilled Manual (for example bakers, bus-drivers, miners, police officers and upholsterers).

Class IV Partly Skilled (for example bar staff and bus-conductors, postal workers and telephone operators).

Class V Unskilled (for example labourers, cleaners, guards and porters).

□ Such a definition of social class inevitably produces a number of anomalies. Try to suggest some.

■ Class and income do not match. For example, aircraft pilots tend to earn more than university teachers. Nor does it make any provision for inherited wealth. Further, the ascription of class is by the occupation of the Head of Household who is generally regarded as the man in a married couple. It makes no adequate provision therefore for changing living patterns, or women's occupations in households where (as is increasingly the norm) both man and woman are in paid labour outside the home.

Because of these anomalies, insofar as a categorization by social class is used in the chapters that follow (mainly in Chapter 7) it is limited to a broad brush division between 'lower' or 'working' class on the one hand and 'upper' (generally called 'middle'!) class on the other.

(e) *Race* The term 'race' presents even more problems than does the term 'sex'. In the latter case there are two distinct terms, sex and gender, to distinguish biological and social ascriptions, but in the case of 'race' the same term is used with several quite distinct meanings, and they often become confused. In everyday language in the UK people speak of 'white' and 'black' races, distinguishing by skin colour; of the 'Jewish race' distinguishing by culture or ethnicity; of 'Asians', distinguishing by continent of birth (or even the continent of parents' or grandparents' birth!); and even of English, Irish and Scottish 'races', distinguishing by nationality. These everyday uses of the term 'race' are very confused. When social scientists and biologists began to attempt to use and therefore define the term more precisely, especially from the 19th century onwards, they at first believed they were both talking about the same thing. That is, they thought that the social label 'race' also describes a biological phenomenon and that there are biologically distinguishing features that separate one socially defined 'race' from another. This belief has deep roots within European and American history, but today finds little scientific support.

Social scientists today recognize that the concept of 'race' fluctuates in different times and cultures. A good example here is how, during the apartheid period, the South African race laws defined Chinese people as 'Coloured' but Japanese people as 'White'). As a result, in a scientific context the term 'race' is replaced with more specific descriptions, such as ethnicity, which make no claims to a parallel biological meaning.

On the other hand, for biologists, 'race' is a technical term derived from population genetics, which may be applied to any species. A race in this sense has a rather precise genetic meaning as a subset of a species within which there is free exchange of genes by interbreeding, and which may be distinguished from other 'races' by differences in the relative frequencies of alleles. Normally such a racial group is formed when there is a barrier (usually geographic, through isolation) over many generations to breeding with other groups. A distinct human race in the biological sense would exist if the frequency with which particular alleles occurred in that group was very different from the frequency with which those same alleles occurred in another group. But when allele frequencies are measured in human populations that are colloquially defined as 'races' (for instance 'English', 'Jews', 'Blacks', 'Whites') it turns out that for nearly all the genes studied the differences between two individuals of different races are no greater on average than those between two individuals of the same race. Over 94% of all the differences are found *within* a given race rather than *between* races. This is because over many centuries there has been a steady movement of people between regions, and very few areas or populations have remained sufficiently genetically isolated to constitute a race in the biological sense. This means that, on average, White or Christian people living in England are likely to be just as similar or different genetically from their White or Christian neighbours as they are from their Black or Jewish neighbours.

Skin colour is a polygenic character; the result of many interacting genes (Book 1, Section 3.5). This characteristic is often used in a political context by racist parties and writers to define race differences. However, the degree of intermixing of Blacks and Whites that has occurred over many generations in most mixed communities in the UK (as elsewhere in Europe and in the USA) means that many 'White' and 'Black' alleles are distributed with rather similar frequencies in both White and Black populations (except, of course, those for skin colour!). Any attempt to use skin colour as some sort of biological marker of race therefore makes no sense. Equally, considerable differences in the distribution of alleles may occur between regions (e.g. between North and South Wales) or even between close villages. However, these are not classified as racial differences because the 'marker' is not visibly expressed, unlike skin colour or religion.

1.4.4 The problems of epidemiology in action

In the field of psychological disorders, one of the earliest epidemiological studies was that by R. E. L. Faris and H. W. Dunham, who looked at the incidence of schizophrenia in Chicago in the 1930s. They observed that the highest occurrence of the condition was in inner city, poor, working-class districts and concluded that there was something about the social conditions of that type of environment which caused or contributed to the schizophrenia.

□ Suggest three other possible explanations for these observations.

■ (a) Schizophrenia could be an infectious disease, and so is more likely to be found in areas of high population density and low amenity, like other infectious diseases.

(b) People who were schizophrenic might be unable to hold down jobs and therefore might tend to drift into poor city areas.

(c) Doctors might disproportionately label poor inner-city people as schizophrenic, while giving other diagnoses to people who show similar behaviour patterns but who are living in more affluent circumstances

Such varying possible explanations show something of the difficulty of interpreting epidemiological evidence. It means that such observations can help define what has to be explained without being able to offer an unambiguous causal explanation in their own right. There are some who would argue, implausible though it seems, that far from smoking causing lung cancer, there is a *covarying* genetic factor which means that people with genes predisposing them to smoke also have genes predisposing them to cancer. Nonetheless, attempts to explain psychiatric disorders ignore epidemiology at their peril, as will become apparent in Chapter 7.

1.4.5 Looking inside the brain

Biological methods for brain investigation will be discussed in this section. If mind disorder means brain disorder, an ideal way to study it would be to look at brain processes directly. It would then be possible to compare aspects of the 'brain state' of people in mental distress with that of people from an otherwise similar (that is matched for age, sex, and so on) but 'normal' group. Alternatively, granted that in some conditions people show periods of distress punctuated by intervals of 'normality' one could ask what is changing in the brain between the normal and disturbed times. A biological measure which is 'normal' in individuals when they are 'normal' but changes when they are distressed is known as a *state marker*, whereas one which is always different in the distressed person compared with a normal person is known as a *trait marker*. Trait markers may imply a predisposition, a vulnerability towards showing particular forms of mental distress, and examples of both trait and state markers will crop up in subsequent chapters. (Note that the term 'marker' is used in a variety of ways; you will also find a different use of the term as a 'genetic marker' later in this chapter (Section 1.4.9) and in Section 4.4).

However, until recent years, the possibility of studying brain phenomena relatively non-invasively did not exist. Perhaps the longest established method derives from the development of the electroencephalograph (EEG) in the 1930s (Book 2, Box 8.2). EEG measurements involve placing disc-shaped electrodes on the surface of the scalp. The electrodes record spontaneous rhythmic fluctuations of voltage, known as EEG waves, in various brain areas. The wave forms are the summed products of the interactions of many millions of neurons, and are obviously most closely related to the activity of the region of cortex closest to the electrode. The precise mechanisms which produce the waves are not known, although most of the electrical activity is believed to originate from the dendrites. The pioneers of the EEG believed that it would provide a 'window' into the brain; wave forms could be correlated with many aspects of cerebral functioning such as sleeping and waking, relaxation and mental concentration and might be expected to show characteristic changes during episodes of brain distress. This is clearly the case for gross brain dysfunctions, such as those resulting from lack of oxygen (hypoxia), or during coma, and for some dramatic neurological conditions, such as epilepsy, when an EEG can be and is used diagnostically. However it is probably fair to say that the EEG has not proved useful so far in distinguishing more subtle aspects of brain and mind disorder. Today researchers are very optimistic about a new form of imaging which, instead of examining the brain's electrical activity, records the fluctuations in the minute magnetic fields which are always associated with an electrical current. Such biomagnetic measurements may indeed provide the type of window into the brain that the EEG promised, though it is still too early to tell.

More direct methods of imaging have become available in the last decades with the development of computerized tomography and its more recent descendant, positron emission tomography (PET; see Book 2, Box 11.1). These scanning techniques provide ways of looking inside the brain directly. They have revolutionized neurology and clinical neuroscience as profoundly as did the development of X-rays for physical medicine at the turn of the century. Their main advantage is that they are relatively simple, non-invasive techniques that can be performed quickly and give immediate results. PET scanning reveals the activity of different parts of the brain in terms of a 'map' of blood flow, receptor activity and neurotransmitter metabolism. The technique can be used to study both normal brain functioning (for example during various forms of mental activity such as listening to music or remembering the route taken on a journey), and also disordered functioning as a result of tumours, strokes or other forms of damage.

PET is much more sensitive than other scanning techniques for detecting and localizing specific molecules in the brain down to a volume of a few cubic millimetres, but one of its major drawbacks is its expense. Another, quicker and cheaper scanning technique is offered by magnetic resonance imaging (MRI) (Book 4, Box 5.1). MRI is not very sensitive to small amounts of any substance so can only detect substances present in large volume in the brain, such as water, but it does enable images to be made of cerebral ventricles and areas of possible damage after disease or injury. Figure 1.2 shows the degree of resolution that can be obtained by MRI scanning to reveal areas of hippocampal damage.

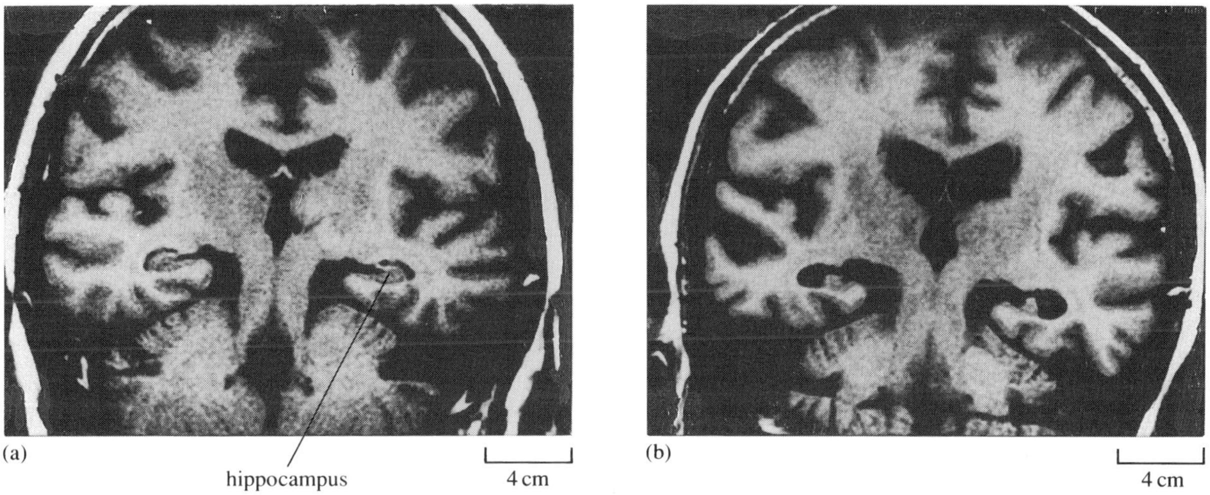

Figure 1.2 MRI scans of (a) normal human brain and (b) the brain of a person with damage to the hippocampus.

1.4.6 Biochemical methods

Before the availability of scanning methods, the only feasible way to study what goes on in the brain as a result of disease or disorder was to examine the brain at post-mortem. Such examination remains important, even today, because of the limitations of just what can be measured by scans. In some disease conditions there are obvious gross abnormalities: tumours are proliferations of glial cells (Section 5.2.5); damage due to stroke or epilepsy often includes scar tissue (Section 5.2.1); Alzheimer's disease produces characteristic plaques and tangles (Section 3.7.1), and in acute alcoholism (Korsakoff's syndrome, Section 5.2.6) the brain becomes dramatically shrunken.

In most types of psychiatric disorder there is no obvious brain abnormality, and this of course forms the basis for the original distinction between neurological and psychiatric conditions. However, in the last few years there have been renewed suggestions that there are some subtle anatomical differences in the brains of sufferers from schizophrenia that are detectable by the new scanning techniques. For instance, it has been claimed that there is a greater degree of hemispheric asymmetry and enlarged ventricles in such brains, as detected by tomography. This claim is considered further in Section 7.6.1.

The following approach has been used in the study of the brains of victims of psychiatric disorder. First, hypotheses are made about which region of the brain might be affected. Second, within that region, a search is made for the biochemical system which might be affected. Finally, the brain samples are analysed and compared with those obtained from 'normal' people. For instance, the limbic system (particularly the amygdala) is considered a likely brain site for biochemical abnormalities in depression and schizophrenia. Tissue from these regions has been tested for disorders of neurotransmitter biochemistry (especially in recent years, the opioid, dopamine and serotonin systems).

□ Why the limbic system?

■ Because the limbic system is associated with motivation and emotion (Book 2, Section 10.4; Book 5, Section 4.4.1) and it is emotional responses which are regarded as being adrift in such psychiatric disorders.

Although it is easy enough to measure the absolute quantity of the transmitter present in a small sample of brain tissue, this is not regarded as a very meaningful figure, because transmitters are rather labile; their amount changes rapidly after death. Indeed, in the living brain transmitter levels show marked circadian fluctuations (Book 5, Section 3.2). It is more useful, therefore, to measure the amount or concentration of the biochemical systems involved in the synthesis, degradation and utilization of transmitter because these are less liable to transient fluctuation.

□ Name some of these biochemical systems (you have met them in previous books).

■ The enzymes of transmitter synthesis and breakdown (for instance, for acetylcholine, they are choline acetylase and acetylcholinesterase), the receptor molecules (for acetylcholine, they are the nicotinic and muscarinic receptors) and the re-uptake systems. (If you do not remember these mechanisms, refer to Book 2, Sections 2.3.3, 4.2, 4.4 and 4.5.)

□ Suppose one finds a difference in the level of some neurotransmitter receptor in a post-mortem sample of the amygdala of a schizophrenic person, compared with that from a normal person of the same age, could one interpret the difference as being the cause of the disorder?

■ There are great problems about drawing such conclusions. Post-mortem samples are not easy to come by, and 'matching' brains is hard, so making comparisons is difficult. If one finds a difference, could this be due to some post-mortem degeneration of the tissue? Or could it be the *result* rather than the *cause* of the disease? (If you are unclear about this, think of having the flu and your nose running. The runny nose is not the cause but the *consequence* of having the flu.) Could the difference have been the consequence of having been hospitalized for many years, or perhaps an effect of the drugs that the schizophrenic person had been taking to treat the condition?

To avoid some of the difficulties with post-mortem samples, a way is really needed to study biochemical changes during life. It is not possible to study brain biochemistry directly. Instead it has become routine to look for biochemical abnormalities in those tissues and body fluids to which it *is* possible to gain access, including urine, cerebrospinal fluid and blood. Although these methods have been standard clinical biochemical practice for many years, there are problems associated with them. One distinguished biochemist compared the task of medical urine analysis to that of trying to identify which way people voted in an election by analysing the contents of their dustbins—you can get some information, but its reliability is not high.

The closest of such body fluids to brain tissue is the cerebrospinal fluid (CSF) that circulates through the brain's ventricles and the spinal canal. Samples of CSF can be withdrawn using a hypodermic syringe by the technique known as lumbar puncture, and such samples can be tested for the presence of various chemicals that might reflect the activity of the nervous system or that are the products of such activity.

Chemicals originating in the brain, or their breakdown products, can also be found in blood and urine. For example, serotonin can be found in blood. However 99% of the body's serotonin content is located in the walls of the intestine where it affects gut movements. Therefore, much of the serotonin measured in blood is derived from the intestine and measuring it provides little useful information about the level of this neurotransmitter in the brain. There is, however, an odd but interesting reason why blood contains a source of material which can provide useful information about brain metabolism. As well as the more numerous red and white blood cells, there is a small class of cells in blood tissue called the platelets (Book 2, Section 5.4.1). These share a common embryological origin with neurons and they carry receptors and re-uptake sites for a number of transmitter systems (notably serotonin) on their membranes. Study of the neurotransmitter metabolism receptors and re-uptake sites of blood platelets can therefore offer a 'surrogate' measure for studying neurons. Many psychotropic drugs, such as anti-depressants, will bind to platelet receptors. There is some evidence that the extent of this binding is affected by a person's emotional state, because, for example, serotonin receptors and re-uptake sites are lowered in blood platelets during states of depression.

Studies on urinary levels of noradrenalin have also proved useful. The body contains a lot of noradrenalin outside the brain itself but it is metabolized differently from central nervous system noradrenalin. Measurement of a breakdown product in urine that comes only from brain noradrenalin can provide clues about brain noradrenergic activity.

□ Suggest one particular reason why, when it is feasible, blood and urine are good sources of material for measurement.

■ Blood and especially urine samples can be taken repeatedly with no deleterious effects on the experimental subject, so that changes in an individual over time can be monitored.

However, such measures may also be misleading, because they can be affected by a variety of other factors. In a notorious case, a biochemical abnormality found in the urine of hospitalized schizophrenic patients was eventually shown to be a metabolic breakdown product derived from tea because the patients were drinking much more tea than were the control subjects from outside the hospital!

1.4.7 Pharmacological methods

Until the last few years, most of the drugs used to treat psychiatric disorders were found using reasoned guesswork; that is, following initial chance or clinical observations about the effect of a particular substance that was perhaps originally used to treat quite another condition (see for example, the discussion about the origins of reserpine in Section 4.3.2). There were no rational ways of developing new drugs based on a clear understanding of the biochemical systems with which they were intended to interact. Once it had been observed that many psychotropic drugs seemed to affect neurotransmitter systems, however, it became possible to test potential new drugs by examining their effect on such systems either in human blood platelets or in experimental animals. The second stage was to make chemical variants of the drugs, by systematically altering their molecular structures and testing each variant in turn to find the most effective one. Today, such studies can be speeded up both by the use of computer technology, which can predict the three-dimensional structure of putative drugs and their interactions with transmitter and receptor systems, and by the genetic engineering techniques which have enabled the actual protein molecules which constitute the receptors to be isolated, cloned (that is, copied to produce bulk quantities using genetic methods) and then studied intensively.

Of course, the developmental path between the identification of a possible new drug and its release into clinical use depends on much more than this, including extensive testing for safety in animals, and then clinical trials, which are not discussed here.

□ Suppose that one discovers that a drug which functions as an antidepressant interacts with a serotonin re-uptake site in blood platelets from human subjects, or with serotonergic systems in the brains of experimental animals. It is then possible to conclude that the cause of depression is some disorder of serotonin metabolism in the brain?

■ No. To see why, consider an analogy. If one has a toothache, aspirin will relieve the pain, but one cannot conclude that the cause of the toothache is too little aspirin in the brain! The aspirin treats the symptoms and some of the consequences of a condition, not its cause, and so may the psychotropic drug. The effects of the depressive condition may be mediated by way of the serotonergic system, and so altering the effectiveness of a serotonergic synapse with a drug may 'mask' the depression without affecting its original cause.

Another problem that frequently arises in interpreting drug action is that the time course and concentration of the drug required to produce a behavioural effect is very different from that which produces the biochemical effect observed in the test tube. Many antidepressants, for example (and also lithium, which alleviates the mood swings in manic depression), need to be taken for several weeks before they are effective in altering mood or behaviour, yet their biochemical effects in the test tube or in experimental animals seem to occur immediately. This tends to complicate any interpretation of the effect of the drug based on a simple correlation between a biochemical and a behavioural response.

1.4.8 Animal models

Many advances in medicine and in the understanding of disease have come about through the study of the analogous condition in experimental animals. A number of diseases are common to non-human and human animals alike, and the effects of drugs and treatments can be studied directly in experimental animals in a way which does not raise the same ethical problems as might their study in humans. However, although the effects of treatments which produce brain disease in humans can be observed in non-human animals (for example, conditions like Parkinson's disease, stroke and alcoholism can be mimicked), when it comes to the psychiatric disorders there are greater problems. These conditions are expressed in humans by subtle changes in behaviour, and above all in people's verbal descriptions of their emotions. Just how is one to recognize a depressed rat or a schizophrenic cat, for instance?

Efforts have nonetheless been made to try to produce such conditions experimentally and to study their effects on brain biochemistry and other measures. Rats have been subjected to conditions in which they are exposed to continued stress, or unavoidable forms of punishment such as electric shock. (Such experiments are generally not permitted under modern animal legislation in the UK, it should be noted.) Repeated exposure of rats to such conditions produces a state of apathetic self-neglect in them which has been described as 'learned helplessness' (Book 5, Section 5.5.3), and may be regarded as analogous to some forms of psychiatric disorder in humans. Rearing social animals like monkeys in isolation (as in the Harlow experiments discussed in Book 1, Section 9.5), results in patterns of behaviour which seem at least superficially to be very similar to those observed in human depression. The animals become withdrawn, rather immobile and huddled, with a lowered appetite and reduced sexual interest. Marmosets have been especially studied in this context, because they generally give birth to twins. If one infant is separated from its family at birth and hand-reared in social isolation, it exhibits behaviour patterns which seem, anthropomorphically, to resemble depression. Aspects of its brain biology can then be compared with those in its normally reared sibling. But can one really say that this is 'the same as' the condition seen in humans? Whether the reward from such experiments in terms of greater understanding of human disorders or better methods of treating them is worth subjecting these animals to such conditions is a question you should consider.

An alternative approach is to use the insights gained from pharmacology. If antidepressants work by *stimulating* serotonin re-uptake systems, then the effect of an agent which *inhibits* such systems in an experimental animal ought to be to produce behaviour analogous to depression, which should in turn be counteracted by the antidepressant. Such behaviour can be quantified. In this way one can observe the behaviour of rats exploring first, an 'open field' which is an arena marked out into squares so that one can count how many squares the rat crosses in a given time, and second, a 'hole board' in which the arena has a number of drilled holes into which an inquisitive rat can poke its muzzle. The behaviour of rats in such arenas is a very sensitive measure of their state of well-being. Rats given tranquillizing or sedative drugs may 'freeze' and fail to explore; other agents may result in hyperactivity. Psychotropic drugs such as lysergic acid diethylamide (LSD) or psilocybins (the active constituent of such recreational agents as 'magic mushrooms') cause hallucinations. Cats fed doses of such drugs which are sufficient to produce hallucinations in humans behave *as if* they are hallucinating, growling or cowering in response to imaginary threats, or hunting imaginary prey. This supports the view that the agents do affect similar classes of behaviour in other mammals to those in humans, and of course makes it much easier to study the intimate details of the biochemical mode of action of the drugs.

1.4.9 Genetic methods

A person's phenotype at any time is the expression of the interplay between genes and environment during the whole of that person's life up to the present time (as discussed in Book 1, Section 3.2.3, and Book 4, Section 2.3.6). When it is said that a brain/mind disorder is genetic, what is meant is that the person suffering from that disorder has inherited a particular combination of alleles which, in the environment in which the person has been reared, predisposes them to express the brain/mind disorder. In that every person's phenotype is uniquely the product of their specific set of genes (their genotype) and their environment, in one sense every aspect of the human condition is genetic. However, while virtually everyone in a human population exposed to an anticholinesterase poison like nerve gas will die, only one in ten thousand newborn children exposed to phenylalanine in their diet will grow up irreversibly mentally retarded. A person's genes are largely irrelevant in the first case, but overwhelmingly relevant in the second, which is why the first is described as an example of environmental poisoning, and the second as an example of a genetic disorder (phenylketonuria, PKU: see Book 1, Section 3.3.3; Book 2, Section 4.5.2; Book 4, Section 4.5.2 and this book, Section 4.1).

So how can one decide whether a particular condition is genetic? One clue is given by the PKU example, in which the brain disorder is associated with a clear metabolic abnormality, that is the inability to metabolize the amino acid phenylalanine because of the non-functioning of a key enzyme. This means that something is biochemically wrong with the phenylketonuric child from birth. This could be a genetic flaw, but there could be other explanations.

□ Try to suggest one.

■ The mother could have had some problem in pregnancy or during the birth process. Defects that are present at birth are generally called *congenital* defects (Section 4.1; Book 4, Section 4.5.2.)

There are many forms that such a problem could take. You will read in Chapter 4 about chromosomal abnormalities such as those which produce Down's syndrome. Severe malnutrition in pregnancy, especially during the period of maximal fetal brain growth (Book 4, Section 4.5.3), can result in a general reduction in brain cell number with variable behavioural consequences. Congenital damage can also result from drugs (like thalidomide) or poisons. Another very common form of brain damage results from a brief period of oxygen deprivation (hypoxia) for the baby during birth (for instance if birth is prolonged or difficult). Hypoxia during birth is one of the commonest causes of infant brain damage and consequent mental retardation or handicap.

□ How might one go about checking whether a condition is really a genetic rather than a congenital disorder?

■ The starting point is epidemiological. First, one studies the incidence of the condition in the population. If the condition is genetic, there will be an increased likelihood that individuals closely related to the person who suffers from the condition will themselves suffer from it.

□ Suppose every individual in a family of two parents and three children is diagnosed as schizophrenic, would that mean that the condition was bound to be genetic?

■ No. If a condition is *familial*, which means that it runs in families, one may suspect a genetic origin but there could be something common about the environment that all members of the family share. They may all get flu, because they live in the same house, but this does not mean the flu is genetic.

In fact, if all the members of the family shared the condition it would be very unlikely to be genetic, because this would imply that the parents, although they are likely to be genetically unrelated, nonetheless by chance, both have the alleles of the gene which predisposes to the condition. This can happen and it is a phenomenon known technically to geneticists as *assortative mating*. An example of such assortative mating is provided by height. A person's height is partly controlled by a rather large number of genes, and there is a tendency for a person to choose a sexual partner who matches them in height (thus tall men tend to pair off with tall women, though the effect is not strong).

To be sure about the genetics of a disorder, it is necessary to look at the condition not merely in members of the immediate family but in their relatives, including grandparents and great-grandparents, if known. One can then construct what is called a *pedigree chart,* describing the incidence of the condition within a family extended in time and space. The extent to which the condition recurs in different members of the same family may provide a clue as to its possible mode of genetic transmission. Figure 1.3 gives a hypothetical pedigree chart for three generations, I, II and III. Each circle represents a female and each square represents a male. Individuals who manifest the condition, or *trait*, are shown as filled circles or squares.

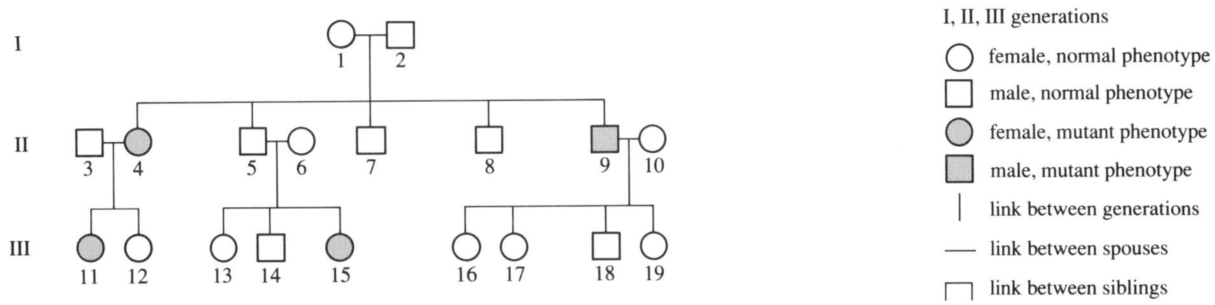

Figure 1.3 A hypothetical pedigree chart.

Study the key to this figure and then try to answer the following questions.

□ How many children and of what sex did individuals 1 and 2 produce?

■ One female (number 4) and four male children (numbers 5, 7, 8 and 9).

□ How many of these five children were affected?

■ Two (numbers 4 and 9).

Often, it is possible to deduce from such family trees whether a disorder is the result of the presence of a recessive or dominant allele (Book 1, Section 3.2.3). The pedigree chart in Figure 1.3 shows the observed pattern of a recessive trait which is only manifest when two mutant alleles of the gene are present. Thus some of the open circles and squares in the pedigree chart represent individuals who are phenotypically normal but who may carry one recessive mutant allele whose presence is masked by the normal allele. On the other hand, people with a dominant disorder need only inherit one mutant allele to manifest the trait.

It is hard to get such pedigree charts running back through three generations because, as you might expect, the criteria used for diagnosis change so radically with advances in biology and medicine. However, even if it is available, a pedigree chart is not often conclusive. Examples of some of the conditions whose genetics are relatively well understood will be discussed in Sections 4.3 and 4.4. However, they are rare, and an alternative approach may then be required.

□ Full siblings have a 50% chance of sharing a particular allele. But there are circumstances in which one individual has exactly the same alleles as another at all its loci. What are they?

■ Identical (monozygotic) twins share an identical set of alleles because they are produced from a single fertilized egg, which then separates into two cells before beginning normal cell division.

By contrast non-identical (dizygotic) twins are the products of the simultaneous fertilization of two eggs, and are no more similar genetically than are ordinary siblings. A particularly simple way of scoring differences between twins is to evaluate traits which are present or absent. Thus, twins may either be *concordant*, that is, both possess or both are free of a particular trait; or *discordant*, in that only one of the pair possesses the trait. Thus if a condition found in one identical twin always or nearly always occurs in the other but not so frequently in non-identical twins or other siblings, it is much more likely to be genetic. However, identical twins also normally share a more identical environment than non-identical twins or siblings (they are always of the same sex and are often dressed alike and treated as 'doubles' by friends and relatives). To rule out a possible environmental effect, much attention has been paid to the very rare and special cases in which identical twins have been separated more or less at birth and reared apart. In this case they have their genes in common but their post-natal environment is different, and it becomes possible to estimate more precisely the relative contributions of genes and environment to the condition. Needless to say this is not a frequent occurrence, and is often associated with special circumstances which make it hard to interpret. (The reverse experimental design involves looking at children with different genetic backgrounds but shared environments. For example, the incidence of a particular diagnosis in adopted children can be compared with that in their biological and social relatives.) Until the development of very modern molecular genetic methods in the last decade or so, studying these rare, separated identical twins was the only way of proceeding in the case of many of the more ambiguous cases of genetically associated conditions (such as schizophrenia).

The newer molecular methods involve more direct examination of the genes and chromosomes themselves (if you are unsure of the meaning of these terms, turn back to Book 1, Section 3.2.3). These techniques rely on the ability to identify the gene as a specific piece of DNA located at a particular site on a chromosome. Although the specific genes associated with a particular condition such as schizophrenia are unknown, many genes have been 'mapped' and their location on particular chromosomes has been established. The presence of such mapped (*marker*) genes may be associated with the production of a particular known protein (often an enzyme) which can be measured. During cell division and replication, genes are shuffled and different sections of paired chromosomes recombine in new patterns (as described in Book 1, Section 3.2.4). Suppose there is an unknown gene, *A*, for a particular condition (say a psychological disorder A) whose location in a particular chromosome is unknown and a marker gene, *M*, which can be identified. Suppose *A* is on the same chromosome as *M*. When recombination occurs, the bit of the parental chromosome carrying *A* will sometimes end up on the same chromosome as *M*, and sometimes not. Because the shuffling is random, the closer together along the parent chromosome that *A* is to *M*, the more chance *A* will have of ending up on the same chromosome in the gamete. One can measure the frequency with which the condition A is associated with the presence of the gene *M*, a measure known as **linkage**. The more frequently *A* and *M* occur together, the more likely it is that A is a genetic condition associated with a gene which lies close to *M* along the chromosome. This likelihood is expressed in a measure known as a *lod score*. While it is not necessary to know in detail how such a score is calculated, you will come back to the principles involved in such linkage studies in Chapters 4 and 7.

Summary of Section 1.4

Studying brain and mind disorders begins with clinical observation and attempts to describe and classify the condition. Classification may be complex and affected by fashion; it may be based on behavioural manifestation; it may be the result of biological investigation; it may be revealed by scanning techniques; or it may even be defined by the response of the person to a drug.

Attempts to explain the origins and perhaps the causes of a condition employ epidemiology, which uses statistical methods to describe the frequency of occurrence of a disease in different groups within the population, classifying by geography, age, sex/gender, race and class, or, seeking for other common features such as whether a person smokes or not. Not all these classifications are without problems, in particular, the definitions of sex/gender, class and race. The first confounds a person's biology with the social ascription of role ('breadwinner' versus 'homemaker' for instance). The second can fail to disentangle wealth and status from occupational group, and also ascribes a married woman's class to her husband's occupation. The third may confuse the very specific biological definition of race as a freely interbreeding population geographically separated from others with social labels such as nationality, religion and ethnicity.

Biological methods of study depend on techniques to examine brain processes. These may include use of the EEG or of modern scanning methods such as MRI or PET. More classical biochemical methods are confined to examining the brain at post-mortem or the analysis of such body fluids as CSF, blood (platelets) and urine. Biochemical analysis in the case of psychiatric disorders has tended to concentrate on transmitters and their receptors, notably dopamine and serotonin. Pharmacological techniques are based on the assumption that if a person's disorder is alleviated by a drug, the biochemical process that the drug interacts with must be the site of the biochemical abnormality in the disorder. Some of the hazards of such interpretations are discussed. Animal models for some types of brain and mind disorder are possible, and the interaction of pharmacological approaches with such animal models can provide a wealth of data, and sometimes enable meaningful conclusions to be drawn.

Finally, genetic methods study the extent to which a condition 'runs in families'. This can happen both as a result of a common environment or shared genes, and the techniques for the teasing out of genetic, congenital and environmental effects include the construction of pedigree charts, the study of twins reared together or apart, adoption studies, and the search for genetic markers for the condition.

Objectives for Chapter 1

When you have completed this chapter, you should be able to:

1.1 Define, use or recognize definitions and applications of each of the terms printed in **bold** in the text.

1.2 Describe the range of conditions embraced within the concepts of brain and mind disorders, and provide examples.

1.3 Distinguish between social, psychiatric and biological modes of explanation of brain and mind disorders.

1.4 Describe critically the methods available for the study of brain and mind disorders, including clinical methods, epidemiology, imaging, biochemical, pharmacological and genetic methods. (*Question 1.1*)

1.5 Distinguish between social and biological ascriptions in the context of sex, gender, class and race.

1.6 Describe the strengths and limitations of animal models. (*Question 1.2*)

1.7 Distinguish between familial and genetic transmission of brain disorder. (*Question 1.3*)

Questions for Chapter 1

Question 1.1 (*Objective 1.4*)
A clinical researcher studies a group of hospitalized patients diagnosed as schizophrenic. Comparing them with a group of 'normal' non-hospitalized controls, matched for age and sex, she finds that there is an abnormal neurotransmitter in the urine of the patient group. The researcher submits a paper to a medical journal announcing the discovery of the metabolite as a probable causative agent in schizophrenia. The paper is rejected. Why?

Question 1.2 (*Objective 1.6*)
Section 1.4.8 describes the Harlow experiments as providing an 'animal model' for depression. How good a model do you judge it to be?

Question 1.3 (*Objective 1.7*)
'He is slow and simple-minded like his father and grandfather. He must have inherited the condition from them.' Must he?

References

Dohrenwend, B. P. *et al.* (1980) *Mental Illness in the USA: Epidemiological Estimates*, New York, Praeger.

CHAPTER 2 PSYCHOLOGICAL ASPECTS OF AGEING

2.1 Introduction

2.1.1 Why study psychological aspects of ageing?

There are plenty of answers to this question. The psychological aspects of an organism cannot be understood by simply describing it at a single point in time without also studying the changes in structure, function and behaviour that occur throughout its life cycle. It is now more necessary than ever before to understand the process of human ageing. Current demographic trends are such that in many developed countries increasing life expectancy and decreasing birth rate have caused a rapid increase in the proportion of elderly people in the population. In 1900 the percentage of Americans over the age of 65 was 4%. By 1980 that figure had risen to 11% and it is set to reach 13% by the end of the century. The trend is similar in Europe. In the UK the proportion of the population who are over 65, expressed as a percentage of those of working age, rose from 11% in 1911 to 28% in 1990. Moreover, the proportion of the population who are over 85 is growing even faster. **Life expectancy** is the *average* number of years a person can expect to live from birth. In 1900 the life expectancy for men and women in the USA was 46 and 49 years respectively. By the 1980s it had risen to 71 and 78 years.

□ Suggest some reasons why this dramatic increase in life expectancy might have occurred?

■ Primarily because of a reduction in infant mortality. Also, many of the diseases like smallpox, tuberculosis and diphtheria that kill young people have now been controlled or eliminated and living conditions have been improved.

As a result, many more people now live on into old age. It is important to note that the statistics do not imply that the ageing process itself has been influenced. Rather, the prevention of premature deaths has allowed more of the population to reach old age.

Figure 2.1 (*overleaf*) shows the contrasting life expectancies for developed and Third World countries. This contrast underlines the importance of environmental factors in determining how long people survive.

The **lifespan** of a species is defined as the length of life of the longest-lived members of that species. For example, the lifespan is around 100 years for humans; 70 years for elephants; 35 years for horses; 20 years for dogs and three years for the house mouse, *Mus musculus*. In contrast to the increase in life expectancy, the lifespan in humans has not changed during the 20th century. Medical science and technological advances have not succeeded in slowing down the ageing process. The fact that the lifespan in different species varies so markedly suggests that there is a strong genetic influence. This conclusion is also supported by studies showing that the lifespan of monozygotic twins is much more similar than the lifespan of dizygotic twins (see Section 1.4.9), and the observation that extreme longevity tends to run in families. The changing age composition of societies in the developed world comes about because many more individuals are living to an age close to the lifespan of the human species.

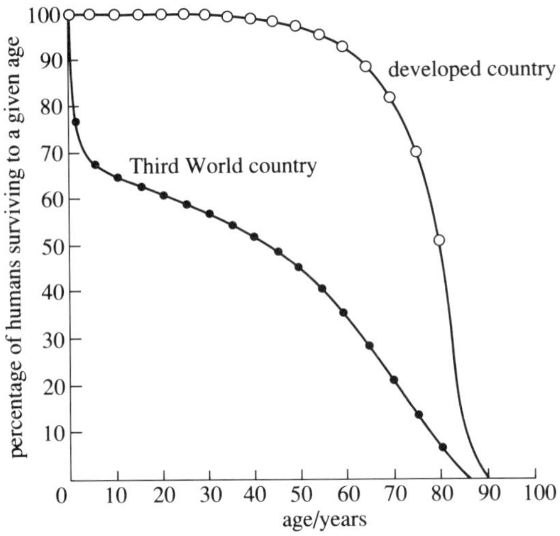

Figure 2.1 Human life expectancies for developed and Third World countries.

□ This shift in the age composition of society raises many social and political issues. Try to list some of them.

■ The allocation of medical resources, financial policy, housing, the provision of care and support services are all affected and making the right decisions depends on being able to assess accurately the effects of ageing on all aspects of physical and mental health, cognition and behaviour.

The study of cognitive function in old age is particularly important for practical issues such as the age of retirement, the relative merits of different types of sheltered accommodation and the provision of leisure and educational opportunities for older people.

2.1.2 What is ageing?

This seems like an easy question to answer: most people would point to birthdays ticking away and visible signs of physical deterioration such as greying hair, wrinkled skin and decline in strength and mobility and perhaps, also, a tendency to be forgetful. In giving such an answer they are actually providing three definitions of age, a chronological one, a physical or biological one, and a psychological one. This chapter is concerned with psychological aspects of ageing, but you need to be aware of the complex relationship between psychological, biological and chronological ageing. It is commonly found that these three indices do not always march in step. For example, one individual of 75 may be physiologically less deteriorated and psychologically less impaired than another individual of 65.

These observations reflect a fundamental problem in the study of ageing, namely the difficulty of distinguishing how far biological and psychological changes result purely from the ageing process and how far they are due to other factors such as ill health or life-style. Figure 2.2 illustrates how the incidence of chronic illnesses increases with age.

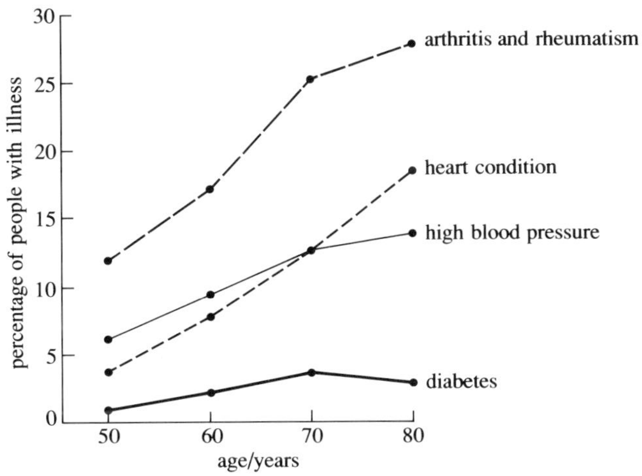

Figure 2.2 Incidence of chronic illnesses in humans from age 50 onwards (in the USA).

Ideally, the researcher aims to distinguish between the effects of 'normal' ageing and the effects of pathological ageing. In practice, this is difficult to achieve. Research on ageing has tended to emphasize the differences between younger and older age groups and has been too ready to conclude that the observed psychological and physiological deficits are age-determined. In the attempt to study normal ageing, researchers have compared groups of different ages and have carefully excluded individuals with identifiable diseases from their samples. However, even when there is no evidence of any disease, there are age-related deficits in hearing, vision, renal function, systolic blood pressure, bone density, the immune system, heart and respiratory function and sympathetic nervous system activity, as well as changes in cognition and behaviour. Some of these deficits are shown in Figure 2.3 (*overleaf*) and the physiological changes associated with ageing are described in more detail in Chapter 3.

These changes mean that it is impossible to study psychological aspects of ageing without a consideration of biological ageing. However, it is very important to remember that there is great variation in the effects of ageing. J. W. Rowe and R. L. Kahn pointed out in 1987 that these findings mask substantial individual differences within the older groups. While group averages may show age-related deficits, there are some older individuals who show no decline at all. Moreover, cross-cultural comparisons have shown that physiological deficits which appear to result from ageing may be due to social and cultural factors. The effects of ageing are less marked in rural agricultural societies, and may therefore be linked to factors such as nutrition, exercise and stress rather than to ageing itself. Within the category of normal ageing, factors such as a healthy diet and lifestyle can reduce the effects of ageing. Similarly, deficits in cognition are much less marked in groups with a high level of education. It has also been established that both biological and psychological decline can be precipitated by adverse changes in social circumstances such as bereavement.

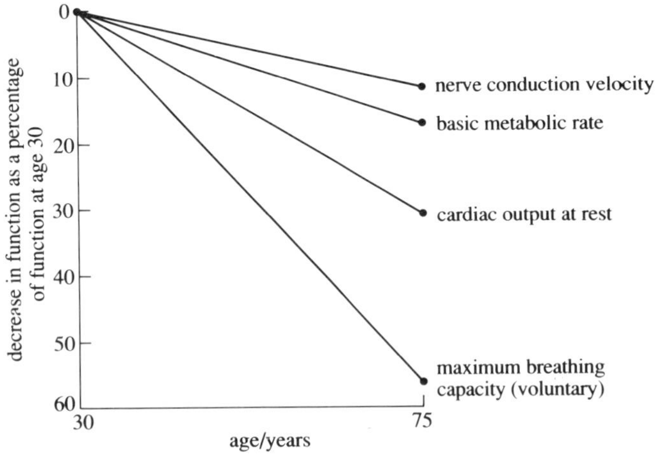

Figure 2.3 Relative loss by age 75 of various biological functions in humans expressed with respect to normal functioning at age 30 (data from a study in the USA).

So far, attention has been drawn to differences between individuals in the effects of ageing, and to the difficulty of separating ageing from the other factors which interact with it. A further problem in the measurement of biological and psychological age arises because different systems within an individual can age independently. The occurrence of greying hair or deafness or absentmindedness does not necessarily correlate with loss of joint mobility or respiratory function. It has proved difficult to identify particular 'biomarkers' that can be used to estimate biological age. Moreover, the rate of ageing within an individual is not constant through the lifespan but accelerates or decelerates at different periods.

This chapter focuses on how ageing affects human mental function. Section 2.2 begins by reviewing the effects of ageing on performance in everyday life situations including work, skills and leisure activities. Many of these observations are familiar from people's own experience. However, the following section outlines methodological issues in the study of ageing and some of the pitfalls and difficulties. The rest of the chapter is concerned with ageing and cognition and describes evidence which shows how different cognitive operations such as perception, memory, and language are affected by ageing. Finally, some of the theoretical explanations of cognitive ageing are considered.

Summary of Section 2.1

Human life expectancy has increased so that the proportion of elderly people in the population of the developed world has increased. Maximum lifespan has not changed. The process of ageing is difficult to separate from accompanying changes in health status. The effects of ageing vary greatly from one individual to another and from one society to another. Both genetic and environmental factors are involved in the process of ageing.

2.2 Behavioural aspects of ageing

This section is concerned with behavioural aspects of ageing in humans, although non-human animals also exhibit age-related changes in social and sexual behaviour. Changes in human behaviour occur with increasing age partly as a result of accompanying changes in physical and cognitive abilities, and partly as a result of conventions imposed by the particular culture. This is true of all the categories of behaviour examined here: work, skills, sports and hobbies, everyday activities and social interactions. A practical way to assess the effects of ageing is to look at changes in the efficiency of performance, as they provide a useful indication of how far performance is affected.

2.2.1 Work

One way to study the effects of ageing on work is to analyse the rate of productivity across the lifespan. H. C. Lehman (1953) calculated for each decade of the lifespan the average number of contributions made in a variety of professions and expressed these as a percentage of the maximum for all ages. As Figure 2.4 shows, the data were remarkably consistent. For practical inventions, literature, music and art, the most productive decade was 30–40 years of age. Although some individuals continue to be active and productive into old age, the rate of productivity declines. Lehman identified reduced physical energy, illness and lowered motivation as factors influencing productivity; alternatively or additionally, there is an age-related reduction in mental ability.

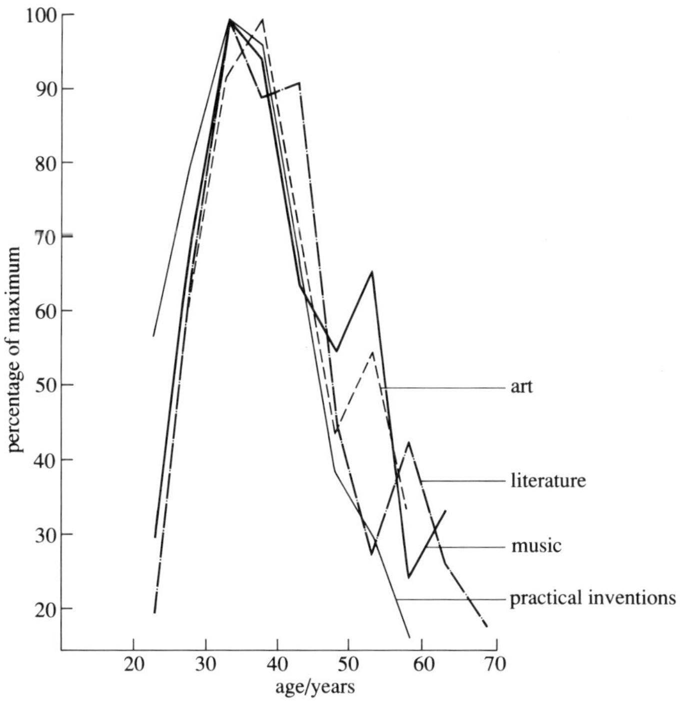

Figure 2.4 Creative contributions at different ages expressed as a percentage of the maximum number of contributions across all ages.

□ Can you think of any other factors which might influence the pattern of productivity across the lifespan?

■ Patterns of employment and career structures: in many professions the pattern is for people to move from creative work into administration as they get older.

When industrial performance in manual jobs is considered some indices show no age differences. Workers in their 50s and 60s are as productive in many jobs as younger workers. Older workers are relatively good at jobs requiring sustained attention. Nevertheless, age deficits are apparent in jobs where performance depends on speed and there is time pressure, and older workers are also slower to learn new tasks. One problem with such comparisons, however, is that older workers transfer out of jobs which they begin to find too demanding. The age distribution of workers in a particular job reflects the effects of ageing on ability to perform that job. Athletes and air traffic controllers, for example, withdraw from their jobs at a relatively early age. In other jobs, such as salespeople and managers, where experience and social skills count for more than speed, older people have been found to be more effective, in some respects, than younger ones. Even so, in general, a graph of performance in skilled and semi-skilled manual jobs shows an inverted U-shape, improving up to the mid 30s or early 40s and then declining gradually.

2.2.2 Skills, sports and hobbies

Many everyday hobbies and activities do not test the limits of physical or mental capacity. Gardening, cooking, attending meetings, writing letters, and so on, can all be carried out competently without imposing excessive demands on memory, speed of response or co-ordination between perceptual and motor systems, so age effects are not very obvious in these activities. However, driving a vehicle is an activity which does make severe demands and does show marked effects of ageing. Statistics indicate that the accident rate per number of miles driven begins to increase for drivers between 45 and 60, and older drivers have even higher accident rates than very young novice drivers. This increase occurs in spite of the fact that older drivers tend to avoid driving in heavy traffic or at night when the chances of an accident are greater.

□ What do you think the causes of car accidents involving young drivers are likely to be, compared with those involving elderly drivers?

■ Young drivers' accidents are typically due to excessive speed, inexperience, drink or fatigue; older drivers' accidents are generally caused by failure to process information, ignoring traffic signs, failing to give way, and improper turns and lane changes.

Typing is another skill which, like driving, depends on perceptuomotor co-ordination (i.e. co-ordination between the eyes and the movement of the limbs). Here, however, a comparison of highly skilled young and older typists showed that the older ones were able to maintain a high speed of copy-typing in spite of a reduction in perceptuomotor speed. They compensated for slower perceptual and motor processes by using a strategy of looking further ahead and so saved time by processing the text in larger 'chunks'. This is a good example of the way that the effects of age can sometimes be offset by compensatory strategies.

In sports and athletics a distinction can be made between habitual performance and maximal performance. Habitual performance is measured in terms of amount of energy expended and it has been estimated that this declines at a rate of 0.5% per year. Maximal performance is measured in terms of speed, strength and stamina and shows a steeper decline. In a study of the effects of age on strength, it was shown that by the age of 70, handgrip strength had declined from its peak value (reached in the early 20s) by 30% in males and by 27% in females, an effect which was attributed primarily to a decrease in muscle mass. Other studies have confirmed the popular view that habitual vigorous exercise offsets the effects of ageing on cognitive abilities such as memory and problem solving, and also reduces anxiety and depression.

Hobbies that are cognitively demanding include playing bridge and chess. Age effects in chess playing have been studied by asking players of different ages and different levels of skill to choose the best move in each of four chess positions. Less skilled players chose poorer moves, but there were no age differences in the value of the chosen moves or in the nature of the search processes preceding the decision. Similarly, older bridge players were able to select an opening bid as accurately as younger players, although they took longer to make the decision. These results indicate that in highly practised activities, like bridge and chess, experience can offset any adverse effects of ageing so that, although performance may be slower, it is not poorer.

2.2.3 Everyday activities

Behavioural effects of ageing are often caused by memory deficits. (Memory mechanisms were described in Book 2, Chapter 11, and Book 4, Chapter 5. Chapter 3 of this book describes the physiological changes in the ageing brain which underlie memory deficits.) A popular myth is that short-term memory (working memory) deteriorates but long-term memory is spared. More careful investigation reveals that this is not strictly accurate. Studies testing people's memory for recent and remote public events have shown a marked decline in older people's ability to recall remote events from long-term memory. And, although it is often anecdotally claimed that older people have excellent recall of events of their own early life, these memories tend to be highly selective ones which have been frequently rehearsed and recounted. It seems likely, therefore that long-term memory is also affected by ageing. Further evidence comes from the finding that elderly people also begin to experience increasing difficulty in retrieving names from long-term memory. Names are not forgotten but are temporarily and frustratingly unavailable when required, although they are usually remembered later. This aspect of memory decline begins relatively early and may be noticeable in the late 50s or early 60s.

Age-related changes also occur in working memory. Working memory stores the current contents of consciousness in that it holds the information that is currently being 'worked on'. An age-related reduction in the capacity of this memory typically begins to be evident in the late 60s or early 70s. This affects some aspects of everyday life such as the ability to remember telephone numbers while dialling, or to carry out mental arithmetic. It can also affect an ability known as prospective memory, that is, memory for plans and actions. Prospective memory is needed to monitor and keep track of ongoing sequences of actions and to carry out planned actions. Absent-minded mistakes such as omission errors (forgetting to put the tea in the pot), repetition errors (putting the tea in the pot twice over), and sequencing errors (putting the water in before the tea), all stem from defective prospective memory. Nevertheless, elderly people can often minimize absent-minded errors by developing effective compensatory strategies, and making increased use of external aids such as diaries, shopping lists and reminders to prop up their memories. They can also opt to concentrate on remembering what is important and not bother about what is unimportant.

The reduction in working memory capacity may affect the elderly person's ability to follow conversation, although deficits in attention and hearing loss are also implicated here. Problems are particularly evident when conversation involves several participants with rapid changes of speakers and topics. The elderly listener may be unable to encode the information sufficiently rapidly and is liable to lose track of who said what. As a reaction to their difficulties some elderly people cease trying to participate whereas others try to control the conversation by long and often inappropriate monologues.

□ Reviewing what you have read so far, list some reasons why ageing does not necessarily involve deterioration of performance.

■ Not all individuals are adversely affected and not all types of performance are affected. People develop compensatory strategies. Deterioration can be offset or cancelled out by practice or training.

Paul Baltes of the Max Planck Institute in Berlin claims that lifespan development is multi-directional. That is, although some abilities decline with age, others continue to develop and improve. He and his colleagues carried out experiments designed to investigate age differences in 'wisdom' — the kind of practical intelligence and good judgement that is based on life experience. Subjects of different ages were asked to suggest solutions to everyday problems of life planning (including, for example, problems about relationships, career choices, personal finance, etc.). Judges who did not know the ages of the subjects rated their proposed solutions. The older subjects were judged to show greater wisdom than the younger ones.

□ Does this finding shed any light on the effects of age on performance in different jobs?

■ Yes, it explains why older people are often better at managerial jobs which require this kind of 'wisdom'.

Baltes claims that lifespan development shows some plasticity, as well as multi-directionality, so that deficits can sometimes be reversed by training. There is some support for this view. Research on cognitive training with the elderly has shown that old/young differences in the ability to memorize lists of words or lists of digits can be substantially reduced by training. The performance of the elderly improved when they were trained to group items into categories. In another programme designed to improve the memory skills required in everyday life, elderly people were taught to use a variety of techniques. They succeeded in improving performance in keeping appointments by the use of written reminders and check lists. Memory for routes was improved by using imagery (mental images) and by techniques such as looking back to memorize the return route to the parked car from the returning viewpoint. Remembering the spatial locations of objects improved if people put objects in places related to their functions. Age differences are not eliminated by these techniques but they are considerably reduced.

2.2.4 Social behaviour

The effects of ageing on social behaviour have been characterized as a process of progressive *disengagement*, resulting in decreased interaction with other people and loss of roles and status. Elderly people retire from work and also gradually withdraw from groups and organizations. This withdrawal is seen as a response to declining energy, but it is also true to say that it is partly imposed by the conventions and stereotypes of developed culture. Elderly people find that their circle of friends and relations diminishes as those of their own age group die, and decreased mobility reduces the possibility of making new friends. Disengagement, however, is considered to be psychologically damaging and the maintenance of social activity is an important element in continued well-being. Poor health, social isolation and cognitive deterioration are all implicated in the psychopathology of ageing. These factors contribute to an age-related increase in the incidence of depression which is epidemic in the older population in developed societies and which is often accompanied by sleep disturbance and by anxiety. The typical form of these disorders is not the same in old age as in youth. Depression in the elderly is characterized by apathy, fatigue and withdrawal, but the feelings of guilt and worthlessness that accompany depression in younger patients are less likely to be present. Anxiety in the elderly rarely takes the form of phobias (such as fear of going out) or obsessive neuroses (such as compulsive washing or tidying). It is more likely to consist of generalized agitation and tension. It is important to note that many of these changes in social behaviour and psychopathology are not necessarily the results of ageing, but are at least partly caused by the cultural conventions which dictate the role and activities of the aged.

□ Suggest some ways whereby, in some societies, cultural factors might act to prevent changes in social behaviour in old age.

■ In societies where elderly people continue to live in extended families, and have a social role and respected status, ageing is not associated with social isolation and disengagement.

Summary of Section 2.2

An age-related decline in the speed of response and in the ability to attend to several things at the same time affects performance in some jobs and in activities like driving. In other jobs, and in activities like bridge or chess, increased wisdom and experience boost the performance of elderly people. Both long-term memory and short-term, working memory tend to decline in old age but, in everyday life, this can be partly offset by compensatory strategies. Age-related changes in social behaviour appear to be largely culturally induced.

2.3 Methodological issues in the study of psychological ageing

Research methods vary with the precise objectives of the researcher. A major distinction lies between descriptive research and explanatory research. Descriptive research aims to *identify* age differences but not to *explain* them. So, for example, descriptive studies may show whether older people are shorter in height, more conservative, better spellers or consume less alcohol but it does not explain why these differences arise. In order to identify age differences, all that is required is that comparisons are made between groups of people of different ages at the same point in time. This is known as the **cross-sectional method** because it compares people from different sections of the population at one time-point. The sample of people who compose each group should be representative of the population in that age band (in terms, for example of gender and class). The characteristic in question (height, conservatism, spelling or alcohol consumption) must be accurately measured and statistical analyses must be performed on the data to find out if there are significant differences between the groups. If the age samples are truly representative, then the mean scores for each group will provide *norms* for that age band. It might be found out, for example, how much alcohol is normally consumed by people in their 20s and people in their 80s.

This kind of research may be quite adequate for the purposes of market research, but it is not explanatory (explanatory research will be discussed in Section 2.3.1). It does not tell researchers what causes the group differences. The differences between the age groups may not be caused by ageing at all. Older people may be shorter because their nutrition was poorer in their youth; they may be better spellers because spelling was formally taught when they were young. These kinds of differences are more properly described as **cohort** differences rather than age differences. Here a cohort refers to a generation, that is, all the people born during a particular period of time. Different age groups come, necessarily, from different generations or cohorts and this means that their life experiences are different. In the case of characteristics which may be influenced by life experience, group differences may be reflecting cohort effects as well as, or instead of, age effects.

□ Suppose it is found that people in their 70s consume less alcohol than those in their 20s and 30s. Could one assume that this was an age effect or might it be partly due to cohort effects?

■ There could be cohort effects here. Those now in their 70s might have formed their alcohol consumption habits at a time when people were more abstemious, less affluent or more influenced by temperance movements.

By contrast with cross-sectional comparisons, the **longitudinal method** involves testing and re-testing the same subjects at different ages over a long time span. The comparisons are therefore free from the effects of factors that are not age-related such as cohort effects. However, there are still problems with longitudinal research.

□ Can you foresee what the problems might be?

- Many subjects drop out over the years and the ones who continue in the study may be those who are least affected by ageing. Another problem with successive re-testing is that the subjects become more and more practised at the tests, so the beneficial effects of practice may mask the detrimental effects of age.

Finally, there are practical and financial difficulties in the way of maintaining a research programme over many years. The longitudinal method is more feasible for the study of ageing in non-human species which have a shorter lifespan.

These considerations about cross-sectional and longitudinal methods are concerned with *who* to compare: different considerations arise in relation to *how* they should be compared. There are three main ways to make age comparisons. These are:

1 the experimental approach;

2 the everyday, naturalistic, approach;

3 the psychometric approach.

2.3.1 The experimental approach

This approach is identified with explanatory research. By contrast with descriptive research where the researcher simply wants to find out what the age-related differences are, in explanatory research the researcher aims to identify what is *causing* the difference between the groups. If the aim is to find out whether the differences are caused by age, the cross-sectional method can be used in an experiment. The different age groups must be carefully matched in terms of any other attributes, apart from age, that might influence their performance. Take, for example, a cross-sectional experiment designed to discover whether there is an age deficit in memory for lists of words. Such an experiment might compare the performance of a young group, aged between 20 and 30 years with an older group aged between 60 and 70 years. The two age groups would need to be matched for verbal intelligence, level of education, health, and possibly also for socio-economic background, ethnic origin and for the gender composition of the groups. By equating the two groups for all these factors the researcher can be reasonably sure that any group differences that emerge are due to age. However, in an experiment of this kind the researcher would also be wise to select the to-be-remembered words carefully. The inclusion of any very modern words or any old-fashioned words might disadvantage one age group. Difficulty in remembering these unfamiliar words would then be a cohort effect rather than an age effect.

The discovery that a group difference is due to age is not always sufficiently informative. Researchers want to know what physiological or cognitive mechanism or process has been impaired by age and so has produced the observed deficit. To move to this level of explanation it is necessary to introduce additional variables into an experiment. Returning to the example of the two age groups being tested for ability to remember lists of words, the researcher might decide to vary the speed of presentation so that one list is presented at a rate of one word every 5 seconds and one list is presented at rate of one word every 10 seconds. Figure 2.5 shows typical data for this kind of experiment.

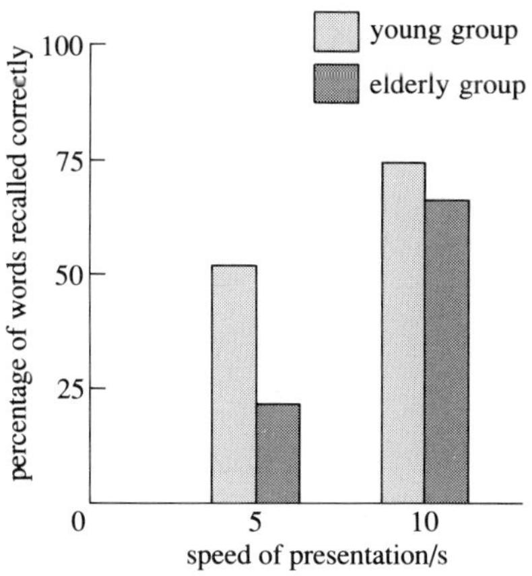

Figure 2.5 Typical data from a memory experiment testing recall of words by young (20–30 years) and elderly (60–70 years) groups, and varying the speed at which the words are presented.

- ☐ What can the researcher conclude from this result?

■ The results show an *interaction* of age and speed of presentation. This means that the older group is much poorer than the young group at the fast rate, but there is little effect of age at the slow rate. Therefore, the age difference in percentage of words recalled is likely to be due to an age-related reduction in the speed of processing.

By manipulating additional variables in this way and observing how they interact with age, it is possible to infer what mechanisms or what component processes have been impaired by ageing.

These principles apply mainly to the experimental approach to the study of psychological ageing, but the experimental method is not always appropriate.

□ Can you suggest some of the advantages and disadvantages of formal laboratory experiments as compared with studying age differences in natural everyday settings?

■ The experimenter can control all the relevant variables rigorously and can measure performance precisely, but the situation and the tasks are unfamiliar and artificial. Elderly people tend to perform much less well in laboratory experiments than they do in everyday life, and age effects that are apparent in the laboratory may be non-existent in real life situations.

2.3.2 The everyday approach

The everyday approach employs a variety of methods including questionnaires and diaries, observation, and informal naturalistic experiments designed to mimic real life conditions.

Questionnaires ask people to rate their own performance. For example, they may be asked to rate how often they forget appointments, or forget shopping, on a 5 point scale ranging from 1 (very rarely) to 5 (very often). Alternatively, they might be asked to rate their own ability as drivers or as bridge-players. Diaries are used to record behaviour over considerable periods of time, with people being asked to record, for example, the frequency and types of memory lapses or of daydreams. There are some obvious objections to be made to these methods.

□ What objections can you think of?

■ These methods rely entirely on introspections or subjective self-assessments which may be inaccurate. Self-ratings are liable to be affected by the memory paradox. That is, subjects with poorer memories forget their errors and so under-report their frequency, so the worse the memory, the better the self-rating. Also, if cross-sectional comparisons are used to compare age groups there are many uncontrolled factors that may contaminate any age effects.

Consider, for example, the self-rated frequency of failures to remember shopping. Older people may make more use of written lists. They may make more effort to remember because failures cause more problems and entail more effort. They may have a less busy life-style so that they have fewer competing demands on memory. They may have organized their shopping so that they only need to follow a familiar and practised routine. All these factors may be enhancing or obscuring the effects of age on performance in real life situations.

Observations of performance in real life situations are more objective when made by the researcher than when they are compiled from self-reports. However, such observations still only yield data which is descriptive rather than explanatory. Everyday life tasks rarely test performance at the limits of capacity, so observed behaviour reflects habits rather than capacities. An elderly person who is in the habit of mislaying objects like keys or a wallet is not necessarily incapable of remembering spatial locations.

Naturalistic experiments offer more possibilities for control of relevant variables and for objective measurement. An example of this method is an experiment in which age differences in prospective memory ability were tested by asking subjects of different ages to telephone the experimenter at designated times. Thus the task is a familiar real life one, but performance can be objectively measured in terms of the accuracy of the timing of the call.

2.3.3 The psychometric approach

The psychometric approach is concerned with the exact quantitative measurement of individual differences in intellectual abilities and personality traits. Instead of doing experiments, the researcher administers standardized tests which yield precise numerical scores. These scores can then be compared with the scores of other individuals or other groups. Psychometric tests are used in this way to compare the scores of groups from different age ranges. One of the main problems here is that, just as it has been claimed that intelligence test material is not 'culture-fair', it is also difficult to ensure that it is 'generation-fair'. The life experience, and particularly the education, of some cultures or races may be such that they are relatively ill-prepared for the test material and, in the same way, older generations may also be ill-prepared for the tests.

The psychometric approach reveals strikingly the extent to which individuals differ from each other. It is particularly evident that the range of individual variation increases with age. It is as if individuality which is partially suppressed in youth re-asserts itself in old age. This finding underlines the fact that generalizations about ageing need to be viewed with caution. When group comparisons show significant age differences it is almost always the case that the young and older groups overlap to some extent, so that some members of the older group are performing as well or better than some members of the younger group. These individuals who are demonstrating 'successful ageing', and show no deterioration, are most likely to be highly educated, healthy and living in their own homes rather than in institutions. A low level of education, poor health and institutionalization represent a cluster of factors often associated with age-impaired performance. The psychometric approach has shown that, because of the great range of individual variation in older people, results expressed in terms of average performance may be misleading. It is important to look carefully at the distribution of scores as well as the averages.

Summary of Section 2.3

Descriptive research identifies age differences whereas explanatory research seeks the causes. Cross-sectional comparisons of different age groups at the same time, or longitudinal comparisons of the same age group at different times, can be used to reveal age effects. The effects of age are confounded with many other factors including cohort effects. Methods of study include formal laboratory experiments, the everyday approach which uses questionnaires, diaries and naturalistic experiments, and the psychometric approach which uses standardized tests. All the methods have some advantages and some disadvantages.

2.4 Age-related changes in cognition

The nature of cognition was discussed in Book 1, Chapter 8. Cognition was identified with holding information in memory over long periods of time and using it flexibly to direct behaviour. This is a very general definition covering both animal and human cognition. In human psychology, cognition is usually defined as thinking, or as the higher mental processes such as language, memory and reasoning that are involved in thinking. However, in practice, it is impossible to separate out thinking from other processes like perception and attention, which feed information to the central thought processes, and from the response systems like speech and action which express the output of the thought processes.

Many psychologists today view human cognition as information processing and divide the cognitive system into a series of stages or component subsystems. Figure 2.6 shows a very simplified version of this information processing model of cognition.

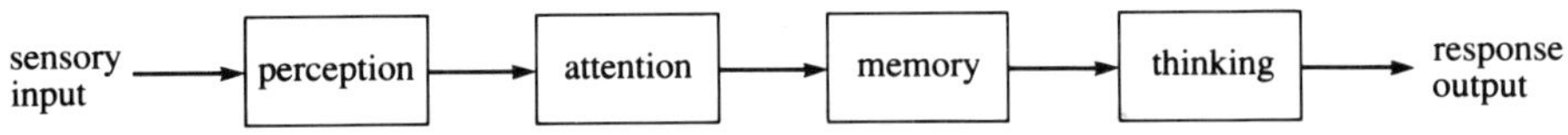

Figure 2.6 A simplified information processing model of the human cognitive system.

In this model, the perception stage is concerned with the acquisition of information from the outside world. The attention process targets and selects some of this information for further processing. The memory stage involves the storage of new incoming information and the retrieval of old information that is relevant to the current situation. The thinking stage involves manipulating this information in reasoning and problem solving. Finally, the response output may take the form of language or of actions.

Numerous flaws in this model have become apparent. In particular, it is now realized that information processing is not a strictly linear one-way series of stages. Most of the processes shown in Figure 2.6 influence each other in an interactive manner. For example, information stored in memory influences the way new inputs are perceived as well as *vice versa*. It is also realized that information processing cannot be considered in isolation from other aspects of human behaviour such as emotions, desires, intentions, beliefs and cultural and evolutionary factors. In 1981, Donald Norman in California incorporated some of these factors into a more complex model of human cognition (Figure 2.7).

The details and terminology of this model need not concern you. The important point to grasp is that the system includes a large number of components which interact with each other. Most psychologists employ the componential (stage by

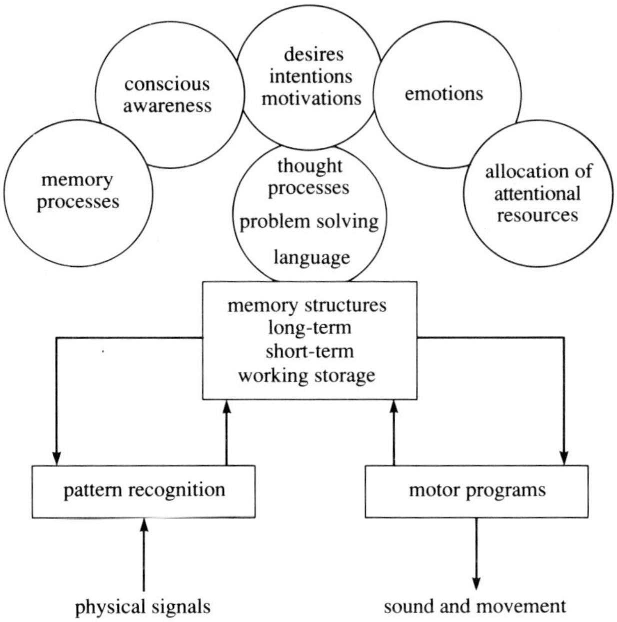

Figure 2.7 A modified version of Norman's model of the human information processing system.

stage) approach which makes experimental analysis of the cognitive system simpler and more feasible than trying to study the system as a whole. This is particularly true when the aim of the investigation is to identify the particular part or parts of the system that are affected by ageing, by injury or by disease. Experimenters work rather like plumbers, testing different parts of the system separately to isolate the faulty stage. And, rather like plumbers or motor mechanics, ageing research usually addresses questions such as: what parts of the system are affected by ageing; in what way are they affected and how does this affect performance?

However, some psychologists have looked for a general factor, like intelligence or speed of processing which might affect all parts of the cognitive system. The sections below outline the age effects that have been found in each component part of the cognitive system (sensory perception, attention, learning and memory, language and thinking) and also examine evidence that general factors are involved.

2.4.1 Age-related changes in sensory perception

Clearly, any deficits in sensory perception have serious consequences since the functioning of the cognitive system is critically dependent on the quality of the information it receives. Age-related changes in sensory perception can be studied at two different levels. The first concerns the efficiency of the sense organs in registering incoming information. The second concerns the higher-level stages of perception which involve the integration and interpretation of stimulus information and require higher-order processes of identification and judgement.

The efficiency of the sense organs is traditionally assessed by psychophysical methods aimed at determining thresholds. Psychophysics is so called because it is the science of measuring psychological responses to physical events. You have already come across the concept of a threshold in Book 2, Section 2.3.3. An *absolute threshold* can be defined as the weakest intensity at which the presence of a particular stimulus can be detected in half of the trials. As the efficiency of sensory perception deteriorates the threshold rises.

□ How does a high auditory threshold affect perception?

■ A sound must be louder before it can be heard.

The *difference threshold*, sometimes known as the JND, or just-noticeable-difference, is the smallest difference between two stimuli which can be detected on half the trials. So, for example, as auditory acuity declines it becomes harder to detect a difference in loudness between two sounds—the JND becomes larger.

Visual sensitivity declines steadily from the age of 40. The absolute threshold for the detection of light declines dramatically. Age differences in visual sensitivity have been examined for dark-adapted vision. When one moves from a light area into a dark area, for example when entering a cinema, vision is very poor. After a while the eyes become dark-adapted and sensitivity increases so that vision improves. The maximum sensitivity of the dark-adapted eye is 200 times greater in adults aged 20 years than in adults aged 80 years. The rate of dark adaptation is also slower in elderly people.

□ How would this affect night driving?

■ Elderly people have poor vision immediately after sudden changes from lit to unlit areas or when the lights of oncoming traffic are passed.

From middle age onwards the loss of elasticity of the lens reduces the ability to change the focus of the eye for near objects (*presbyopia*) and there is also a shrinkage in the effective field of vision. A reduction in the size of the pupil and increased opacity of the fluid within the eye result in less light reaching the retina. These changes occur in the course of normal ageing, but there is also increased likelihood among elderly people of pathological conditions such as glaucoma (a build-up of pressure within the eye which causes damage), cataract (an increased opacity of the lens which reduces the amount of light reaching the retina) and retinal degeneration (deterioration of the light-sensitive cells of the retina).

In hearing there is a progressive decline in sensitivity to high frequency sounds known as *presbycusis*. Figure 2.8 shows the steep decline in hearing for sounds with frequencies higher than 1 000 Hz. This is why older people cannot hear very high frequency sounds like the cries of bats while many children and young people can.

Although sensitivity to low frequency sounds is fairly well preserved, the high frequency loss distorts perception of music and spoken words. A similar decline is evident in the difference thresholds for discrimination between sounds of different pitch. There is a substantial increase in the JND in pitch between two high frequency tones.

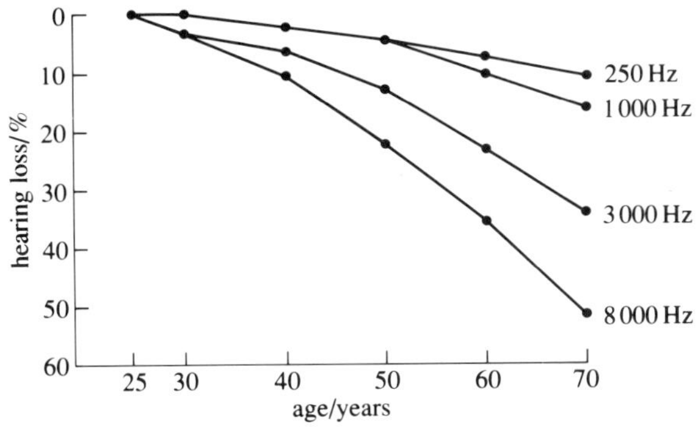

Figure 2.8 Percentage decrease in auditory sensitivity (measured as the absolute threshold) for sounds of varying frequencies with increasing age.

At the higher level of perceptual judgements (as outlined in the first paragraph of this section) age differences are well documented. Accuracy in colour identification is reduced by about 25% between the ages of 20 and 70, and several aspects of spatial perception are impaired. These include depth perception and judging the position of a moving target. Deficits in both spatial and temporal judgements can be explained in terms of a phenomenon known as *stimulus persistence*. It is claimed that in old age there is an increase in the length of time for which neural activity persists after the termination of the external stimulus. In consequence, two stimuli occurring in rapid succession are perceptually fused; they are not perceived as separate events. For example, a light repeatedly turned on and off will, if this is done quickly enough, appear to be on all the time. A measure known as the critical flicker fusion frequency identifies the precise rate of alternation between a light going on and off, at which the on/off flicker fuses and is seen as a persisting light. For older people, fusion occurs at a slower rate because, perceptually, the light is 'on' for longer. This increase in stimulus persistence can also affect processes such as speech perception if separate sounds are perceptually fused. In addition, there is evidence of a general decline in the rate of processing stimulus information so that elderly people take longer on average to identify visual and auditory stimuli. Note, however, that these data are based on group averages, and there is actually considerable variation from one individual to another.

2.4.2 Age-related changes in attention

Attention is the mechanism that governs the amount and type of information reaching conscious awareness. The cognitive system has a limited capacity and cannot process all the information that is present in the environment. A mechanism is required, therefore, to select the most important or relevant source of stimulus information and to allocate processing capacity to it. This mechanism, commonly known as attention, takes three different forms:

vigilance (or sustained attention)

selective (or focused attention)

divided attention.

Vigilance refers to the ability to maintain attention and detect any changes. Book 1, Chapter 10 discusses the kind of vigilance animals employ to detect the approach of a predator. During experiments on human vigilance, people are commonly asked to detect a change in a long sequence of stimuli as, for example, in monitoring an assembly line for defective items or a radar screen for warning blips. In this kind of task, age differences are evident if the rate of presentation is fast, if the task is prolonged to the point of fatigue, or if it involves a memory component in keeping track of preceding states.

Selective attention involves focusing on a selected input and screening out other irrelevant inputs (Book 1, Chapter 2). The most familiar example in everyday life is listening to one speaker in a crowded room and ignoring other speakers. There is evidence from numerous experimental situations that elderly people are more easily distracted by irrelevant stimuli and that this impairs their performance. However, age changes are most marked in divided attention tasks where the subject needs to process more than one source of stimulus information, like the air traffic controllers mentioned in Section 2.2.1 who must monitor many incoming messages simultaneously. An experimental method often used to assess ability to divide attention is called dichotic listening. A series of items (such as letters, digits or words) are presented over earphones in such a way that each ear receives a different item at exactly the same time, as shown in Figure 2.9.

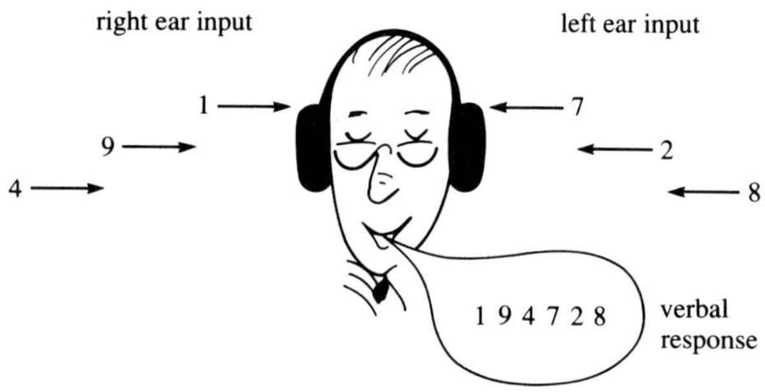

Figure 2.9 A dichotic listening experiment. See text for details.

Subjects asked to report the items from both ears usually recall all the items from one ear (1—9—4) followed by all the items from the other ear (7—2—8). Age differences are most marked in recall from the second ear. Since the input to this ear has to be held in memory while the first ear input is being reported, it seems likely that the age deficit is largely due to a memory deficit rather than to inability to divide attention. This method has yielded another interesting finding. Subjects normally exhibit a 'right ear advantage', that is, superior recall of inputs to the right ear, reflecting the fact that right ear inputs are transmitted directly to the left hemisphere which, in right-handers, is usually specialized for language processing. (See Book 2, Chapter 11 and Chapter 6 of this book for more details of this organization.) Left ear inputs are transmitted initially to the right hemisphere and are recalled less well. The age deficit in dichotic listening is particularly marked for left ear inputs.

□ It has been suggested that, in old age, neural degeneration is greater in the right hemisphere. Suggest what effect this will have on dichotic listening.

■ Older people will have relatively greater difficulty with left ear inputs than will younger people.

Ability to divide attention is also tested in dual tasks. For example, how well can people do reasoning problems while simultaneously holding in memory a list of digits? One study used reasoning problems of the form

'Is A preceded by B?'—AB—Yes/No

'Is B followed by A?'—AB—Yes/No

Older people responded more slowly to the problems, recalled fewer digits, and showed a larger decrement when performance on the dual task was compared with performance on the reasoning task alone. These results appear to show that elderly people have difficulty in dividing attention between two tasks. In everyday life, dual tasks can often be combined successfully when they are highly practised because some components of the tasks become automatic and do not require the conscious monitoring that uses up attentional resources. Age deficits in divided attention show up when conscious attention must be paid to several things at once.

The mechanism underlying age differences in divided attention is not clear. One widely accepted explanation is that, with increasing age, there is a diminution in the limited capacity or amount of attentional processing resources available in the cognitive system. A second possibility is that the observed effects are due to a reduction in the rate of processing with the result that each component stage takes longer to complete. When multiple inputs have to be processed, or multiple tasks performed simultaneously, the effects of slowing accumulate and spoil performance. Both these factors, reduced attentional resources and slower processing, are general factors which would have widespread effects at all stages of cognitive processing and in many different tasks.

2.4.3 Age-related changes in learning and memory

Learning and memory are so closely related that it is difficult to distinguish between them. Roughly speaking, memory involves the retention of information, whereas learning, as defined in Book 1, Chapter 6, is the process by which memory is acquired. Both learning and memory are affected by ageing. Age differences have been found in both classical and operant conditioning (Book 1, Section 6.3). In a standard classical conditioning experiment, a puff of air to the eye is the unconditional stimulus, the eye-blink being the unconditional response. A change in the brightness of a light then serves as the conditional stimulus.

□ What does the graph in Figure 2.10 (*overleaf*) show about the rate of conditioning in elderly adults as compared with young adults?

■ Elderly people are conditioned more slowly than young adults. The older adults show very little conditioning even after 8 blocks of trials.

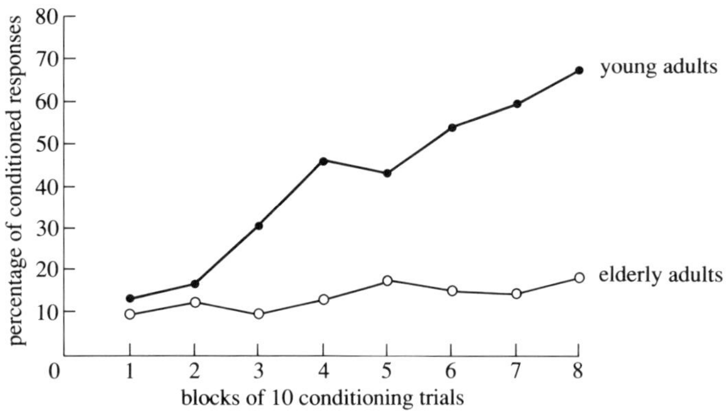

Figure 2.10 Age differences in the rate of classical conditioning (eye-blink response) for young adults (in their 20s) and elderly adults (in their 70s).

In elderly subjects the conditional response has also been found to extinguish more slowly. In operant conditioning there is less evidence of any age-related changes. Much of the research in this area has been carried out on rats rather than humans. Negative reinforcement (also called avoidance learning) is a type of operant conditioning in which, for example, the organism learns to respond to a warning signal so as to avoid a harmful stimulus (Book 1, Section 6.3.4). There is evidence that avoidance learning is poorer in old rats as compared with young rats, especially in complex avoidance tasks, but whether such findings can be generalized to human learners is questionable.

Studies of skill learning have been of two contrasting kinds. In the laboratory, comparisons of elderly and young subjects learning unfamiliar and rather artificial tasks, like tracing patterns which are mirror-reversed, have generally shown age deficits. In contrast, real life studies have produced different results. A group of elderly people aged 63 to 91 were given one German lesson a week for 3 months. At the end of this period over half the group passed an examination at a standard which school-children required three years to attain. In the Netherlands in 1990, Janet Jackson compared the progress of young and elderly students learning to play the violin. The elderly did as well or better than the young. In both these studies, however, there was no control over motivation and the amount of practice. The elderly were especially keen students and spent much more time on their studies. The same is often true of older Open University students: there are no differences in grades achieved when over 60s are compared with under 60s, but the older group spend more time on assignments and course work.

□ What can be concluded from the contrasting findings of laboratory and real life studies?

■ Age differences in learning occur when amount of practice is equated, but older people are able to learn and acquire new skills if they can devote extra time and effort to the task.

Older people often complain that their memory has deteriorated, but believe that this effect is not across-the-board. Some aspects of memory seem to be impaired while other aspects continue to function reasonably well. These introspections are consistent with research findings. Researchers have sought to distinguish between different kinds of memory and to identify the precise location of age-related deficits. The capacity of working memory is reflected in the span of immediate memory—the longest string of digits, letters or words that can be reproduced in the correct serial order immediately after presentation. When the items are numbers this is known as *forward digit span*. It is usually found that this span declines slightly from around 6.7 items for people in their 20s to about 5.5 in the late 60s or early 70s.

Instead of being simply a storage space, working memory is thought of as a limited capacity work-space or 'desktop', corresponding to the current contents of consciousness, and used for both temporary storage of information and also for re-organizing, rehearsing and manipulating that information. Whereas experiments such as those measuring forward digit span showed only slight age differences in storage capacity, substantial age effects are evident when information has to be re-organized rather than just stored. For example, *backward digit span*, when the digits have to be reported in the reverse order, declines markedly with age. Similarly, a group of researchers in the USA recently found no difference between young and elderly subjects on simple forward digit span, but they then gave their subjects a 'loaded word span' test. In this test, subjects had to listen to a set of spoken sentences, decide if each one was true or false, and at the same time remember the last word of each sentence. At the end of the set they were asked to recall all the last words. In this task there was a large age decrement. Young subjects recalled on average the last words of 4 sentences whereas elderly subjects only managed an average of 2.25.

□ Why is the loaded word span task especially difficult for elderly people?

■ Memory, attention and speed of processing are involved in the following ways. First, the limited capacity of working memory is overloaded. Second, subjects have to divide attention between understanding the sentences and remembering the last words. Third, subjects cannot control the speed of the spoken input and have to keep pace with the speaker.

Working memory is a key element in the cognitive system, contributing to many cognitive tasks such as reasoning, calculation, speaking, listening, reading and writing. Consequently, an age deficit in working memory has knock-on effects in all these tasks.

Age differences in long-term memory are also evident in many other different tasks. Much of the research has used verbal learning tasks in which memory for lists of words is tested. In these tasks, age differences in memory could stem from inadequate encoding of the material, poor retention, or difficulty in retrieving items that are stored in memory. These three processes are difficult to disentangle experimentally. It has been shown that age differences in verbal memory tend to diminish if the elderly are instructed to use more effective strategies of organizing items into groups or using imagery (as described in Section 2.2.3).

□ What does this indicate?

■ It shows that older people do not normally use these more effective strategies for encoding, but can do so if instructed. It also shows that the age differences originally obtained were partly due to strategy differences, as well as to differences in the efficiency of the memory mechanism.

Difficulties in retrieval appear to be the major source of the age deficit. Elderly people appear to be more vulnerable to interference from unwanted items so that their recall of designated material is, typically, cluttered with irrelevant items. They are also prone to *tip-of-the-tongue* states when retrieval is temporarily blocked even though the to-be-remembered item (such as a person's name), may be very familiar, and is usually recalled later. These effects of age on memory are found in formal laboratory experiments, but they are also apparent in real world situations. There has been some recent research into the accuracy of people's reporting of events which they have witnessed.

□ Would you expect to find an age difference in the accuracy of reports of eyewitnesses?

■ Yes. Perceptual and attentional deficits would tend to affect elderly witnesses' original perception of the event. Difficulties in retrieval would increase the inaccuracy of the report. Memory might be distorted by interference from irrelevant material.

Currently, memory theorists tend to divide long-term memory into two memory systems, *episodic* and *semantic* memory, each containing different kinds of information (See Book 2, Section 11.4.4, where this distinction was introduced). Episodic memory consists of personally experienced events which occurred at a specific time and place. Semantic memory stores general world knowledge (such as the facts that 2 + 2 = 4; that lemons are sour; that ice is cold), concepts and language. The two systems are not completely independent. Episodic memory is functionally dependent on semantic memory because encoding and interpretation of an experience draws on stored semantic knowledge. Memory of what you ate for breakfast today is episodic, but is based on semantic knowledge about eggs and high-fibre cereals, for example. Nevertheless, the effects of ageing reflect the partial independence of the two systems. Episodic memory tends to be impaired by normal ageing but the general knowledge and vocabulary stored in semantic memory are relatively unaffected, even though their retrieval may slow down.

□ Just to recapitulate, which aspects of memory are most impaired by ageing?

■ Working memory, episodic memory and retrieval from long-term semantic memory.

2.4.4 Age-related changes in language

Studies of language ability distinguish between spoken and written language and between production and comprehension. In a study of spoken language production, normal healthy adults aged 50 to 90 years were asked to give an oral narrative account of a significant event in their own lives. The researchers analysed a 20-minute sample of speech from the middle of the narrative. The 50- and 60-year-olds produced a greater variety of grammatical forms and used more complex constructions such as subordinate clauses (e.g. 'Because Bill left the party without his coat, John was upset'). Although the 70-, 80- and 90-year-olds used simpler constructions and made more grammatical errors, there were no age differences in fluency or vocabulary.

- □ Why does the speech of older people show a reduction in syntactic complexity?

- ■ The production of complex constructions and especially embedded clauses, places a heavy load on working memory. Reduction in the capacity of working memory enforces a reduction in grammatical complexity.

This conclusion is reinforced by the finding that there was a high positive correlation between forward digit span (which is a measure of working memory capacity), and the mean number of subordinate clauses used. Thus those with larger digit spans produced more sentences with subordinate clauses, and those with smaller spans used simpler constructions.

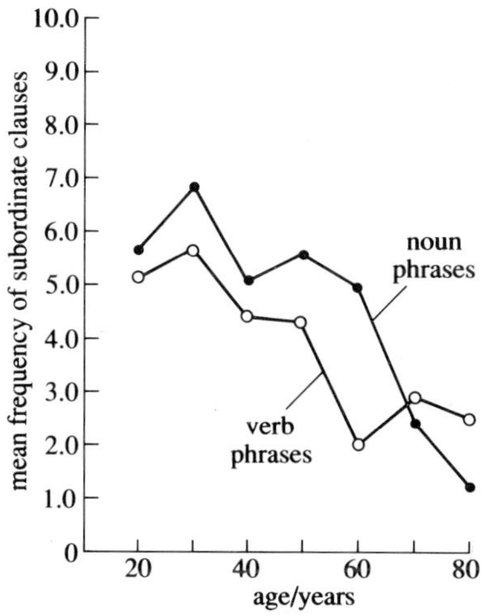

Figure 2.11 The mean frequency of subordinate clauses of two different kinds in samples of writing by writers of different ages.

The same research group also analysed written diaries kept by people of different ages over many decades. This study included a longitudinal examination of the output of the same writers over seven decades. Sentence complexity, especially use of subordinate clauses, again declined with age. As shown in Figure 2.11, the incidence of two kinds of phrase, 'noun phrases' and 'verb phrases', in subordinate clauses is reduced with age. (Note, however, that clarity of expression is not necessarily adversely affected by this change.)

In the comprehension of spoken language the performance of elderly people is affected by perceptual deficits and by slowing of the rate of processing. Presbycusis (Section 2.4.1) makes speech perception more of an effort and may force the listener to rely on context-based guesses to fill in words that cannot be identified, as in 'The child was bitten by the ...'.

Context-based guessing works well enough provided the listener hears enough of the speech to guess the rest. If they only hear 'The child was ... by the ...', then they are in trouble. They do not have enough context on which to base a guess. Experiments varying the rate of speech and the type of material show that the older listener also has problems when the speech rate is fast and the topic unfamiliar. The combined effect of limitations on speed of processing and working memory capacity is that comprehension may be superficial or fragmented. The older listener is unable to carry out the deeper levels of analysis necessary to extract fully the meanings that are implied, but not actually stated. Because older people have difficulty in simultaneously holding in mind the phrases they are currently processing and linking them to what has gone before, they may be unable to integrate different pieces of information that are not adjacent in the discourse. Finally, there may be difficulty in separating out the gist, or main ideas, from less important details.

These problems are much less evident in the comprehension of written material.

- □ Why should ageing affect the comprehension of spoken material more than written material?

■ Reading is self-paced so the reader can take all the time that is needed. Also the reader can back-track and re-read if necessary. Listeners have to keep pace with speakers and cannot always ask for speech to be repeated.

Age differences in comprehension of written material are slight or non-existent. This confirms that comprehension processes are intact, but that problems arise from slower processing and reduced memory capacity.

2.4.5 Age-related changes in thinking

Higher-level cognitive processes such as thinking, reasoning and problem solving involve both memory and language, and so are likely to reflect any deficits in these systems. Many kinds of reasoning are also influenced strongly by practice, familiarity and education, and these factors have to be taken into account when assessing age differences. Age differences in reasoning are often due to differences in the way a problem is tackled. This was apparent when middle-aged and older subjects were tested on a version of Twenty Questions. They were shown an array of 42 pictures of common objects and the task was to discover which object was the target by asking the smallest possible number of questions. The most effective strategy is to frame questions that will eliminate a whole class of objects (such as 'is it synthetic?'). Fifty eight per cent of the questions asked by the middle-aged, but only 3% of the questions asked by the elderly, were of this kind. For the most part, the older people used the poor strategy of asking questions relating to only one object at a time (e.g. 'does it tell the time?').

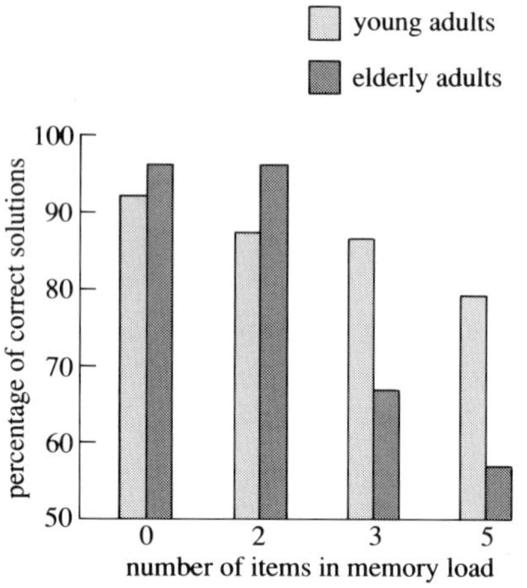

Figure 2.12 Age differences in ability to solve arithmetic problems correctly as a function of memory load in young adults (18–22 years) and elderly adults (68–70 years).

Other studies have revealed age differences in capacity rather than in strategy. Ruth Wright (1981) gave young and elderly subjects mental arithmetic problems (adding a two-digit number to a three-digit number). The memory load was varied by changing the number of digits that remained visually on display and the number that had to be held in memory. Have a look at Figure 2.12.

□ What does the histogram in Figure 2.12 show?

■ It shows that, although the elderly perform better than the young when up to two of the digits have to be held in memory, their performance declines steeply as the size of the memory load increases.

In a study of verbal reasoning, young and older subjects were given problems in the following form.

The boys measured their heights.

David was taller than Bob.

Bob was taller than James.

James was taller than Ron.

Is Bob shorter than Ron?

In this task the load on memory can be varied by changing the order of the sentences so that the information has to be mentally re-arranged. The age difference was again found to increase with the memory load. The researchers concluded that the reduced capacity of working memory of the older subjects was used up by the need to store all the data, leaving insufficient capacity for manipulating it.

❑ Do these studies show an age deficit in reasoning ability?

■ No. Age differences can be attributed to faulty strategies or to impairment of working memory rather than inability to reason.

Summary of Section 2.4

The cognitive system can be divided into several different stages or components. Several of these are affected by ageing. At the perception stage there is a loss of sensitivity in both vision and hearing. At the attention stage, the ability to focus on one input and screen out others and the ability to divide attention between different tasks are both impaired. There is a reduction in the capacity of working memory, and learning (acquiring new skills and new knowledge) is slower. These changes in the cognitive system affect processes like language production and comprehension and reasoning. Although they are detectable experimentally, it is important to note that they may not make very much difference to performance in everyday life situations. People can adapt effectively by working a little slower, doing one thing at a time, using calculators and writing down things they need to remember.

2.5 Psychometrics and theories

2.5.1 Age-related changes in intelligence: the psychometric approach

The psychometric approach differs from the experimental approach in that the researcher does not introduce or manipulate additional variables, but simply records the scores obtained on standardized tests, and compares these with the norms for the group to which the individual belongs, or the norms for other groups. Scores on intelligence tests should have predictive validity, that is, they should correlate with performance on real-life tasks, such as exam grades or job-related performance. However, these predicted correlations are not always found.

The definition of intelligence is controversial. It has been said sceptically that 'intelligence is whatever intelligence tests measure'. However, one definition proposed by D. Wechsler in 1944 is 'the aggregate or global capacity of the individual to act purposefully, to think rationally and to deal effectively with his environment'. This definition implies that there is some kind of general ability that affects performance on all cognitive tasks.

The measurement of intelligence is traditionally carried out by administering a battery of sub-tests of different kinds and deriving a composite score which is used as an index of intelligence. Scores on individual sub-tests are usually highly correlated (i.e. good scores on one test go with good scores on other tests; poor scores on one test go with poor scores on other tests). This supports the idea that there is a common factor of general intelligence. However, when intelligence is tested across the lifespan, the sub-tests show quite different trends. Scores on some tests decline with age and scores on others do not. This finding casts doubt on the concept of intelligence as a general unitary ability. Whatever conclusions are reached about the nature of intelligence, anyone interested in the effects of ageing must consider the different trends of the various sub-tests separately.

A negative relationship between age and intelligence scores was first noted during World War I when a battery of intelligence tests called the Army Alpha test was administered to large numbers of soldiers and further studies have since confirmed these trends.

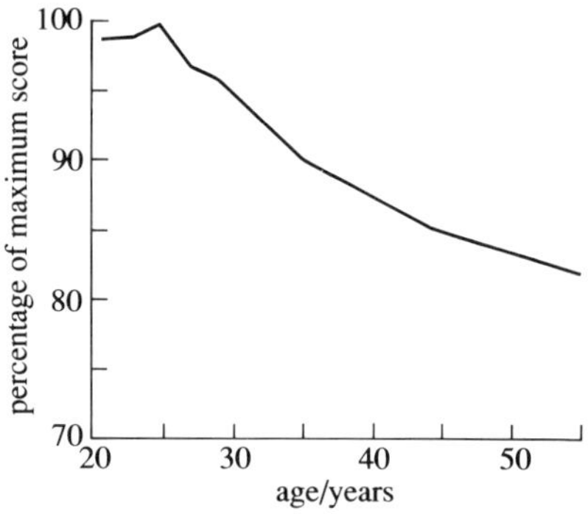

Figure 2.13 Scores on the Army Alpha intelligence test as a function of age.

□ Look at Figure 2.13, which shows the average scores on the Army Alpha intelligence test at different ages. What is the decrease in scores over each ten-year period after the age of 25?

■ After the age of 25 the decline is around 5–10% per decade.

The Wechsler Adult Intelligence Scale (WAIS), discussed in Book 2, Chapter 11, is the most widely used intelligence test. It consists of 11 sub-tests. Four of these are considered to measure verbal ability. In the Vocabulary test the subject is asked to give definitions of 40 words. The Information test consists of general knowledge questions. The Comprehension test asks for interpretation of proverbs and the Similarities test consists of verbal analogies (the subject is given two words and asked to explain how they are similar). The first three of these tests show no decline with age and there is only a slight reduction in the Similarities test. So, in general, there is relatively little evidence of age-related decline in verbal ability.

Other tests in the WAIS battery are classed as measures of performance. The Arithmetic test and a combined forward and backward digit span show a moderate decline in performance after age 50. Four other tests measure spatial ability. Object Assembly and Block Design are both types of jigsaw; Picture Completion requires spotting the missing elements in a picture and Picture Arrangement requires a series of pictures to be arranged in a meaningful sequence. All these spatial tests exhibit a steady decline as shown by scores on the Picture Arrangement test presented in Figure 2.14. In the speeded Digit-Symbol Substitution Test (also shown in Figure 2.14) strings of single-digit numbers written in rows are presented. The task is to substitute a specified abstract symbol for each digit (e.g. writing in > for each number 9), and the score is the number of substitutions made in 90 seconds. Scores on this test fall off even more sharply with age.

The WAIS battery does not include tests of reasoning, but Raven's Progressive Matrices are often used to test reasoning ability. An example is shown in Figure 2.15. (They are called 'progressive' because they get progressively harder.)

This test consists of a large number of such problems increasing in difficulty. Performance is measured in terms of the number of problems solved correctly within a fixed time. Scores remain stable up to age 40, but thereafter decline at a rate of 10% per decade.

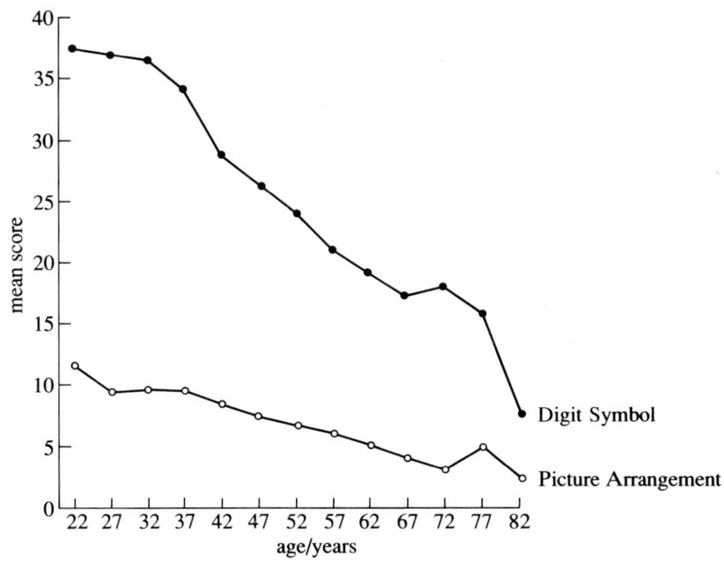

Figure 2.14 Scores on the Picture Arrangement and Digit-Symbol tests as a function of age.

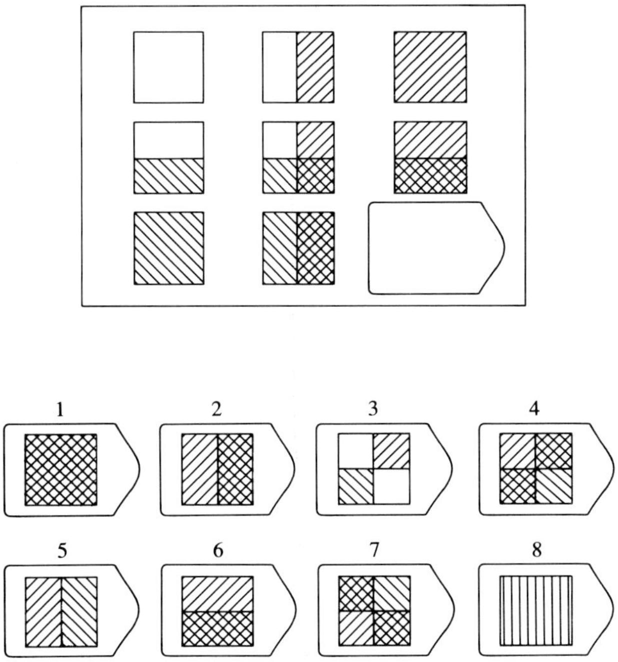

Figure 2.15 An example of a problem from Raven's Progressive Matrices. The subject must select one of the lower patterns to fit the space in the upper array. (The answer is given in the Summary of Section 2.5.)

Thus the overall picture is of stable verbal abilities, but declining spatial and performance abilities. An alternative way of describing these trends is to say that the ability to acquire new information or to use it to solve problems is impaired, but the utilization of old information (such as language or general knowledge) is preserved.

The use of intelligence tests in assessing age-related changes in intelligence has been criticized on the grounds that the tests are not valid for older people. It is argued that the elderly are disadvantaged in that the tests are unfamiliar to them; some of the tests are speeded; the test content is too abstract and performance is strongly influenced by level of education. It is also argued that the tests fail to reveal subtle age differences that are found in studies which are designed to be more discriminating. For example, psychometric tests show no age effects on verbal ability but experimental studies of language use (Section 2.4.4) have revealed some deficits. While these considerations do suggest that intelligence test results need cautious interpretation, the psychometric approach to ageing supports the view that age changes in cognition originate in specific components of the system rather than unitary general factors.

2.5.2 Theoretical explanations of cognitive ageing

This section reviews some of the explanations that have been offered for the effects of ageing on cognitive processes which have been outlined above. It was pointed out that some approaches presuppose that the cognitive system is unitary and look for a single general factor which governs the efficiency of all the sub-systems. Other approaches presuppose that the sub-systems are modular, so that some may be independently impaired while others remain intact. Some of the theoretical explanations that have been proposed are outlined below.

1 *Reduced processing capacity* This is a general factor which can explain many of the observed age effects. Support for this idea comes from numerous experiments showing that the age deficit increases as the task demands are increased. The effects of task complexity occur across a wide range of different tasks, so it appears that the important factor is not *which* processes are involved, but *how much* processing is required. The explanation suffers, however, from lack of any agreed definition of what exactly constitutes processing capacity. A further problem is the finding of considerable plasticity in processing resources rather than a fixed limit. An extreme example of this plasticity is the case of an elderly woman who, with very prolonged training, increased her digit span to 120 items.

2 *Reduced speed of processing* This is a single general factor which is less ambiguous than 'processing resources', and which J. L. Fozard in 1981 identified as 'the quality of mental performance that most consistently distinguishes young and elderly adults'. It has been suggested that slowing results from damage or disorder of the basal ganglia. Nevertheless, age differences do appear in some tasks which are self-paced (as, for example, in the diary writing described in Section 2.4.4 and in retrieving information from long-term memory), so that speed cannot be the only factor involved.

3 *Reduced working memory capacity* This is a less general explanation in that the locus of impairment is identified with one component of the system, rather

than a factor affecting the whole system. However, it is difficult to think of any cognitive tasks in which working memory is not involved (except for purely automatic processes), so knock-on effects of a decrement in working memory are bound to be widespread.

4 *A production deficiency* This explanation attributes age deficits to faulty strategies, called production deficiencies, so age differences are seen as being due to differences in practice, experience or training which bias elderly people to approach tasks in a different way. For example, when they are asked to read and remember a passage of text, older people typically try to memorize it verbatim; younger people try to pick out the important points and remember those. It is said that such differences in strategies stem from educational practices. The older people were trained as youngsters to learn by heart, but young people did not have this experience. It follows from the production deficiency explanation that age deficits can, in principle, be overcome. However, this explanation fails to give a complete account of age effects since, although it has been shown that the performance of the elderly can be substantially improved if they are guided to use optimal strategies, age differences still persist in many tasks. Moreover, the use of non-optimal strategies may be induced by limited processing resources or limited rate of processing.

□ How might limited resources affect the choice of strategy?

■ Because they process more slowly, older people might not have enough time to employ time-consuming strategies. For example, it has been shown that using imagery (forming mental images) can improve memory performance, but imagery takes time to generate.

So which of the above explanations is correct? They are not mutually exclusive. Each of these factors probably contributes to age deficits in performance. Depending on the nature of the task, one factor may be more important for some tasks and less important for others.

These explanations of cognitive ageing are all couched in terms of the information processing framework. The drawback with this approach is that it does not map readily onto a biological explanation. However, these explanations have recently been re-formulated in terms of network models. According to these models, information is represented in memory as sets of nodes and links forming a network. Each node represents a concept and is interconnected by linking pathways to other associated nodes. Links are made during learning and retrieval, and utilization of information takes place by activation of the relevant nodes and links. Within this model, age deficits can be attributed to any of several different changes in the efficiency of the network. These include a reduction in the total amount of activation (corresponding to lower processing capacity); a reduction in the number of nodes that can be simultaneously active (corresponding to a reduction in working memory); a reduction in the speed of the spread of activation (the rate of processing) or a faster rate of decay of activation. This re-formulation has the advantage of seeming somewhat easier to map onto the neurophysiology of neural networks. However, it will be some time before it is possible to map the accounts of psychological ageing in this chapter onto the accounts of biological ageing in the next chapter.

Summary of Section 2.5

The psychometric approach has revealed large individual differences between older people. The variation in test scores is greater for the old than for the young. Verbal tests show very little effect of ageing, but spatial tests and speeded tests show much larger effects. Theoretical explanations of age deficits in cognition are based on information processing models and include reduced processing capacity; reduced speed of processing; reduced capacity of working memory and poor processing strategies. More recently, explanations of ageing are being framed in terms of neural networks rather than information processing models.

The answer to Figure 2.15 is (1).

Summary of Chapter 2

The increase in life expectancy in developed countries has changed the age composition of these societies, so that there are many more elderly people in the population. In consequence, it is important to understand the physical, behavioural and cognitive changes that occur as a result of ageing.

An accurate assessment of the level of cognitive competence that can be expected in elderly people is necessary for many aspects of social policy, environmental design and marketing. It is also important in providing a base-line against which to assess pathological changes and shows what level of performance should be aimed for in remedial therapy. The effects of ageing vary greatly from one individual to another and it is often difficult to disentangle physiological and cultural factors. Social behaviour may be affected by the loss of a defined role and status and by the social isolation that occurs in old age. Older people may withdraw from social contact and become depressed, but these changes are caused by the cultural conventions of developed societies and by physical ill health, rather than by ageing itself. This chapter has examined some of the changes that occur in the course of normal ageing. It has not included the effects of pathological ageing, such as Alzheimer's disease, which is described in Chapter 3.

Studies of normal people have shown that older people are at a disadvantage in occupations and activities which require the individual to work fast or which need physical strength or optimal eye-hand co-ordination. However, increased experience and expertise may compensate in some jobs and activities. In everyday life, absentmindedness and other memory deficits are associated with normal ageing, but can be offset by the use of memory aids and appropriate strategies.

This chapter has concentrated on the effects of ageing upon cognition. In order to explain these effects, various research methods have been briefly reviewed and a model of the cognitive system was introduced. Perceptual acuity declines, with the result that the quality of the information reaching the cognitive system is impaired. Components of the cognitive system that process the information are also impaired. Both short-term and long-term memory are affected. In short-term, or working memory, the amount of information that can be held in consciousness is reduced. Information is not lost from long-term memory, but becomes more difficult to retrieve when it is wanted. There is a decline in the speed at which information can be processed which can cause a back-up of information in the system. There is also a decline in the ability to attend to several things at once which causes a deterioration in activities like driving, and in the ability to perform different operations (like reasoning and remembering), simultaneously. Although these changes are clearly apparent in experimental or psychometric tests, they have relatively little effect in everyday life because most everyday activities are not sufficiently demanding to expose the underlying impairment.

Theoretical explanations of age-related changes in cognitive processes have sought to identify the component parts or processes of the cognitive system which are affected. These explanations have been expressed in terms of an information processing model of cognition. However, recent developments have begun to formulate explanations in terms of neural networks. With further progress along these lines it should become easier to map the cognitive effects of ageing onto the neurophysiological effects to be described in Chapter 3.

Objectives for Chapter 2

When you have read this chapter you should be able to:

2.1 Define and use, or recognize definitions and applications of each of the terms printed in **bold** in the text. (*Question 2.1*)

2.2 Describe how the age composition of the human population is changing and explain why. (*Question 2.1*)

2.3 Explain the distinctions between normal and pathological ageing, and between chronological, biological and psychological ageing. (*Question 2.2*)

2.4 List some ways in which job performance and skills are impaired by ageing, and some ways in which performance is unaffected by ageing. (*Question 2.3*)

2.5 Understand the strengths and weaknesses of different approaches to and methods of studying ageing. (*Question 2.4*)

2.6 Describe those aspects of cognitive processing that are impaired by ageing. (*Question 2.5*)

2.7 Recall how the impairment of cognitive processes affects performance in different tasks. (*Questions 2.5 and 2.6*)

2.8 Be able to assess critically the value of psychometric intelligence tests to reveal the effects of ageing. (*Question 2.7*)

2.9 Discuss and evaluate theories about the effects of ageing on cognition. (*Question 2.8*)

Questions for Chapter 2

Question 2.1 (*Objectives 2.1 and 2.2*)
What is the difference between life expectancy and lifespan?

Question 2.2 (*Objective 2.3*)
Why is it difficult to study 'normal' ageing? Suggest two reasons.

Question 2.3 (*Objective 2.4*)
Explain how the effects of ageing on performance can sometimes be masked or counteracted.

Question 2.4 (*Objective 2.5*)
Outline the advantages and disadvantages of the experimental method of studying ageing, as compared with the everyday approach which studies age effects in real life situations.

Question 2.5 (*Objectives 2.6 and 2.7*)
Which of the following list of activities and abilities are not affected by ageing and which are likely to be impaired by ageing?

playing chess

memory of routes

vocabulary

wisdom

judgements of colour and space

ability to do two or more things at once

Question 2.6 (*Objective 2.7*)
Outline the factors which can operate to make performance on memory tasks less difficult for older people.

Question 2.7 (*Objective 2.8*)
Why are intelligence tests said to be 'unfair' to the elderly?

Question 2.8 (*Objective 2.8*)
What are the advantages of network models in explaining age effects?

References

Wright, R. E. (1981) Ageing, divided attention and processing capacity, *Journal of Gerontology*, **36**, pp. 605–614.

Lehman, H. C. (1953) *Age and Achievement*, Princeton University Press.

Further reading

Kausler, D. J. (1982) *Experimental Psychology and Human Ageing*, John Wiley & Sons. This is a very comprehensive and detailed text-book.

Rowe, J. W. and Kahn, R. L. (1987) Human ageing: usual and successful, *Science*, **237**, pp. 143–149.

Salthouse, T. A. (1985) *A Theory of Cognitive Ageing*, Elsevier/North Holland.

CHAPTER 3
BIOLOGICAL ASPECTS OF AGEING

3.1 Introduction

During this century, progress in medicine and improved social conditions in the more developed countries has increased life expectancy and gone some way towards fulfilling the dream of humans to live in eternal youth. However to become older and older often means that people live in bodily and mental infirmity over a longer period of time and problems of ageing often involve problems of the ageing nervous system. Chapter 2 discussed the psychological aspects of ageing. This chapter considers biological ageing. It looks at morphological, biochemical and physiological aspects and their interrelationships in both the normally and the pathologically ageing brain. Historically, psychological and biological approaches to the study of ageing have progressed separately. This is in part due to the inherent complexity of nervous tissue and in part due to the fact that some biological measurements can only be made after death. Consequently it is not always possible to link which biological changes are associated with which psychological ones, though the chapter will give pointers where such an explanation seems possible.

Suppose that a certain psychological function is thought to be associated with a particular brain region and that, with advancing age, the psychological function deteriorates and degeneration is seen in the brain region. It might well be that loss of the psychological function is caused by the observed brain degeneration.

- □ Why is caution in order in concluding this?
- ■ It might be that (a) the degeneration of the brain region under consideration is secondary to brain degeneration elsewhere, or that (b) changes in behaviour contribute in some way to changes in brain structure.

Such caution is necessary in interpreting the observations but it does not preclude tentative theorizing in an attempt to relate behavioural change to changes in brain structure and physiology.

The present chapter concentrates on the changes that occur in brain structure and function with age. When the nervous system was introduced in Book 2, its so-called 'normal' function was described. For example, the transmission of action potentials along myelinated axons was described in terms of intact neurons with intact myelin wrapped around their axons. Dopaminergic transmission was described in terms of neurons able to synthesize adequate amounts of dopamine. By implication, the nervous systems under consideration were not those of aged, malfunctioning subjects. The present chapter compares such normal nervous systems with those which have changed their function with ageing (e.g. by loss of neurons or by loss of the ability to synthesize dopamine). In addition, it also compares the normal aged nervous system with the pathological nervous system.

When studying this chapter it is important to bear in mind these two comparisons. Is pathological ageing simply a consequence of a speeding up of those processes that lead to normal ageing or are different processes involved? An understanding of the changes in the normal ageing brain can lead to a better understanding of what goes wrong in the case of pathological ageing.

Figure 3.1 shows the difference between normal and pathological ageing of the nervous system in humans. Normal ageing is generally regarded as involving a loss of neurons (see Section 3.4) which leads to 'healthy' senescence, basically the simple slowing down of the normal function of the nervous system. The outcome of pathological ageing is the development of clinical syndromes—medically recognized conditions which include the group of mental disorders of middle and late life known as the **dementias**, along with various subgroups of nervous system disease.

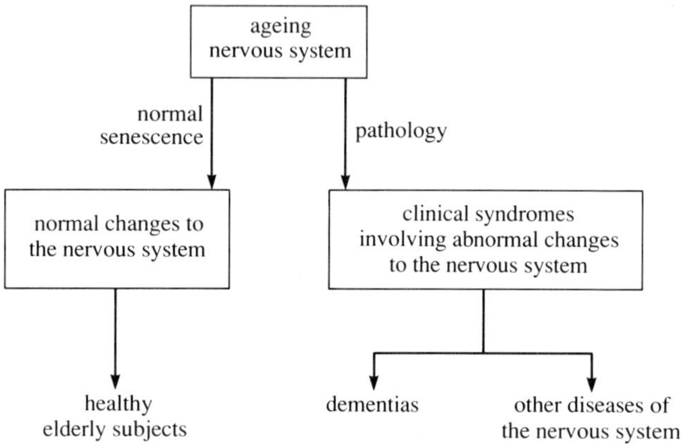

Figure 3.1 Normal and pathological ageing of the nervous system. See text for details.

The most common of the pathological diseases associated with ageing are grouped under the 'umbrella' term of dementias (Figure 3.1). Dementias are commonly defined as progressive loss of cognitive and other intellectual functions. The damage to the nervous system involved in dementias can be transient but is usually permanent, and far exceeds the rate and extent of deterioration seen with normal ageing. There are many causes of dementias, some associated with diseases occurring primarily outside the nervous system (e.g. in blood vessels) and others occurring within the brain. The chapter ends with a review of the dementias, looking, in particular at the commonest and most severe form, called Alzheimer's disease (Section 3.7).

Changes that occur during ageing can be considered at a number of levels. Section 3.4 examines the morphological changes in the brain as a whole as well as at the neuron level, while Section 3.5 looks in detail at physiological and biochemical aspects of the ageing brain.

The nervous system, like other organs and tissues, has been shown to possess some powers of plasticity that may help to compensate for changes associated with ageing. Nevertheless the capacity of these compensatory mechanisms may not be adequate to overcome all the factors which can lead to the development of neurodegenerative disease such as the dementias. External and internal influences may upset the brain's homeostatic mechanisms and lead to a breakdown of normal neural functioning.

Degeneration in the elderly is not confined to the nervous system. The immune system has a number of features in common with the nervous system (Book 2, Chapter 5). The changes that it undergoes with ageing are considered in Section 3.6.

It is hoped that in time biological investigations of the ageing brain will open the way for preventative therapy that will forestall the onset of diseases, such as dementias, relieve their symptoms, or arrest their relentless progression and lead to more effective care and counselling. For these reasons, much present-day neurobiological research is directed towards studying the functioning of the ageing brain in both normal and diseased states.

However, the inherent complexity of neural tissues in terms of variety of cell types and intricate intercellular connections poses a formidable obstacle to biological investigation, which generally relies on tissue sampling and the interpretation of biochemical analyses. As was discussed in Section 1.4 of this book, chemical findings may reflect the progression of a disease and thus represent the pathological *consequences* rather than the fundamental *cause* of the illness. For obvious reasons, most research using human brain samples must be carried out on samples obtained from volunteers after their death. As will be discussed in more detail in Section 3.3, general deterioration after death must be borne in mind when interpreting data from such studies. Furthermore, because of the difficulties involved in obtaining and using human tissue, some of the information contained in this chapter comes from research on experimental animals.

The next section begins with a discussion of what normal biological ageing is and its possible causes.

3.2 What is normal biological ageing?

As discussed in Section 2.1.2, biological ageing may be defined as a process involving normal, inherent and progressive change of physiological function over the entire lifespan of an organism. Ageing is generally thought to be, to a greater or lesser extent, a genetically programmed process but also, as explained in Chapter 2, is affected by environmental factors.

The ability of an organism to adapt to changes in its normal environment gradually declines following the onset of maturity. With age, there is a reduction in the body's ability to recognize as abnormal the range of substances with which it comes into contact and a consequent decrease in their rate of breakdown. This means that the mature central nervous system, along with other body systems, may become more vulnerable to neurotoxic substances (for example, alcohol or barbiturates).

□ Why is this particularly important in the case of the nervous system?

■ Neurons do not replicate, and thus regeneration within the central nervous system is very limited.

The causes of the degenerative processes of ageing are still unclear. Ageing seems to be due to the breakdown of cellular mechanisms which may be mediated through internal genetically-controlled events or through external influences. In the former case, there may be genes that switch off essential controlling factors at a certain age. This might explain the restricted and characteristic lifetimes of particular species. which could therefore be regarded as a part of an animal's phenotype (Book 1, Section 3.2.3).

However, there is no firm evidence for the existence of particular genes associated with senescence, with the exception of *progeria*, a rare condition that begins in childhood and which is characterized by retarded growth and accelerated ageing. Some human genetic diseases do manifest themselves in older but not in younger people. An example is Huntington's disease (discussed in Section 4.3.1) which develops in people carrying a particular dominant allele.

Various theories have been put forward to explain changes in gene expression with age. For example, the *somatic mutation theory* proposes that ageing is due to random somatic mutations, i.e. mutations that affect somatic (non-gamete) cells (Book 1, Section 3.2.5), some of which result in chromosomal damage in cells including those of the central nervous system. Some of these mutations escape the repair mechanisms of the cell and so accumulate with age.

3.3 Problems of studying biological ageing

Research in gerontology—the study of ageing—is impeded by a number of difficulties and as a result the literature on ageing is quite contradictory and confusing. Much of the available data that relate to biochemical measurements are made on the brains of laboratory animals, particularly rats, and the applicability of such information to humans is often very limited (Section 1.4.8).

Another factor to be considered is that the maintenance of colonies of old animals in the laboratory in order to study the effects of ageing requires considerable long-term financial investment. This is particularly true in the case of non-human primates: apart from the ethical and conservation reasons against maintaining such colonies, the costs are prohibitive.

The rate of ageing is usually expressed chronologically, yet many species, and even individuals, age at different rates. Variations in rates of ageing with respect to humans were discussed in Section 2.1.1. Some investigators in the field have claimed that one- to two-year-old rats may be regarded as elderly, even though some animals can survive for up to three years. Are these very elderly rats a superior group who have survived disease and therefore do not accurately represent a normal sample? In order to overcome the effects of infection, some laboratory animals are reared under special pathogen-free conditions and thus have the potential to live to great ages. At what age can they be described as old?

External factors have a significant effect on lifespan. Of particular importance is nutrition. As mentioned in Section 2.1.2, the state of nutrition of human beings significantly affects their lifespan. This also appears to be true for laboratory animals since a restriction of the amount of food eaten can significantly extend the lifespan. For example, mice given free access to food have a mean survival time of 24.9 months compared to 29.9 months for those on a carefully controlled diet. This suggests that longevity may be greatly modified not only by type but also by availability of food.

□ What other possible variables in the environment of an animal may affect its lifespan?

■ Other possible variables include amounts of stress, stimulation and exercise.

It was pointed out in Section 2.1.2 how important such factors are in determining psychological ageing in humans.

Although it is possible to make biochemical measurements on the brain of elderly animals, it is more difficult to design suitable behavioural tests in order to correlate the two types of data. This makes it difficult to establish relationships between metabolic activity at the cellular level and function at the level of the whole organism. In humans, psychometric assessment (Section 2.3.3) is easier than in experimental animals, but, as you learnt in Section 2.1.2, in the elderly, impaired hearing or eyesight have to be allowed for.

The chemical composition of tissues and neuropathological changes are affected by conditions both before and after death. For example, the brain of someone maintained on a life-support machine or that of a person with a protracted illness, may undergo non-specific degenerative damage. After death, changes in cell biochemistry and morphology take place very quickly and it has been shown that comparisons between post-mortem brain samples are only valid when the corpses are refrigerated immediately after death and the brain specimens taken within one day.

Added to these difficulties are problems that are not specific to the study of ageing but which arise because of the complexity of the structure and function of the brain. For example, there are immense problems in identifying cell types and in determining the physiological functions of each type (see Book 2, Section 2.4). Such analyses are crucial to a full understanding, not only of the interrelationships between different types of cells, but also of how the nervous system works as a whole. It is possible to count cells automatically by machine to analyse brain tissue sections but this relies mainly on cell morphology, so that if cell shrinkage occurs neurons may be miscounted as the smaller glial cells. This difficulty of identifying particular cell types has recently been overcome with the development of monoclonal antibodies (described in Book 4, Box 2.3). Recall that antibodies recognize specific antigens and thus can be used to 'label' particular cell components. Staining techniques that combine stains with these antibodies can be used to highlight particular cell types within a preparation of brain tissue. Figure 3.2 (*overleaf*) illustrates the staining of astrocytes (a kind of glial cell) in a preparation of human brain.

The problem of identifying the specific neurotransmitters used by particular cells in the brain, although not specific to the study of ageing, impedes research into the ageing brain. Identification of catecholaminergic neurons (Book 2, Section 4.5.2) and their connections has been made possible by methods that again utilize antibodies. However, for other neurotransmitters, the instability of the transmitter molecules has necessitated the use of indirect methods to detect their presence (Section 1.4.6). One way is to measure an enzyme involved in the manufacture of the transmitter (for example, choline acetyltransferase for cholinergic neurons). Where amino acids such as glutamate (see Book 2, Section 4.2.5) are thought to be transmitters, their identification is much more complicated. An estimate of the density of receptors specific to the transmitter concerned may be the only way of obtaining an indication of the substance involved. Such techniques have been successfully applied to the identification of synapses that use a peptide transmitter (Book 2, Section 4.5.6).

You should bear in mind the difficulties of collecting and interpreting data in studies on the ageing nervous system when reading the rest of this chapter.

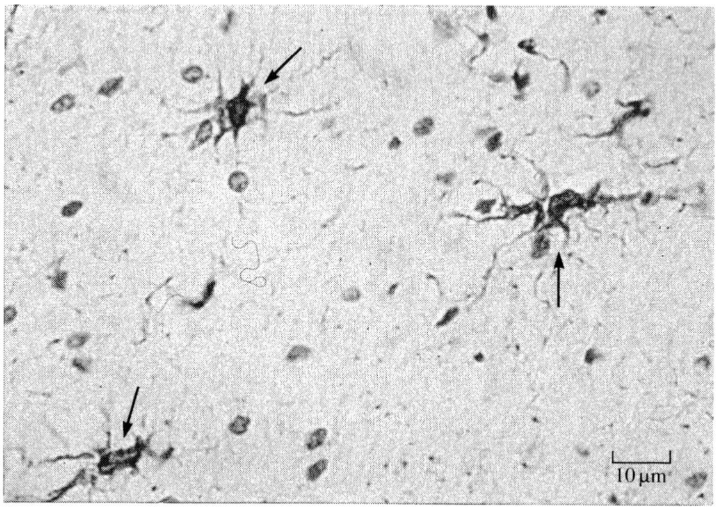

Figure 3.2 Astrocytes (arrowed) in the human brain detected by a technique that uses horseradish peroxidase to bind to antibody, which gives a coloured, and therefore visible, product (see Book 4, Box 2.3).

Summary of Section 3.3

This section has reviewed the problems of studying biological ageing. One problem is that of defining what constitutes an elderly animal because different strains have variable longevity. In addition ageing is affected by external factors such as nutrition and stress. But the biggest problems, although not specific to the study of ageing, are at the biochemical and cellular levels in identifying particular transmitter substances and cell types, in making comparisons between different animals and the applicability of findings from animals to humans.

3.4 Gross structural changes in the ageing brain

The brain plays a critical integrative role in the functioning of the whole organism. The most striking change to occur in the brain with increasing age is the decrease in its water content. It is important to note that this decrease is in the *intracellular* content and is not due to reduction in the fluid surrounding the brain. This is one of the most obvious markers of change occurring in the neurons of the brain.

After the age of about 50 years the wet weight (i.e. the weight of the tissue in its natural state, including its water content) of the human brain slowly decreases, though there is considerable variation between individuals. Thus, the average weight of the normal young adult male brain is about 1 400 g, which by the time he is 65, may drop to 1 200 g. The average female brain weight is roughly 100 g lower than that of males of the same age throughout adult life. This difference in brain size is related to body size and not to sex. A larger than average woman will have a larger brain than a man who is smaller than her.

This decrease in wet weight with age also occurs in laboratory animals. For example, in young rats the intracellular water content of the brain is about 79% of the total cell weight but this value decreases to about 72% by the age of 30 months.

□ Why must this age-dependent decrease in water content of the brain cells be taken into consideration when interpreting physiological measurements?

■ If the proportion of the tissue weight that is due to water is decreasing, any age-dependent reduction in rates of synthesis or activity would be much more pronounced if expressed relative to wet tissue weight than if expressed relative to dry tissue weight (i.e. the weight of tissue that has been dried to remove all its water content).

Table 3.1 illustrates this point. It gives a comparison of data showing protein content of the brain in organisms of different ages. The apparent increases in protein content with age, which are statistically highly significant, are principally due to decreases in tissue water content.

Table 3.1 Average protein content of the brains of normal rats and humans of different ages.

Tissue	Age	Protein content/ mg g^{-1} wet weight tissue
Rat brain	2–4 months	117.2
	6–12 months	116.1
	about 2 years	141.8
Human brain	16 years	43.0
	90 years	51.0

So far, the wet weight of the normal aged brain has been considered. Now examine Table 3.2, which compares brain weight in normal elderly males with that in men with senile dementia and Alzheimer's disease.

Table 3.2 Average brain weights in normal elderly male subjects and male individuals (of similar body size range) with senile dementia or Alzheimer's disease.

Sample	Mean age/years	Brain weight/g
Controls (normal elderly individuals)	75.0	1 200
Senile dementia	80.5	1 101
Alzheimer's disease	65.4	1 150

□ From an examination of Table 3.2, what is the striking feature about brain weight?

■ The diseased brains are lighter than the normal elderly controls.

Computerized tomography (described in Section 1.4.5) allows the structure and function of the brain to be investigated simultaneously. This technique has revolutionized methods of assessment of neurological ageing and disease.

□ What are the advantages of brain scanning studies? (You may need to check back to Section 1.4.5 to answer this.)

■ The techniques are non-invasive and simple, can be performed quickly and give immediate results.

Brain scans have revealed that sulci get wider and the ventricles get larger with increasing age. (Sulci are the deep furrows formed by the infolding of the cerebral cortex, allowing for a large increase in the surface area of the human brain; Book 2, Section 1.1, Figure 1.1.) Tomography reveals that from about 40 years of age onwards there is a small enlargement of the third ventricle and alterations in sulci. However, in common with most such phenomena, there is marked variation between individuals; there is also a sex difference, men showing greater changes in the fourth decade compared with women. In addition, tomography has proved an invaluable aid to diagnosis for physicians who can use it to detect causes of brain malfunctions such as tumours.

The next two sections go on to look at the morphological changes that occur in the ageing brain at a different level, the cellular level.

3.4.1 Changes at the cellular level in the ageing brain

The survival of healthy neurons depends on continued interactions between them within the neuronal network. The presence of growth factors of various kinds (for example, chemotrophic factors, see Book 4, Section 3.2.2) and stimulation by neurotransmitters are essential for normal function. Normal ageing in the central nervous system is a process of deterioration that involves both the death of some neurons in specific regions of the brain and also the regression of dendrites and axons which in turn lead to cell death. In the cerebral cortex of some elderly people this regression of dendrites progresses until in some cells only stubs remain. Since replication of neurons does not occur, the central nervous system is especially vulnerable to degenerative change.

Selective degeneration of neurons occurs in some parts of the cortex. For many years it was thought that there is a trend throughout life for the whole of the neuronal population throughout the brain to decrease. This was largely based on a pioneering study carried out by Harold Brody in the mid-1950s, who claimed that there was a general loss of cortical neurons varying from 15% to 35% between early and late adult life. This dubious conclusion, which was based on the percentage reduction of brain weight, was questioned, however with the advent of more accurate techniques. But Brody did identify loss of neurons in the primary sensory and motor cortical areas and this may possibly account for the motor changes seen in the elderly.

There is in fact no substantial evidence that neuronal loss occurs on a large scale as a universal effect throughout the whole brain. It is now estimated that cell loss in the cortex does occur every day of adult life at the staggering rate of more than 100 000 neurons per day. However, somewhat reassuringly, it is also estimated that given the total number of neurons in the cortex, this daily loss is insignificant and does not drastically affect cortical function. But, given time, this relentless daily attrition in the number of cells in the cortex does contribute to the process of normal ageing of the brain.

Within the brain and brain stem the percentage loss of cells varies widely between regions. Counts of neuronal number in the ventral cochlear nucleus of the auditory area have clearly shown that a reduction in brain cell number does not occur here. However, measurable neuronal loss occurs in the cerebellum. Although the number of Purkinje cells (Book 2, Section 8.8.8) varies greatly from one person to another, there is a detectable reduction of about 2% of neurons per decade up to the age of 60 years, after which the loss is greater. Another region particularly prone to neuron loss is the substantia nigra of the basal ganglia.

So far in this section, neuron loss in normal elderly individuals has been reviewed. Consider now neuron loss in the pathological ageing brain.

Substantial cell loss occurs in specific regions of the brain in elderly individuals with some types of brain disease. Neurodegenerative diseases are among the most common conditions affecting the mature nervous system and yet their cause remains unknown. The rate of loss is greatest in motor neuron disease where there is progressive loss of motor neurons in the spinal cord and brain stem and loss of pyramidal cells in the motor cortex.

One pathological feature common to all the dementias is degeneration of the cerebral cortex, with other brain structures affected to various degrees. Given the evidence for localization of cognitive functions in the cortex (Book 2, Chapter 11), it is not surprising that dementias are associated with reduction in intellect and loss of awareness.

In the case of Parkinson's disease (discussed in detail in Section 4.3.1), selective cell loss is confined in the early stages of the disease to the substantia nigra of the basal ganglia (Table 3.3).

Table 3.3 Cell losses in the substantia nigra in normal elderly subjects and individuals with Parkinson's disease, senile dementia or Alzheimer's disease (males and females).

Samples	Mean age/years	Neuronal loss in substantia nigra
Controls (normal elderly individuals)	65	21%
Parkinson's disease	63	81%
Senile dementia	65	22%
Alzheimer's disease	65	26%

- □ From an examination of Table 3.3, what is the striking feature about percentage cell loss in the substantia nigra?
- ■ Although there is loss of cells in all individuals, the percentage loss is exceptionally high in individuals with Parkinson's disease.

There is a further important point to understand about the functioning of the nervous system. Damage or degeneration of the type that is frequently seen in the ageing brain, that is sustained within one region, can have profound effects upon the function of other brain regions. This is best illustrated with a rather dramatic example. In a wide variety of experimental animals, following the amputation of a limb, a drastic form of lesioning, degenerative changes and neuronal loss can be clearly seen in certain brain regions. This emphasizes the fact that all parts of the nervous system interact with each other as a functional unit and the same is true for different regions of the brain. It is important to bear this in mind when looking at the specific changes associated with particular disease states such as Parkinson's disease.

Despite the changes in the normal ageing brain, much of the essential circuitry remains unaffected. In affected regions such as the cortex some of the structural losses may be compensated for by intrinsic adaptive mechanisms. Evidence presented in Book 4, Chapter 3, showed that the dynamic interactions between axons and targets during development do continue in the mature adult to a limited extent. Loss of neurons in the mature brain causes their axons to degenerate, resulting in loss of inputs to downstream cells. In turn this triggers a chemotrophic factor such as nerve growth factor to be released which causes axons of surviving neurons to sprout extra branches and re-innervate. This process enables the brain to compensate at the structural level but it is difficult to determine to what extent it compensates at the functional level for loss of neurons. This process of axon sprouting has been demonstrated only in certain regions of the damaged brain (this is described in more detail in Section 5.4.2).

Dendrites are also affected by age. The results of some research on dendrite lengths (Buell and Coleman, 1979) in New York are shown in Figure 3.3. These reveal that in normal elderly subjects (average age 79 years), dendritic trees were more extensive than in the brains of middle-aged subjects (average age 50 years), with most of the difference being due to increases in the number and average length of terminal segments of the dendritic tree. These results provide morphological evidence for plasticity in the mature and old human brain.

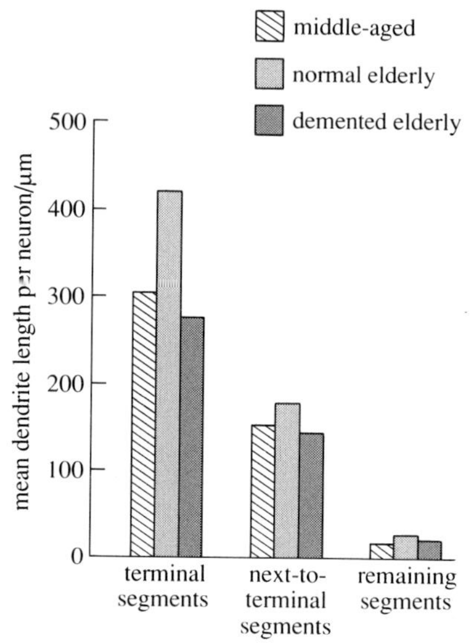

Figure 3.3 Mean length of dendrites in the brains of normal middle-aged subjects (average age 50 years), normal elderly subjects (average age 79 years), and subjects who had suffered from senile dementia.

Importantly, these dendritic trees continue to grow between middle and old age (between 50 and 70 years of age) but not thereafter. This emphasizes that mechanisms of plasticity also have a finite life.

□ How do the data for normal elderly subjects in Figure 3.3 compare with those for subjects who suffered from senile dementia?

■ The dendritic trees have shorter segments in the subjects who suffered from senile dementia.

In fact not only are they shorter but counts reveal that they also have fewer branches.

These compensatory mechanisms of axon growth and extension of dendrites may be part of an extended programme to maintain and adapt brain structure. Buell and Coleman suggested that there may be one population of neurons that dies and regresses (described at the beginning of this section) and another that survives and grows (supported by the data in Figure 3.3).

□ From an examination of the results presented in Figure 3.3, which of the two populations of neurons in the Buell and Coleman model appears to be predominant in the normal elderly aged brain?

■ The population of neurons that survives and grows seems to be predominant in the normal elderly brain.

In marked contrast, it seems that the population that regresses and dies predominates in the demented brain.

3.4.2 Accumulation of pigment in the ageing brain

Other changes beside the loss of neurons occur at the cellular level. One of the most notable is the accumulation of insoluble pigments in certain neurons as well as in many other cells in the body. The highest proportion of heavily pigmented cells are large neurons that occur in the motor nuclei of the brain and spinal cord, whereas small neurons such as those in the brain stem accumulate much less of this pigmentation.

There are two different types of insoluble pigments that accumulate with age. A dark brown coloration, seen specifically in catecholaminergic neurons (for example, in the substantia nigra), is due to the presence of the pigment *neuromelanin* and has been shown to be more prevalent in older subjects. Thus, in the substantia nigra, a region rich in dopamine-containing neurons, increased darkening of this area occurs in the ageing brain. One possibility is that the pigment may accumulate as a result of prolonged functional demand for transmitter molecules. There is some evidence to support this. For example, there is less pigment in the relatively unused lateral geniculate nucleus of blind people in comparison with the constantly used lateral geniculate nucleus of sighted individuals. However, it is not clear whether the pigment is in any way linked to neuronal death. A study carried out on the dopaminergic system in ageing rats (21–26 months old) showed that the accumulation of neuromelanin with age did not significantly impair neuronal function.

More widely dispersed is *lipofuscin*, a yellow, insoluble fatty pigment containing a protein. It has been suggested that accumulation of lipofuscin leads to the death of neurons, but there is still no actual evidence that it is harmful to cells and it could just be related to some other unknown event. For example, in certain areas of the brain, cells accumulate large amounts of lipofuscin with age but loss of neurons is minimal. Furthermore, the fact that there is no loss with age in the number of the very heavily pigmented neurons in the auditory pathway suggests that this is not always the case.

Thus, although various changes that take place in neurons of the ageing brain are well documented, the specific effect of these changes on neuronal loss with age remains a matter of conjecture.

3.4.3 Changes in the very old human brain

The number of very old people—those aged 80 years and more—is increasing in Western society. Further changes occur in the normal very old brain in addition to those already seen up to the age of about 80. These cellular changes, which occur *throughout* the brain, have been described by studies using the technique of Golgi-impregnated brain sections (see Book 2, Box 2.3) but have not so far been quantified.

The changes include continued and extensive loss of dendritic spines, appearance of varicosities (swellings) and distortions of horizontal dendritic branches. These are followed by progressive enlargement of the neuronal cell body and eventually a total loss of dendrites. Finally, as these degenerative changes progress, the cell body disappears, resulting in cell loss. Relevant here is the fact that increase in dendritic length of surviving neurons (described in Section 3.4.1) no longer occurs in individuals after about 70 years of age.

Summary of Section 3.4

This section looked at the structural changes in the normal and pathological ageing brain. The most notable change is the decrease in wet weight with age which is even greater in the pathological elderly brain. Selective degeneration of neurons occurs in some parts of the cortex and non cortical areas of the brain. This cell loss is more extensive in individuals with neurodegenerative diseases.

There is evidence that loss of neurons may be compensated for at the structural level by adaptive mechanisms in surviving neurons such as axon sprouting and extensions of dendrites, at least in the normal elderly brain up to the age of about 70 years old. But in the demented brain these compensatory mechanisms are less evident.

The relationship between cell loss and the accumulation of the pigments neuromelanin and lipofuscin in neurons in the normal elderly brain is not clear. Further structural changes occur in the brain in very old people, aged 80 years and more, which result in further cell loss.

3.5 Physiological and biochemical changes in the ageing brain

Having considered the gross and cellular changes that occur in the ageing brain, this section outlines some of the underlying physiological and biochemical aspects of these changes. It looks at how the brain's electrical activity, energy metabolism and its complement of neurotransmitters change with age.

3.5.1 Electrical activity in the ageing brain

The electrical activity of the brain can be measured by electroencephalography (Section 1.4.5 and Book 2, Box 8.2). This technique, which provides an indication of electrical events related to general activity in a brain area rather than the activity of a single cell, can be used routinely in fully conscious human subjects or in animals under a variety of experimental conditions. It is believed that an EEG of a resting human monitors the maintenance activity of the underlying brain cortex (Figure 3.4a).

Some researchers report that, with increasing age, there is a difference in the energy (electrical activity) of the EEG. Figure 3.4b shows a comparison of the frequency distribution of EEG energy between elderly (75–95 years) and young adult (20–43 years) subjects.

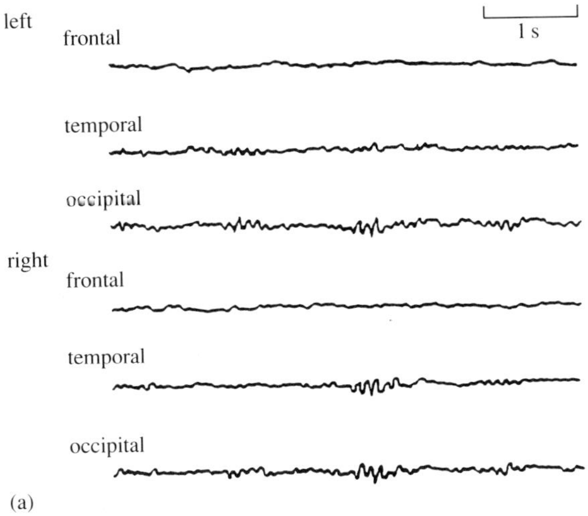

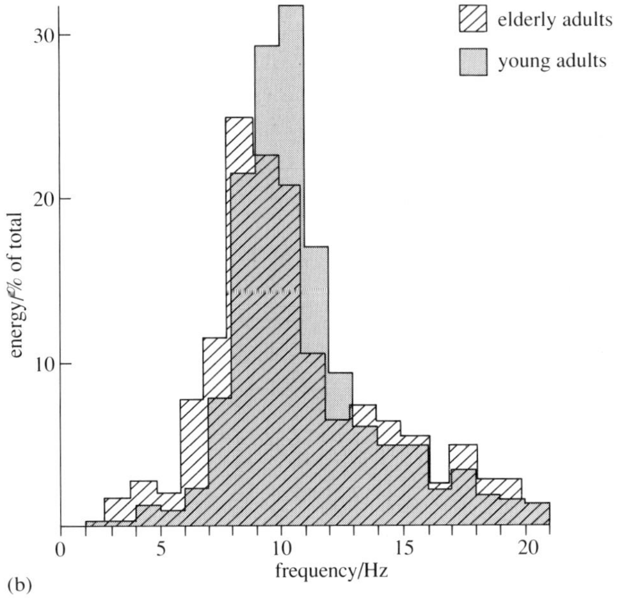

Figure 3.4 (a) An EEG recorded from the scalp surface of a normal individual at various points over the left and right hemispheres. (b) EEG energy (electrical activity) expressed as a percentage of the total in elderly (75–95 years) and young adult (20–43 years) subjects.

- □ From an examination of Figure 3.4b, does the range of frequencies represented in the EEG change with age?
- ■ No. Roughly the same frequency distribution is seen.

□ Does the percentage of electrical activity at some frequencies change with increasing age?

■ Yes. At around 9 Hz and above there is a decrease, and below 9 Hz there is an increase.

It is generally agreed among researchers that EEG traces in about 24% of apparently normal old people show abnormalities. In individuals with neurodegenerative diseases, this proportion is much higher, 50% or more. The abnormalities may arise as a secondary effect of the development of clinical symptoms of, for example, hypertension (high blood pressure) or arteriosclerosis (hardening of the arteries) in some individuals in both the normal elderly group and the pathological elderly group. But why these anomalies are present in those individuals not manifesting these circulatory symptoms in both groups has yet to be explained.

3.5.2 Evoked potentials

An alternative technique to measuring the spontaneous electrical activity of the brain as described above, is to record EEGs while subjects are presented with particular stimuli. Stimuli commonly used are flashes of light and a rotating black and white chequerboard which specifically stimulate the visual system. The multitude of action potentials generated in the sensory system following stimulation travels to the sensory areas of the cortex where it sets off responses in a large number of cortical brain cells. The synchronized activity of these cortical cells—the *evoked potential*—can be measured by electroencephalography. However, evoked potentials are highly sensitive to the conditions of stimulation and to the demands placed upon the subjects being tested. Small changes in these variables (e.g. light intensity) can be sufficient to abolish any age-dependent effects that exist. Great care must therefore be taken in carrying out experiments and in interpreting the resulting data.

Such evoked potentials can be compared on the basis of three features: the early component (the period before the potential becomes negative), the wave height (amplitude) and the late component (the period after the potential first becomes negative). These three features are illustrated in Figure 3.5.

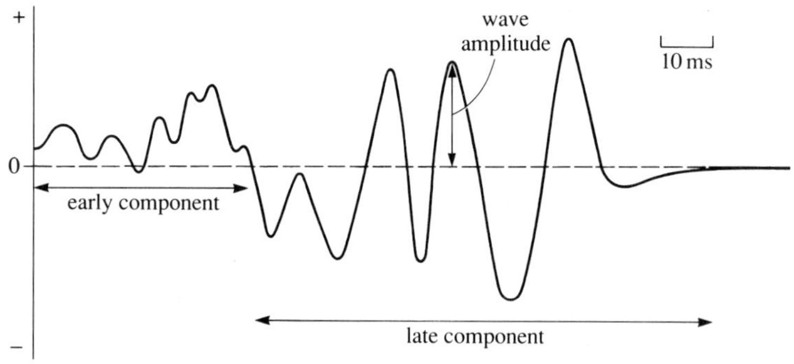

Figure 3.5 The early component, late component and wave height in an evoked potential. See text for details.

Visual evoked potentials in normal elderly subjects differ from those in younger individuals in a number of ways. The early component lasts longer, the wave has a higher amplitude and there is a longer late component. Individuals with dementias show even longer components, particularly the late component.

Difference in evoked potentials with age is not confined to the visual system. For example, somatosensory evoked potentials can be measured on the scalp following stimulation of the nerves of the skin, usually that of the hands, and show a similar pattern of differences between young and elderly subjects.

These differences between young and elderly subjects and between normal elderly and demented subjects may possibly be due to slower conduction of action potentials along the nerves with age. It is not easy to extrapolate from changes in electrical activity at the cellular level to changes in psychological skills that accompany ageing. However, it is not difficult to appreciate that a slowing down of transmission in neurons in a sensory pathway and a change in cortical responses *might* in some way underlie, for example, slower reaction times and changes in cognitive skills as a function of ageing.

3.5.3 Energy metabolism in the human brain

Even though the brain makes up only about 2% of the total body mass of the adult, it consumes about 20% of the blood glucose and oxygen. The brain is totally dependent on adequate supplies of oxygen and glucose if it is to continue functioning. Its high rate of utilization is maintained, not because the capacity of blood to carry these substances in the brain is greater but because the *rate* of flow through the brain is faster than in any other organ. Regional cerebral activity can be assessed by measuring cerebral blood flow.

Ageing involves changes in the blood circulatory system including that which supplies the brain. These changes sometimes lead to hypertension and hardening of the arteries (see Section 3.5.1) which in themselves can contribute to large changes in brain function. These particular problems are not considered further here. Instead this section concentrates on the reduced rate of blood flow to the brain which is a normal occurrence in the ageing brain.

Studies of the effect of ageing on cerebral circulation and metabolism started shortly after the introduction in the late 1940s of a technique developed to measure average cerebral blood flow (CBF) and metabolism. CBF in the brain of old but healthy individuals, in whom blood circulatory disease is absent, is slightly reduced compared with that in young individuals. The CBF in young subjects (mean age 21) averages 6.2 microlitres (μl) mg^{-1} dry tissue h^{-1} and is reduced in normal old subjects (mean age 71) to 5.7 μl mg^{-1} dry tissue h^{-1}.

Use of positron emission tomography (PET) scans (Section 1.4.5 and Book 2, Box 11.1) has enabled the measurement of regional CBF relatively non-invasively in humans. Investigators have used this approach to map cerebral metabolism and function in response to a variety of sensory stimuli. In the normal human brain, the local rate of supply of nutrients, as measured by CBF, is closely related to the local oxygen or glucose utilization rate. So measuring the rate of glucose or oxygen utilization is a reliable, though indirect, way of estimating CBF.

Results using this technique have confirmed the slight difference between young and old subjects. In addition, there are significant differences between the results obtained from normal old people and those suffering from dementias. Table 3.4 gives the average cerebral metabolic rates and performance on certain psychologicai tests in elderly normal and demented people.

Table 3.4 Average cerebral metabolic rates and performance on psychological tests in normal elderly and demented individuals.

Subjects	Global rating *	Mean metabolic rate of glucose used/mg^{-1} l^{-1} min^{-1}
Controls (normal elderly)	1	57
Senile dementia:		
Group A	3	40
Group B	5	35

*Based on psychological testing, each subject was assigned a 'global rating' of 1 to 5 as follows: 1 = normal; 2 = very mild dementia; 3 = mild dementia; 4 = moderate dementia; 5 = moderately severe dementia.

□ Examine the data in Table 3.4. Is there a correlation between the rate of glucose utilization and the degree of dementia?

■ Yes, there is a negative correlation—the rate of glucose utilization decreases as the degree of dementia increases.

Thus it seems that blood flow declines slightly but measurably with age but this reduction is of greater magnitude when there is intellectual deterioration. It has been suggested that the small decline in cerebral metabolic rate that occurs with age, and the further decline that occurs with dementia, may be due to similar underlying morphological changes but that more dramatic alterations occur in patients with dementia. However, it is difficult to determine whether the decline in cerebral metabolic rate is the cause or an effect of behavioural changes.

3.5.4 Energy metabolism in animal brains under experimental conditions

Well-controlled experimental studies of the brain are obviously more feasible in animals than in humans. This section is mainly concerned with experimental observations of laboratory animals. It considers not only glucose metabolism but also oxygen consumption in isolated tissue slices of the brain.

Studies of brain tissue from rats and mice in tissue culture confirm the observations made on living humans that changes in glucose consumption occur with age. Data on the regional utilization of glucose within the rat brain have been obtained using 2-deoxyglucose autoradiography (described in Book 3, Box 4.1). In this technique, the use of radioactively labelled deoxyglucose allows investigators to determine which parts of the brain are active at any particular time and enables them to distinguish between discrete regions of the brain in different states of functional activity. The labelled glucose accumulates in active cells, and after appropriate treatment and autoradiography of the tissue, these active regions can be distinguished under the microscope. Although different studies by different research groups show wide variations in the amount of reductions, there is a consistent drop in the glucose consumption with an average drop of about 20% in the brains of 2–3-year-old rats compared with those of 4–6-month-old rats. Interestingly, reductions are greatest in structures associated with visual and auditory function. There seems, therefore, to be a measurable decline in metabolic activity within certain regions of the central nervous system with increasing age in rats.

The patterns of glucose metabolism show a similar trend in mice. The rate of glucose utilization in brain slices of two different strains of mice decreases with increasing age (Table 3.5).

Table 3.5 Rate of utilization of glucose in brain slices taken from two different strains of mice at 3 different ages, measured as a percentage of the rate for 3-month-old mice.

Mouse strain	Age		
	3 months	10 months	30 months
Strain 1	100%	77%	58%
Strain 2	100%	81%	65%

□ What is the advantage of using two different strains of mice in such a study?

■ The use of two different strains minimizes the possibility of the changes being strain-specific and not applicable to the species as a whole.

Intriguingly, these changes in glucose utilization are correlated closely with the development of progressive behavioural deficits observed in the mice. These findings parallel those found in humans (Table 3.4) but again it is difficult to determine whether the decline in glucose utilization is the cause or the effect of the observed behavioural changes.

Metabolic rate can also be estimated by measuring oxygen consumption. Studies on rat brain slices showed that oxygen consumption in the cortex falls between the ages of 4 and 24 months from 5.51 to 4.82 $\mu l\, mg^{-1}$ dry tissue h^{-1}, a reduction of 14%. The amount by which it decreases varies slightly between regions of the brain. For example, oxygen consumption in the hippocampus decreases by as much as 21%.

Thus studies on oxygen and glucose utilization in isolated animal brain tissue show similar reductions to those found from studies of living humans.

3.5.5 Neurotransmitter alterations with age

Ageing is often associated with physiological changes such as disturbed sleep, increased anxiety, decreased motor activity, and declining or altered mental activity (Section 2.2). As a result, it might be anticipated that there could be parallel changes in the ability of the brain to synthesize and/or respond to certain neurotransmitters. In fact what is observed are changes relating to a few specific neurotransmitters.

Unexpectedly, the activity of the enzyme monoamine oxidase (MAO) is increased in the ageing brain. MAO can inactivate catecholamine neurotransmitters (dopamine, adrenalin and noradrenalin). In keeping with this, small reductions are seen in the total catecholamine concentration of the brain with age. The two regions of the brain where there is a notable catecholamine decrease are the striatum in the basal ganglia where dopamine is reduced, and the hippocampus where the level of noradrenalin is affected.

The losses in the dopamine and noradrenalin systems possibly correlate with altered sleeping habits, depressive illness and dyskinesia. Dyskinesia is a loss of the sense of awareness of the relative movements of the joints, muscles and limbs of the body in space. It is estimated that there is a reduction of about 50% in the efficiency of this process throughout the average human lifespan.

It has been shown that there is an age-related reduction in the concentration of the postsynaptic muscarinic cholinergic receptor protein in the hippocampus (Book 2, Section 8.8.7).

□ Can you suggest a possible consequence of such a reduction? (You will have to recall details of the function of these receptor proteins from Book 4, Section 5.7.4).

■ A loss of hippocampal cholinergic receptor proteins may possibly explain the reduction in efficiency of short-term, or working, memory often associated with the elderly.

In contrast to increased activity of MAO, normal ageing seems to produce little or no change in the activity of choline acetyltransferase, the enzyme involved in the formation of the neurotransmitter acetylcholine. Intriguingly, in patients suffering from Alzheimer's disease, the cholinergic system is particularly affected (see Section 3.7.1).

Summary of Section 3.5

During normal ageing there is a slight reduction in cerebral blood flow, rate of glucose metabolism and oxygen utilization. Electrical activity in the brain changes with age and this may be due to slower conduction of action potentials. The most important change in neurotransmitters is in catecholamines, whereas cholinergic neurons retain most of their original activity.

3.6 Ageing and immunity

Having reviewed the changes in the nervous system with age, it is of interest to look at changes in the immune system which has a number of features in common with the nervous system and with which it interacts in a number of important ways. The details of the immune response and its regulation were discussed in Book 2, Chapter 5.

There is a very gradual decline with age in the levels of activity displayed by white cells involved in the immune response to an infectious organism. When young people are compared with elderly people in good health, the older subjects show marked reductions in many measures of immune function, for example, the rate at which white cells replicate in response to appropriate signals, the effectiveness of cytotoxic (cell-killing) white cells in destroying infected cells, the amounts of lymphokines (signalling molecules) secreted by activated white cells and the number of receptors for lymphokines and neurotransmitters found on the surface of white cells. Some studies have shown that immune activity in healthy elderly people can be less than 30% of that in young people.

However, comparing young adults with elderly people obscures age-related changes that begin in childhood. The thymus gland—a vital site for the maturation of an important group of white cells, the T cells—is at its largest at about one year of age, begins to shrink slowly until puberty, and then deteriorates very rapidly until it is barely detectable as a small nodule from mid-life onwards. This organ secretes *thymic hormones* which help to activate the T cells when an infection is detected. The output of these hormones falls sharply as the organ shrinks and by the age of 60 they are no longer detectable in the bloodstream.

Some aspects of immune function do not appear to deteriorate with age. For example, the total number of white cells in the circulation remains the same as long as the person is in good health and so does the 2 : 1 ratio of helper T cells to suppressor T cells. Problems arise when a novel infection, that is, one to which the elderly person has had no previous exposure, proliferates in the body. The reduced capability of the immune system results in a slower and less effective response, which may allow the infection to cause more severe or more persistent symptoms than in someone younger. Infection in elderly people is often prolonged—coughs and colds seem to hang around for much longer than they did in their youth. Respiratory infections such as bronchitis and pneumonia account for about 80 000 deaths a year in England and Wales (approximately 15% of all deaths), and the great majority of these deaths are among people over retirement age.

Another aspect of the relationship between ageing and immunity is the gradual increase in the incidence of the *autoimmune diseases* (anti-self diseases; Book 2, Section 5.6.2). In these diseases, the immune system mistakes normal body cells or chemicals for foreign infectious organisms and attacks the organs and tissues of its own body, causing very serious damage which may prove fatal. A number of prevalent diseases have been identified as having either an autoimmune component in the pathology of the condition, or of being entirely due to such anti-self attacks. Table 3.6 (*overleaf*) lists some of the more widely known autoimmune diseases. (You are not expected to remember these details.)

Myasthenia gravis is an autoimmune disease in which the immune response is inappropriately directed against proteins in motor end plates (see Book 2, Section 2.4.2).

□ What effect do you think such an attack might have on the affected person?

■ It causes severe disturbance to transmission of signals between motor neurons and the postsynaptic muscle. People afflicted with myasthenia gravis have great difficulty in co-ordinating muscular movements and may exhibit muscle tremors as a result of the damaged end plates spontaneously generating muscle potentials.

Table 3.6 Some of the autoimmune diseases affecting humans.

Insulin-dependent diabetes	autoimmune destruction of the cells that produce insulin results in failure to control sugar levels in the blood
Hashimoto's thyroiditis	inflammation of the thyroid gland, leading to low output of thyroid hormones and disturbances to body metabolism
Coeliac disease and ulcerative colitis	inflammation and destruction of the gut lining, leading to haemorrhage and disturbances to body metabolism
Myasthenia gravis	destruction of the motor end plate at neuromuscular junctions
Rheumatoid arthritis	inflammation of the joints, leading to deformation and chronic pain
Ankylosing spondylitis	progressive deformity of the spine
Psoriasis	a chronic inflammation of the skin
Addison's disease	destruction of the adrenal cortex, leading to low output of corticosteroids and metabolic disturbances
Pernicious anaemia	destruction of red blood cells

There are a number of so-called demyelinating diseases, including multiple sclerosis, some of which may have an autoimmune component in the mysterious destruction of myelin sheaths around neurons (Book 2, Section 2.4.4).

□ What is likely to be the consequence of myelin loss?

■ It leads to profound disruption of the normal pattern of action potentials travelling along the nerve axon. The speed of propagation of the action potentials is also greatly reduced. This may lead to relapses and remissions of neurological damage as in blurring of vision or difficulty in muscular coordination.

Autoimmune diseases are normally treated by drugs that suppress immunity and by anti-inflammatory drugs such as steroids. Both these methods have serious side-effects, including an increase in infections resulting from immuno-suppression. Book 2 (Section 5.6.2), discussed some tentative methods for controlling autoimmune diseases by classical conditioning, which have been demonstrated in animals. It remains to be seen whether humans can be conditioned to suppress their immune reactivity to self tissues.

The relationship between age and the onset of autoimmunity is not clear-cut. Some of the conditions listed in Table 3.6 arise primarily in the young (for example, insulin-dependent diabetes), and others show a peak of incidence in mid-life, with a gradual decline in the rate of onset among older people. But as a very broad generalization, autoimmune reactions are associated with increasing age. There are two main types of explanation for this phenomenon. The first assumes that the normal immune system does not contain any cells that could recognize and attack 'self' tissues, but a new population of 'anti-self' white cells may arise later in life as a result of mutation in the somatic cells. There is evidence that the rate at which mutations occur increases and the efficiency with which mutations are corrected falls gradually with age. The second type of explanation assumes that the immune system always has the capacity to recognize and attack 'self' tissues, but this potentially damaging outcome is normally inhibited.

□ Can you suggest how anti-self white cells might be inhibited? (You will have to recall details of the control of the immune system from Book 2, Chapter 5.)

■ The major control mechanism is supplied by the activity of helper T cells and suppressor T cells—so inhibition of anti-self white cells could either be achieved by the *absence* of the correct activating signals from the helper T cells, or by the permanent *presence* of inhibitory signals from the suppressor T cells.

The breakdown of either of these inhibitory mechanisms will lead to autoimmune diseases and the suggestion is that such an eventuality becomes more likely with increasing age. The nervous system contributes levels of fine-tuning to the control of immune responses (Book 2, Section 5.6), so it too may become less efficient at preventing autoimmunity as it ages.

In conclusion, it should be emphasized that even though the incidence of many autoimmune diseases increases with age, these are still relatively rare conditions that do not affect the great majority of elderly people.

3.7 Neurological disease of the elderly

Neurodegenerative diseases are among the most common conditions affecting the mature nervous system. They are characterized clinically by a slow but relentless deterioration of intellectual faculties resulting eventually in death. The most common of the neurodegenerative diseases associated with ageing are the dementias. Alzheimer's disease, Pick's disease, Huntington's disease, Creutzfeldt–Jakob disease, and senile and pre-senile dementia are members of the group of diseases that are termed *primary dementias*. Alzheimer's disease is discussed in Section 3.7.1; the rest are considered in Chapter 5.

In contrast, the *secondary dementias* occur in association with disorders that include syphilitic infection, blood vessel disease, brain tumours, ~~myxoedema~~ (a macrocytic anaemia condition that results from a deficiency of vitamin B12), alcoholism, nutritional deficiencies, encephalitis, and trauma.

Senile dementia is generally defined as a dementing process originating in the brain of a person over the age of 65. The term pre-senile dementia has been used when the disease has an earlier age of onset, which can be as young as 45. Onset of pre-senile dementia is characteristically slow and fitful but eventually it develops into full-blown dementia. These definitions apply to clinical examination. Pathological examination after death reveals that most cases of both senile and pre-senile dementia show structural changes in the brain which are especially characteristic of Alzheimer's disease. Strict criteria used to diagnose Alzheimer's disease (see Section 3.7.1) were introduced by Larry Embree and his colleagues in the USA but unfortunately they can only be confirmed after the sufferer's death.

In order to consider the relative importance of each of the primary dementias, it is necessary to have some idea of their relative incidences. Pick's disease,

Creutzfeldt–Jakob disease and Huntington's disease are rather rare and occur in less than 1% of the elderly population but Alzheimer's disease and senile dementia are relatively common and together occur in about 15% of the population over 65 years of age.

Alzheimer's disease is undoubtedly the most common of the primary dementias occurring in the pre-senile period of life. Although its incidence does increase with each decade of lifespan, the disease is nevertheless distinct from the process of ageing. It is now generally agreed that Alzheimer's disease is essentially identical to the most common form of senile dementia and the term is now used to include elderly as well as pre-senile people.

This high incidence highlights the enormous impact that the primary dementias have upon modern society. Whilst tranquillisers may be used to calm agitated behaviour in demented patients, generally the most important consideration is to find the most suitable environment for that person to live in. Family and social pressures often dictate that this means hospitalization.

Rather than producing a lengthy documentation of all the many possible neurological diseases of the elderly, the following section is confined to a more detailed examination of Alzheimer's disease which, as described above, is the most common of the dementias.

3.7.1 Alzheimer's disease

In 1906 a German physician called Alois Alzheimer reported what was then recognized as a new kind of disease but which now bears his name. A 51-year-old woman became ill in the spring of 1901, had an overwhelming jealousy of her husband and above all developed increasing memory impairment. She was unable to find her way about in the house, dragged objects to and fro, hid herself and thought that people wanted to kill her. At this time she would scream loudly for no apparent reason and suffered from serious perception disorders. When the doctor showed her familiar objects, she gave the correct name for each, but immediately afterwards had completely forgotten everything that she had been shown. While reading, she would omit sentences or spell every word and read without expression. In a writing test, she often repeated the same syllables while omitting others and became confused and absent-minded. By late November of that year, she was admitted to hospital, where she was permanently disorientated as to time and place. Erroneously, she thought her daughter was married and could not remember her husband's name. In 1902 she became completely impossible to manage, screaming for hour after hour and stamping her feet. In the end she was completely apathetic, confined to bed in the fetal position (arms hugged to her chest and her legs drawn up) and incontinent.

After her death her brain was found to be atrophic, that is, greatly reduced in size. When sections of the cortex were treated with a silver-based stain (Book 2, Box 2.3), Alzheimer found neurons containing tangled strands or fibrils; in some cases the neurons were completely replaced by such tangles. He concluded that these tangle 'tombstones' must be composed of a very inert fibrous material which survived even though the cells had been destroyed. Alzheimer concluded that this was a peculiar and little-known disease process. In addition he found scattered deposits of silver-staining masses of ill-defined material throughout the entire cortex.

The importance of this discovery was recognized by the distinguished contemporary psychiatrist Dr Emil Kraepelin who in 1910 emphasized that Alzheimer's disease was a particularly serious form of senile dementia. For many years subsequently it was thought that Alzheimer's disease was a condition affecting relatively young individuals, up to the age of about 65 years. It was also considered that the disease was due to arteriosclerosis, that is, hardening and eventual blockage of the blood vessels. Amongst others, the British neuropathologists Bernard Tomlinson and Nicholas Corsellis, in very extensive studies carried out in the early 1970s, showed that the histological changes in both elderly and young Alzheimer's patients were indistinguishable. Moreover, disease of the blood vessels of the brain was absent in more than half the cases that they observed.

It is now known that the main behavioural symptoms characteristic of the disease are increasing memory impairment and perception disorder. They also include bouts of jealousy, violent reactions to everyday events and incontinence. All the observations of Alzheimer's disease that have been carried out reveal the same striking structural abnormalities that accompany progressive cerebral disorder in patients with the disease. The brain is generally atrophied (see Table 3.2) with the frontal and temporal lobes showing the most wasting away. Tissue staining and examination under the light microscope show that the histological hallmarks of this disease are *senile plaques, neurofibrillary tangles* and, in the hippocampus, *granulovacuolar degeneration*. These conditions (explained below) are all indicative of the chronic degeneration of neurons.

Senile plaques consist of an ill-defined mass of abnormal unmyelinated neuronal processes, thickened axons, degenerating dendritic processes, distended neuronal processes, abnormal axon terminals, and enlarged synaptic endings (Figures 3.6 and 3.7, *overleaf*). Lipofuscin granules (see Section 3.4.2) are also present and, often, the senile plaque has a more or less dense core of fibrillar material. Electron microscopic studies have shown that plaques contain contributions from many different cell types and that no single cell or portion of a cell is completely responsible for their existence.

The neurofibrillary tangles seen in Alzheimer's brain tissue are made up of many closely packed neurofilaments that displace other cytoplasmic organelles within the neurons. The individual neurofilaments are generally tubular in structure (Figure 3.8, *overleaf*). They vary in diameter from 10 to 20 nm. There seems to be no continuity between the closely packed fibrous bundles of these neurofibrillary tangles and normal cellular organelles. Although lipofuscin bodies are often present, the amount of lipofuscin is not correlated with the degree of degeneration seen.

Small vacuoles or holes may appear in the cytoplasm of the neuron; this is called granulovacuolar degeneration. The vacuoles are up to 5 μm in diameter, each containing a small granule some 0.5–1.5 μm across.

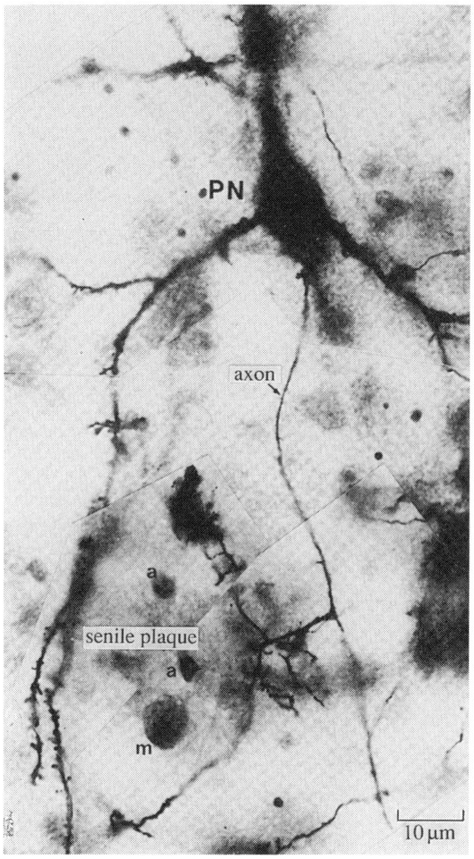

Figure 3.6 Photograph of a Golgi-stained cortical neuron which has axon branches involved in a senile plaque. (Ignore other labels which were references used by the original researcher.)

Histochemical techniques, the staining of tissue to reveal the localization of particular chemicals, have been extensively applied to samples of Alzheimer's tissue over many years and, indeed, provided some of the earliest clues as to the biochemical nature of this disease. Electron microscopy has been particularly valuable in examining the detailed structure of senile plaques. Histochemical methods have shown that senile plaques display a striking increase in the activity of certain enzymes, and electron microscope studies have revealed that there is an increase in the numbers of mitochondria present. It has been suggested that a local increase in metabolic processes, as reflected by increased enzyme activity, may play an important role in the formation of senile plaques. One important point to note is that the development of plaques is not confined to neurons that respond to specific neurotransmitters—they form in neurons of many types.

One notable exception to this increase in the activity of enzymes is monoamine oxidase (MAO). This is intriguing since this is the one enzyme that shows an identifiable increase in the normal ageing brain (see Section 3.5.4) and emphasizes the fact that Alzheimer's disease is not simply a speeded up form of the normal ageing process. However, the cholinergic system is less efficient in people with Alzheimer's disease. It is thought that the reduced synthesis of acetylcholine in the brain may be associated with the severe memory loss from which they suffer.

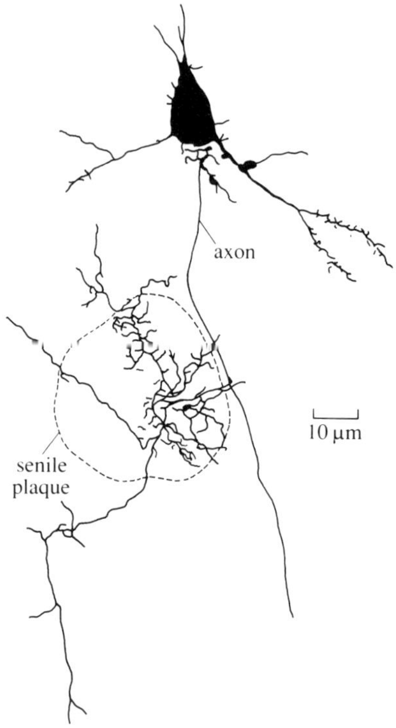

Figure 3.7 Drawing of the same neuron as photographed in Figure 3.6 showing a more complete picture of the axonal branching in the senile plaque.

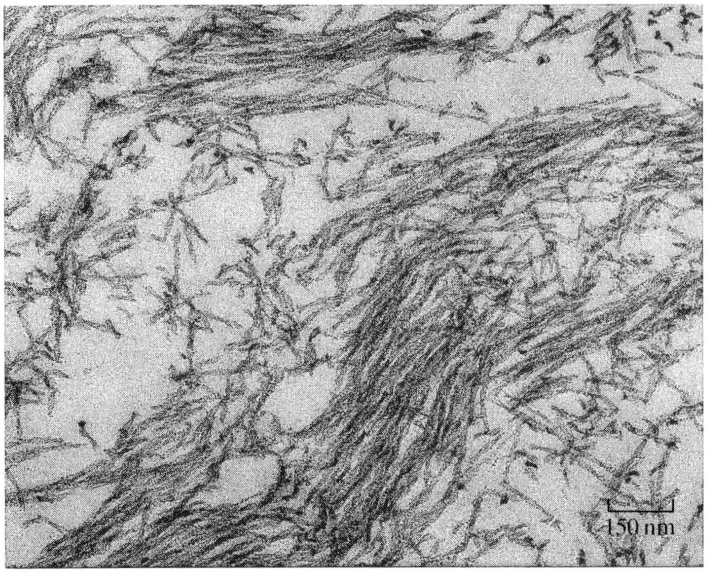

Figure 3.8 An electron micrograph of a neurofibrillary tangle.

Unlike senile plaques, neurofibrillary tangles show a marked lack of enzyme activity thus providing evidence for separate, though unknown, causes of these two manifestations of the condition.

Until the 1990s, no genetic basis of Alzheimer's disease had been demonstrated, but some investigators believe that a more detailed examination of families in which a number of cases appear might reveal that hereditary factors are involved. A family was identified in Nottingham that suggested there was a pattern of inheritance of an 'Alzheimer's gene'. Further studies provided evidence that about 10% of cases of the disease are inherited. The remaining cases are presumably due to environmental factors though whether such environmental influences and the supposed 'Alzheimer's gene' interact at all is unknown. In the late 1980s the idea that dietary intake of aluminium, for example in drinking water, was a factor in causing Alzheimer's disease generated much excitement. However, at the time of writing (1992) this hypothesis remains to be proven.

Several avenues of investigation are likely to occupy the attention of researchers into Alzheimer's disease in the years to come, including attempts to locate the gene or genes that are involved in at least some cases of the disease. However, attempts to investigate the disease on the basis of its behavioural and physical symptoms are equally important. Here, one starts with the most consistent and specific abnormalities and traces their development backwards, with the aim of eventually identifying those molecular changes that may be involved in the development of the disease. Much research effort will continue to be directed towards studying the formation of senile plaques because of the fact that they involve neurons of many types. The growing opportunities, in terms of more sophisticated technology and deeper background knowledge, for understanding the biochemical events underlying the structural changes that take place in the Alzheimer brain bode well for unravelling its causes, and thus ultimately preventing, this common and tragic disorder.

Summary of Chapter 3

Biological ageing is a process that involves normal, progressive and inherent change of biochemical and physiological function as well as structural cellular change over the lifetime of an organism. Among the many problems that beset the study of biological ageing is the difficulty of applying findings from experimental animals to the human situation.

The wet weight of the brain progressively decreases with age as a result of intracellular water loss. The issue of how much neuronal loss occurs during ageing is still somewhat controversial but it seems that the amount varies between different regions of the brain. Brains of very old patients show evidence of progressive enlargement of neuronal cell bodies, continued loss of dendrites and the development of varicosities.

In healthy old people the regional cerebral blood flow (CBF) is slightly reduced compared with young subjects but in those with dementia it is reduced even more. In animals, the rate of glucose and oxygen utilization by the brain decreases with age. On the other hand, except for a decrease in levels of dopamine and noradrenalin, ageing does not seem to be accompanied by marked changes in neurotransmitter metabolism. Both electrical activity in the brain and electrical responses to stimuli change with age. Given all these biological changes it is perhaps not surprising that there is loss of some cognitive functions and motor skills with increasing age.

The level of immune response to infection gradually declines with age; some studies have shown immune activity in elderly people to be less than 30% of that in younger subjects.

In some elderly people, neurodegenerative diseases lead to deterioration in intellectual faculties and dementia. Primary dementias are characterized by degeneration of the cerebral cortex and varying amounts of involvement of other brain structures. Secondary dementias occur in association with other disorders. Alzheimer's disease is the most common of the primary dementias occurring in the pre-senile period of life. In Alzheimer's disease, chronic degeneration of neurons occurs and the disease is characterized by the presence of senile plaques, neurofibrillary tangles and granulovacuolar degeneration.

You have seen in this chapter a number of examples of where changes in the structure and performance of neurons and groups of neurons occur as a function of ageing. In some cases, it is not difficult to appreciate how the change in structure would relate to a change in performance of the neurons in question. Somewhat depressingly for those getting older, the changes tend to be those characterized by either a *reduction* or *loss* of function. For example, loss of myelin leads to slower action potential transmission, loss of dendrites prevents normal presynaptic inputs from affecting a neuron and reduction in the synthesis of dopamine lowers transmission in dopaminergic pathways. In Chapter 4 the discussion will return to such changes and, for example, it will show how loss of dopamine from certain neurons affects the neural systems of which they form a part. In some cases, such changes at a neural level can be further mapped onto behavioural changes. For example, a reduction of dopamine in people with Parkinson's disease is associated with reduced motor skills. In other cases, more work is still needed to relate the psychological and biological aspects of ageing.

Objectives for Chapter 3

When you have completed this chapter, you should be able to:

3.1 Define and use, or recognize, definitions and applications of each of the terms printed in **bold** in the text. (*Question 3.7*)

3.2 Describe some of the difficulties associated with the study of ageing, and some of the pitfalls in trying to relate physiological to biological factors of ageing. (*Question 3.1*)

3.3 Distinguish between normal and pathological biological ageing at the cellular, biochemical and behavioural levels. (*Question 3.2*)

3.4 Discuss the evidence for neuronal cell loss taking place in specific regions of the normally ageing brain. (*Question 3.3*)

3.5 Explain how regional cerebral blood flow is used as an experimental measure of energy metabolism in the brain. (*Question 3.4*)

3.6 Describe how imaging techniques have contributed to the study of the ageing brain. (*Questions 3.4 and 3.5*)

3.7 Describe those aspects of immune function that are affected by age. (*Question 3.6*)

3.8 Describe the main behavioural manifestations of Alzheimer's disease.(*Question 3.8*)

3.9 Name and briefly describe the structural changes in the nervous system that accompany Alzheimer's disease. (*Question 3.9*)

Questions for Chapter 3

Question 3.1 (*Objective 3.2*)
Why must great care be taken in making conclusion about the ageing brain from studies on post-mortem tissue?

Question 3.2 (*Objective 3.3*)
(a) Which of the following increase, decrease, or remain about the same in the normally ageing brain?

1 neurons in the cortex

2 cerebral blood flow

3 latency of evoked potentials to stimuli

4 level of monoamine oxidase (MAO), the enzyme that can inactivate catecholamine

(b) How do the above differ between the normally ageing brain and the pathologically ageing brain?

Question 3.3 (*Objective 3.4*)
Describe the model of ageing in the central nervous system that proposes that one population of neurons dies and regresses whereas another population survives and grows.

Question 3.4 (*Objectives 3.5 and 3.6*)
What has the use of PET scanning revealed about changes in cerebral blood flow (CBF) that occur with age?

Question 3.5 (*Objective 3.6*)
Describe the main advantages of using scanning techniques to study brain function.

Question 3.6 (*Objective 3.7*)
What are the two main types of explanation for the fact that autoimmune reactions become more prevalent with increasing age?

Question 3.7 (*Objective 3.1*)
What is meant by the term senile dementia?

Question 3.8 (*Objective 3.8*)
What are typical behavioural traits of a person suffering from Alzheimer's disease?

Question 3.9 (*Objective 3.9*)
What are senile plaques and neurofibrillary tangles as described in people with Alzheimer's disease?

References

Buell, S. J. and Coleman, P. D. (1979) Dendrite growth in the aged human brain and failure of growth in senile dementia, *Science*, **206,** pp. 854–856.

Further reading

Davies, D. C. (ed.) (1989) *Alzheimer's Disease: Towards an Understanding of the Aetiology and Pathogenesis*, J. Libbey, London.

CHAPTER 4
BIOLOGICAL APPROACHES TO THE STUDY OF NEUROLOGICAL DISEASES

4.1 Genes and environment

This chapter explores the possible causes of neurological diseases and disorders. It focuses on one or two examples that have been chosen because a great deal is known about them and because they highlight the known causes of the disorder—environmental factors, genetic influences or an interaction between these two.

Diseases or disorders can be classified in a number of ways. One way is to consider them as being on a spectrum. The environmentally determined conditions caused by nutrition factors (for example, the intake of abnormally high doses of heavy metals) or infectious diseases (such as poliomyelitis), at one end of this spectrum, and genetically determined diseases such as Huntington's disease and Lesch-Nyhan's disease at the other end. In these two examples the sufferers are ill because in each case they were born with a defective allele of a single gene, that is, a gene mutation (Book 1, Section 3.2.5). The sequence of bases in the allele is altered in some way so it no longer provides the blueprint for the correct structure of a protein or for a functioning enzyme, and there is no known way of modifying the environment so as to alleviate the effect of the mutation. Some congenital diseases such as Down's Syndrome (see Book 4, Section 4.5.2), are associated with chromosomal abnormalities that involve a number of genes.

Between these two ends of the spectrum lie many disorders in which environmental and genetic factors both play a substantial part. For example, there are nervous disorders in which genetic influences may be important, such as epilepsy, multiple sclerosis and Parkinson's disease. These disorders, like many others, show no clear genetic association even though there is some evidence that genetic inheritance is one important factor determining the likelihood that a person will develop one of them. It is difficult, however, to measure the genetic effect in these disorders. Clearly, the environmental influences on a person during development are important in determining whether symptoms will develop or not. If more members of a family are affected than can be accounted for by chance, it may be because they share similar environments, or, because they have certain alleles in common or both.

This picture is further complicated by the fact that, in general, no two genotypes respond to the same environment in the same way. The diverse phenotypes that may arise from the interplay between a given genotype and the various environments in which this genotype may live constitute the **norm of reaction** of that genotype. It is evident that the entire norm of reaction of any genotype can never be known. Indeed, to know it one would have to expose different individuals with this genotype to all environments and to study the resulting phenotypes. This is impossible, because the number of environments is infinite and because large numbers of individuals with identical genotypes do not exist among humans.

A further complicating factor is that genetically determined diseases produce different clinical symptoms in different individuals. This is because their expression (their phenotype) is also affected by the expression of other alleles present at other gene loci. This is sometimes referred to as the genetic background. For example, enzyme activity, receptor binding activity, or structural protein functions may vary slightly, albeit still within the normal range and these variations may be sufficient to affect the expression of the disease.

Molecular techniques are used to investigate the molecular components of single genes or alleles. The disease phenylketonuria (PKU; see Book 1, Section 3.3), has been understood much better since these techniques have made possible the identification of the gene at the DNA level. The phenotypic expression of mutant alleles can be understood in terms of the biochemical defect, such as the absence of a normally functioning enzyme which disrupts a critical biochemical pathway. This in turn results from a change in the structure of the DNA. Recently, the gene for Lesch-Nyhan syndrome has also been identified, but how this links to the phenotype of the individual remains a mystery. Other genes, such as the one associated with Huntington's disease, defy isolation, although their approximate location on a particular section of a specific chromosome is known.

It is important to remember that geneticists often name genes after the *abnormalities* that result from the gene mutation. Thus the name usually reflects the associated abnormal phenotype rather than the normal gene product. In fact a name such as 'Huntington's disease' tells us nothing about the normal gene product.

Some words of warning about the terminology sometimes used to describe the relationship between genes and the environment are appropriate here. Phrases such as 'the gene for Huntington's disease' are often used but may be misleading in that they suggest a direct causal relationship between a sequence of DNA and the phenotype. It is important to remember that the phenotype of an organism results from a complex and poorly understood series of interactions between genetic and environmental factors in a dynamic relationship. This was exemplified in Book 1 by the disorder phenylketonuria. Patients suffering from the disease are unable to convert phenylalanine into tyrosine but may be freed from symptoms by a change in the environment.

□ How is phenylketonuria treated?

■ By control of the dietary intake of phenylalanine and tyrosine so that normal blood levels can be maintained.

There are many genetic diseases for which no important environmental factor has yet been identified. However, there are other diseases where the association between environmental factors and the effects on health are convincing, as in the case of heavy metal poisoning and phenylketonuria. It is this group of diseases that are discussed in the next section.

There is one final point to bear in mind. The nervous system of an individual cannot be considered in isolation from other systems. For example, an adequate supply of oxygen and nutrients from the respiratory and circulatory systems are essential for the correct functioning of the nervous system. Similarly, diseases cannot be considered in isolation; each patient who suffers from the disease shows a personal and individual reaction.

This chapter first examines poisons and their effects on the nervous system, mainly through the examples of lead and copper. Second, it turns to a group of complex movement disorders exemplified by Huntington's disease and Parkinson's disease. In contrast to Parkinson's disease, Huntington's disease is associated with the inheritance of a single defective allele. The final section considers genetic approaches to understanding neurobiological disease and in particular Lesch-Nyhan disease.

4.2 Neurotoxic disorders

The nervous system has been shown to be especially vulnerable to chemical poisons. *Neurotoxicology* studies the action of chemical, biological and certain physical agents that produce adverse effects on the nervous system and/or behaviour.

You may not be surprised to learn that responses to poisons vary from individual to individual.

□ Why should this be so?

■ Because of the interplay between the poisons, other environmental factors and genetic factors.

Genetic factors causing increased susceptibility to neurotoxic disorders may play a role in both the metabolism of the poisons and their interaction with the nervous system. One of the best understood examples where genetic factors are involved in the metabolism of poisons is that of the breakdown of alcohol.

The major product of alcohol digestion and degradation is a compound called acetaldehyde. It is responsible for a number of symptoms occurring during intoxication and its after-effects. Alcohol is broken down in the liver by two enzymes: alcohol dehydrogenase (ADH, not to be confused with a hormone having the same initials!) converts alcohol to acetaldehyde and acetaldehyde dehydrogenase (ALDH) breaks down the acetaldehyde. ADH exists in two alternative forms, ADH_1 and ADH_2, each being the product of a different allele of the same gene. ADH_2 is many times more active than ADH_1. Thus acetaldehyde is formed at a different rate depending on the alleles carried by the individual.

ALDH also exists in different forms, and again each is the product of a different allele, ALDHI and ALDHII. The former has a much higher affinity for acetaldehyde and breaks it down quickly, thus preventing high concentrations forming after alcohol intake. Individuals who do not have the ALDHI allele tend to experience discomfort and symptoms of intoxication even at low alcohol intake levels.

The alternative alleles of both genes differ remarkably in their distribution between races. More than 90% of white Europeans have ADH_1 only, whereas 85% of Chinese have ADH_2. On the other hand, virtually all white Europeans have both ALDHI and ALDHII while about 50% of Chinese have only ALDHII. As a result people vary, both in the time they take to become intoxicated and for the length of time they are affected. Thus, there are clear genetic variations involved in the process of metabolism of alcohol. The same is almost certainly true for other poisons, and this must be borne in mind when interpreting available data. (You do not need to remember the details of this or other examples given in this chapter; rather, examples are intended to illustrate the important principles that you are expected to learn.)

Disorders of the nervous system may occur following exposure to a wide variety of substances, some examples of which are shown in Table 4.1.

Table 4.1 Substances poisonous to the nervous system.

Abused substances (e.g. alcohol, inhalants and narcotics)
Therapeutic drugs (e.g. anti-epileptic, anti-cancer drugs)
Chemicals designed to affect certain non-human organisms (e.g. pest control products)
Industrial chemicals and chemical warfare agents (e.g. trichloroethylene used in drycleaning; mustard gas)
Some food additives and some natural components of food (e.g. copper is a natural component of some foods)
Products of certain organisms (e.g. snake venom, poison from marine animals such as jellyfish)

In the last few years, more and more chemicals have been recognized as potential poisons, because of their ability to cause neurophysiological alterations, behavioural abnormalities, or disruption of neurochemical activities. This is partly because since the 1980's there has been an increased interest by research workers. In addition there has been enhanced public awareness of the major problems caused by deliberate or unintentional exposure to hazardous chemicals. This section focuses on one of these, the environmental chemical lead.

4.2.1 The poison lead

Metals can have both therapeutic and poisonous effects which arise from their physical and chemical properties. 'Heavy metals' such as lead and mercury can form complexes with proteins. It is for this reason that they have become of toxicological interest. Iron is an important constituent of haemoglobin (the oxygen-carrying pigment in vertebrate blood), and so is required in relatively *large* amounts by the body. Other metals such as copper and manganese are constituents of enzymes and are thus essential trace elements, which means they are necessary in *small* quantities in the diet.

This section provides a description of both the clinical and experimental neurotoxicity of lead. The general hazards of lead have been known for centuries. Historically lead was used as solder, as 'sugar of lead' for sweetening wine, as a constituent of ceramic glazes, cosmetics and water pipes, in paints, and as a fuel additive in petrol for internal combustion engines. The Romans, at least the emperors and the aristocrats, almost certainly suffered from lead poisoning. One result of such poisoning is gout and there is considerable evidence to show that gout was common in Rome. This was a result of Roman drinking and eating habits, particularly of a substance called sapa. This was a boiled down grape syrup used as a sweetener and preservative both for wine and food stuffs. The boiling was done in lead kettles and consequently sapa was a rich source of lead.

The effects of lead are not confined to the human population. Earthworms have a system for detoxifying ingested lead whereby it becomes bound within cells surrounding the gut. It is released into the soil when the worm dies. But other organisms have not evolved such a system and are dramatically affected by lead in the environment. One such example in the UK is the mute swan, *Cygnus olor*. In 1979 it was estimated that as many as 3 000 swans a year could be dying from lead poisoning as a result of eating lead shot used as weights and carelessly discarded by anglers. Swans swallowed the lead, either accidentally or mistaking it for the grit they need to aid digestion. Once swallowed, the lead lodges in the gizzard and slowly permeates the blood. The lead inhibits peristalsis, which is the process that propels food through the gut by means of rhythmic muscular contractions. The end result is that the swans die of starvation even though their guts may be full of food. There is some evidence that the swan population is increasing now that alternatives to lead weights are in use.

There is a wealth of sources of lead in the environment. Lead exists in two forms, organic (as in petrol) and inorganic (as in paint). The difference here does not concern us so much as the effect once inside the body. In 1983 a nationwide study of 53 towns and villages throughout Britain pinpointed lead in house dust and school playgrounds as a potentially important hazard to young children. Lead dust is present in the air and in the soil. From there it contaminates food and water. However, practically all the lead found in the atmosphere comes from car exhausts. Lead is put into petrol in the organic form to improve the octane quality, or anti-knock performance of the fuel, so that the engine runs better. It also lubricates the valves in the engine. Lead enters the atmosphere from exhaust pipes mostly in the form of inorganic lead, which is formed by the action of other petrol additives. Nevertheless some organic lead does get through, particularly when an engine starts from cold. Other major sources of inorganic lead are paint and the pipes that carry water in many older houses.

Organic lead is far more difficult to detect in humans than inorganic lead and is easily absorbed by the skin. This chapter is concerned with inorganic lead, which enters the body mainly through the mouth in food, water, dust and dirt and which is mainly held by the red blood cells. Analysis of blood is the usual method of testing for lead in the body. But it is important to bear in mind that the human body also carries an additional lead burden, which is difficult to detect, in the form of organic lead.

Figure 4.1 (*overleaf*) shows the correlation between the amount of lead in petrol and the average levels of lead in people's blood between 1976 and 1980 in the USA. Notice the strong correlation between the amount of lead in the environment and the amount in people's blood. In 1975 the USA government began to phase out the use of lead in petrol. This was subsequently matched by a fall in the levels of lead in people's blood. Such a correlation, however, cannot prove that one of the effects is caused by the other. For example, there may have been a decrease in the use of other lead-containing substances over the same period which could account for the decrease in blood levels of lead.

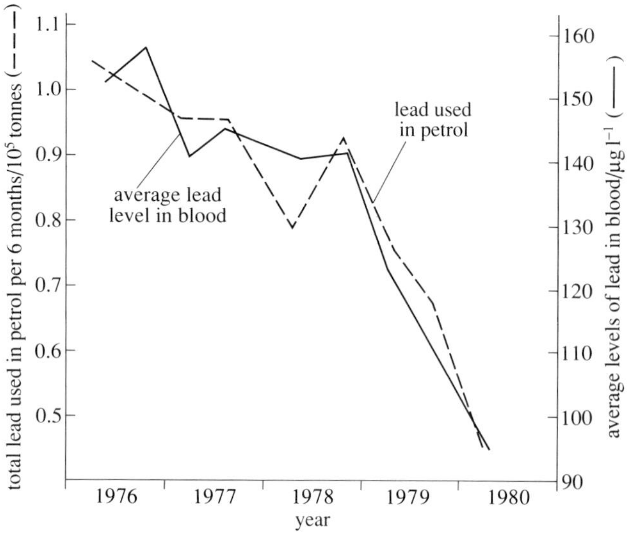

Figure 4.1 A possible relationship between the lead in petrol and lead in the blood.

However, an Italian study conducted in Turin between 1977 and 1979, using special *isotopes* of lead in petrol, provided evidence for a causal relationship. An isotope of a particular substance has identical chemical properties to the usual form of the same substance but has a slightly different atomic structure sufficient for it to be readily identified and separated from it. The experiment involved substituting a special isotope of lead for the normal additive in all petrol sold in the area. A mass screening of residents then assessed what proportion of the lead in their blood consisted of this isotope. This experiment found that 25% of lead in the blood of Turin residents came from petrol. This was a minimal figure because the quantity of lead in the blood was still rising 18 months later when the experiment ended. By 1990 most countries had placed some limits on the maximum amount of lead in petrol, although most still do not use exclusively lead-free petrol.

It is difficult to determine a maximum safe daily intake of lead because other factors affect the amount of lead taken up by the body. For example, people with diets low in vitamin D and iron take up lead more readily. Other factors that increase the uptake of lead include the acidity of the gut, a lack of calcium in the diet, the amount of alcohol consumed, and cigarette smoking.

Quantifying the risks from lead is not easy. Data from epidemiological studies (Section 1.4.2) in many countries, and from experimental studies in which animals are given lead in their diet, demonstrate psychological impairment at blood level concentrations as low as 150 μg^{-1}. In the USA in 1988 it was estimated that 17% of children exceeded this level.

4.2.2 Clinical toxicology of lead exposure

The evidence that lead is poisonous to the nervous system is very strong. Unfortunately, the neurotoxic effects are not readily reversible when external exposure stops, because lead is stored in the body. Neurotoxic effects are induced prior to the onset of toxicity to any other organ systems.

High-dose exposures, although relatively rare, do still occur. They are associated with so called lead encephalopathy, the symptoms of which include lethargy, seizures, brain oedema (swelling), and eventual coma and death. At lower doses, the signs of poison are not always readily observed, but there is a wealth of evidence, as shown below, to suggest that they are real. Furthermore, there appear to be long-term effects even from low level exposure.

The special sensitivity of children was not appreciated until this century. Children, particularly those under 5 years of age, demonstrate higher mean blood level values than adults given the same environmental exposure.

□ Suggest possible reasons for this observation.

■ Apart from the obvious difference in body size between adults and children, there may be enhanced rate of absorption in children, and/or toddlers may eat non-food products that are more likely to contain lead.

Lead absorption from the gut appears to differ with age. In adults approximately 10% of ingested lead is absorbed, whereas this can be as high as 50% in children. Furthermore, lead can pass through the placenta to reach the developing fetus.

A large number of studies on the behaviour and intelligence of lead-exposed children compared with children with no unusual exposure to lead have been carried out. Apart from the amount of lead in the body, other factors need to be taken into account in such studies. These include the source and duration of the original exposure; the fact that behaviour and intelligence are affected by genetic, perinatal, nutritional and socioeconomic variables; the difficulty in measuring certain features of behaviour such as attention span; and the problem that children may have been exposed to other poisons. Furthermore, the two groups studied, lead-exposed and those with no unusual exposure to lead, must each consist of a representative sample. It is important to note that lead exposure is highest in inner cities where the inhabitants may also suffer from poor housing and poor nutrition. These and other factors, such as the use of different techniques for measuring the quantity of lead in the body, mean that the interpretation of some of the results from such studies is not straightforward.

Even so there is general consensus among researchers that the performance of lead-exposed children is impaired. The debate is mainly centred around the level, in terms of blood lead, at which adverse effects on the nervous system can occur. Some of these adverse effects are listed in Table 4.2 (*overleaf*).

Table 4.2 Effects of lead exposure in children.

(a)	Impaired reaction time
(b)	Decreased mean IQ score
(c)	Impaired attentiveness
(d)	Lowered visual and auditory sensitivity
(e)	Hyperactivity
(f)	Impulsiveness
(g)	Decreased learning in both acquisition and retention
(h)	Slow nerve conduction velocity

4.2.3 Nature and mechanism of lead toxicity

To understand the mechanism of lead neurotoxicity, it is necessary to turn to neurochemistry. Even at low levels of exposure, lead produces extensive neurochemical changes. In order to investigate the mechanisms of toxicity at the neurochemical level, animals, particularly developing rodents, have been used. It is important to bear this in mind when examining data from such investigations, since any conclusions drawn may not be directly applicable to humans. Some of the consistent findings from work carried out in living animals are summarized in Table 4.3.

Table 4.3 Some neurochemical effects of lead exposure in rodents.

(a)	Changes in amounts of neurotransmitter release
(b)	Increase in the density of cerebellar GABA receptors
(c)	Blocking of the release of acetylcholine at neuromuscular junctions
(d)	Interference with ion movements at neural synapses
(e)	Inhibition of the action of enzymes associated with the neural membrane

□ Examine the neurochemical effects of lead exposure in rodents given in Table 4.3. Are both presynaptic and postsynaptic aspects of neurotransmission affected?

■ Both are affected. For example (a) is a presynaptic change and (b) is postsynaptic.

□ Is the functioning of the peripheral nervous system affected by lead exposure?

■ Yes, as demonstrated by (c).

At the cellular level the nature of lead neurotoxicity is not fully understood. Several studies have been undertaken to determine the localization of lead inside cells of the nervous system. There is general agreement that the site of lead's action is on cellular membranes and particularly within mitochondria, the organelles which are crucial for energy production in the cell.

At present there is no satisfactory unifying hypothesis to account for the clinical and experimental neurotoxicity of lead. Nevertheless, among the neurotoxins that play a significant role in clinical disease, sufficient is known about lead to guide decisions necessary to reduce exposure to this poison. The most important aspects of treatment are preventative because lead accumulates in the body and because the associated effects are not readily reversible when external exposure stops.

4.2.4. Copper and the brain

In marked contrast to lead, copper is an essential element for life. It is an important constituent of enzymes, an important component of the energy-producing system in the mitochondria of cells and also essential in the CNS for myelin formation.

Copper absorbed from the gut is taken up by tissues, mainly the liver. It is excreted in urine and bile fluid (a substance produced by the liver and released into the gut). Copper deficiency can result from malnutrition but an excess of copper is poisonous. Both these conditions can arise in people as genetically determined 'inborn errors of metabolism'. Given the importance of copper it is not surprising that, in contrast to lead, there are genes specifically involved in the metabolism of copper in humans. A study of some of these genes provides an introduction to the importance of genetics in neurology.

Wilson's disease is an inherited recessive disorder. This means that it is caused by a recessive allele of a particular gene (see Book 1, Section 3.2.3). People suffering from the disease have an excess deposit of copper in their tissues, particularly the liver and the brain. The administration of an isotope of copper in both normal and affected individuals allows the movement of labelled copper to be monitored over a number of days. In contrast to normal individuals, little of the labelled copper is released from the liver in affected individuals. This is probably due to defective bile excretion. The total body retention of labelled copper is also markedly prolonged in patients. Not surprisingly, affected individuals suffer from liver disease. But they also suffer from destruction of neurons in the brain, particularly in the basal ganglia.

□ What would be a possible treatment?

■ Remove the excess copper.

The excess copper can be removed by oral administration of certain chemicals (either penicillamine or triethylene tetramine). These drugs increase the excretion of copper in the urine.

□ Removal of the copper results in reversal of the symptoms except where destruction of cells has occured. How does this result compare with similar observations for lead poisoning?

■ Unlike lead, at least some of the damage caused by excess copper is reversible.

As long as a correct copper balance is maintained, a remarkable degree of recovery follows. Thus Wilson's disease, like phenylketonuria, is a striking example of a reversible biochemical abnormality.

In contrast to those who have Wilson's disease, individuals suffering from Menke's disease (also called kinky hair disease) are unable to absorb copper from the gut. Menke's disease is also an inherited recessive disorder. Infants, if left untreated, fail to thrive and suffer extensive neuronal degeneration. The hair is characteristically fragile, sparse, white and kinky, hence the name of the disease. The profound copper deficiency can be corrected by administering copper other than via the gut, generally by injection.

Rarely, individuals who have not inherited the defective allele manifest the same symptoms as people with Menke's disease. Such individuals might be described as phenocopies (Book 1, Section 3.6.4), since their symptoms mimic those of an individual with a genetic defect.

□ Suggest how a phenocopy of this disease might arise?

■ Copper deficiency could occur as a consequence of malnutrition.

Interestingly, it has long been known to the veterinary profession that lambs born of ewes grazed on copper-deficient pastures develop the demyelinating disease, *sway back*. The symptoms of these copper-deficient newborn lambs are similar to those of infants suffering from Menke's disease.

Summary of Section 4.2

This section has considered the interaction of specific metal poisons in the environment with the genotype of an organism and showed how these can cause disorder of the nervous system. One way of approaching the study of a disease is to begin with the suspected causative agent and identify the associated phenotypic effects. Lead is an example of an environmental poison. When inside the body it is a neurotoxin. It has profound effects on the neurochemistry of rodents and is associated with behavioural changes in humans, particularly in children. There is no known gene involved in the metabolism of lead. In the case of copper, however, which is an essential element for life, there are a number of genes involved in its metabolism. Abnormalities in these genes give rise to Wilson's disease or Menke's disease. Phenocopies of these phenotypes can be produced by environmental factors. Wilson's disease and Menke's disease illustrate the importance of genetics in neurology.

4.3 Basal ganglia disorders

In trying to understand any disease the primary aim is to determine the link between cause and the clinical symptoms or expression of the disease. The link between copper metabolism (cause) and the clinical symptoms of Menke's disease and Wilson's disease is clear. The identification of that link gives a better understanding of the normal functioning of the organism as well as providing a basis for treatment.

More frequently the study of diseases begins at the other end of the link, with the presentation of a constellation of symptoms from which the nature and cause of the disease has to be diagnosed. Too often it is not possible to establish a connection with any degree of certainty because of the richness of symptoms which in turn reflect the complexity of the functioning of the body and its interaction with the environment. Another way of attempting to learn about neurological function is to investigate one particular region of the brain through the study of disorders associated with it. Best understood in contemporary neuroscience is the close link between the function of the basal ganglia and the clinical symptoms of a number of crippling movement disorders. The study of these disorders has had a strong influence on our understanding of basal ganglia function.

Two movement disorders are considered here, Parkinson's disease and Huntington's disease. In each case the smooth coordination of muscles is lost and in each something goes wrong with the basal ganglia. In addition, they are both progressive diseases that develop in later life, usually in the fourth to the sixth decade. But there the similarities end. The development of Huntington's disease, in contrast to Parkinson's disease, is associated with the inheritance of a mutation in a particular gene. In addition the clinical symptoms shown by sufferers are markedly contrasting; those with Parkinson's disease suffer from slowness and poverty of movement whereas those with Huntington's disease show excessive movement of the limbs, head and trunk.

Following a description of the clinical symptoms of people with these two movement disorders (Section 4.3.1), the neuropathology and the nature of their disorders at the neurological level is considered (Section 4.3.2). The final section (4.3.3) considers the reasons for the development of these diseases (their aetiology) by exploring the possible causative factors, including genetic and environmental influences.

4.3.1 Clinical symptoms of movement disorders

Parkinson's disease

Box 4.1 A case study of Parkinson's disease

Bill is 65 years old. He has a tremor in his right hand and right leg due to alternating contraction of opposing groups of muscles. The symptom is worse when he walks. He is unable to write, feed himself or fasten his shoe laces. He is unable to rise from a chair without help and is always fatigued. A second feature of his disease is muscle rigidity. When his joints are passively moved and the affected muscles stretched, increased resistance can be detected. But the most incapacitating feature of his disease is slowness and difficulty in initiating movements. This not only affects his ability to write and feed himself, but also causes facial immobility and slow and quiet speech. His posture and balance are affected and he walks with a hurrying gait (as shown in Figure 4.2, *overleaf*). He suffers from periods of depression, fading concentration and inattentiveness. These change to disorientation and confusion as the disease progresses. Bill suffers from Parkinson's disease.

Figure 4.2 The posture and hurrying gait of advanced Parkinson's disease.

The disease gets its name from James Parkinson, an English doctor, who in 1817 wrote a detailed description of it in his *Essay on the Shaking Palsy*. By accurate observation he was able to separate the illness from other similar movement disorders.

One in every thousand people in the UK develops Parkinson's disease, usually between the ages of 40 and 70. One in every hundred people over 60 years of age suffers from it, the illness affecting slightly more men than women. Bill (Box 4.1) shows the three striking features of the disease (see also Book 2, Section 9.4.3). These are tremor (involuntary shaking movements), rigidity (stiffness of muscles) and slowness and poverty of movement. From this triad of fundamental signs stem a wide variety of disabilities. Parkinson's disease falls into the general category of *hypokinetic* disorders in which there is too *little* motor activity.

Huntington's disease

Clinically, Parkinsonism is the opposite of another category of movement disorders, *hyperkinetic* disorders where there is an *excess* of motor activity. These are characterized by an excess of movement with uncontrollable and relatively rapid motor activity. Chorea (derived from the Greek word for 'dance') is the most common of these disorders. Affected individuals suffer from rapid movements of the trunk, head, face and limbs that interrupt normal movement. Huntington's disease (also called Huntington's chorea), was first described by George Huntington in 1872 and afflicts people in this way.

Huntington's disease is progressive and usually appears between the ages of 30 and 50, although some people are stricken in childhood and others not until after the age of 70. At the beginning of the disease the involuntary movements (chorea) may be brief, but as it progresses, the movements become more pronounced until the sufferer is constantly in motion. Psychological manifestations may include mood shifts, impulsive behaviour, chronic depression and mental deterioration.

Traditionally, movement disorders were classified simply on the basis of their clinical appearance. However, with increased understanding of their neuropathology, a number of movement disorders have been identified that fit into the category of hypokinetic disorders along with Parkinson's disease or that of hyperkinetic disorders along with Huntington's disease. So far the clinical symptoms have been described. The next section goes on to consider their neuropathology, that is the cause and the nature of the diseases at the neurological level.

4.3.2 The neuropathology of basal ganglia disorders

In both Parkinson's and Huntington's disease the smooth co-ordination of muscles is lost and in both of them something goes wrong with the basal ganglia, the structures that are crucial to regulating movement. People suffering from these diseases have other complicated symptoms including behavioural changes but this chapter concentrates on their abnormal movements.

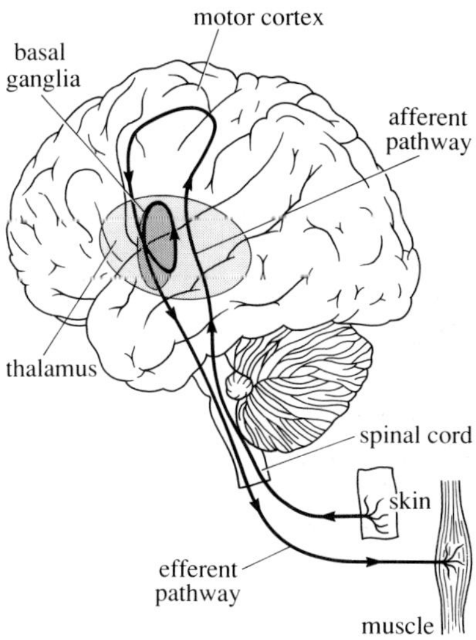

Figure 4.3 Schematic drawing showing the position of the basal ganglia within the brain and the circuit from the cortex to the spinal cord via the basal ganglia.

The various parts of the basal ganglia, their connections and their inputs and outputs were described in Book 2, Section 9.4.3. The various parts are briefly reviewed here before examining the connections between them and the influence they have, both on each other and on the overall motor output from the brain.

In the past ten years, substantial advances have been made in understanding the functioning of the basal ganglia. These have come from a study of detailed anatomy, including sensitive methods for tracing nerve tracts, physiological studies of the effects of neurotransmitters, pharmacological data and post-mortem studies of the anatomical changes associated with human movement disorders. The results of these studies have enabled researchers to make correlations between anatomy or biochemistry and the clinical features of the disease, as well as increasing understanding of how the basal ganglia are involved in movements.

Complex movements are generated in the motor area of the cerebral cortex, in response to information from the environment and all parts of the body. The cortex sends instructions direct to the spinal cord which in turn controls movement. It is this information that is modified by a circuit to the basal ganglia which feeds back to the cortex (see Figure 4.3). The signals along this loop ensure that movements are smooth and coherent.

The basal ganglia are complex structures made up of a number of regions:

- the caudate nucleus and putamen, collectively called the striatum
- the globus pallidus, which literally means 'pale globe'
- the substantia nigra, which literally means 'black substance'

The physical relationship between these regions and the thalamus is shown in Figure 4.4 (*overleaf*). The known connections between the parts of the basal ganglia are very complex and research continues to reveal new connections. Looking at the circuitry of even the most complex computer is nursery work by comparison. This chapter deals only with the major pathways in the basal ganglia, sufficient to make the overall function clear. The details of the circuitry described are not so important as an understanding of the general principles of function of the normal and diseased basal ganglia.

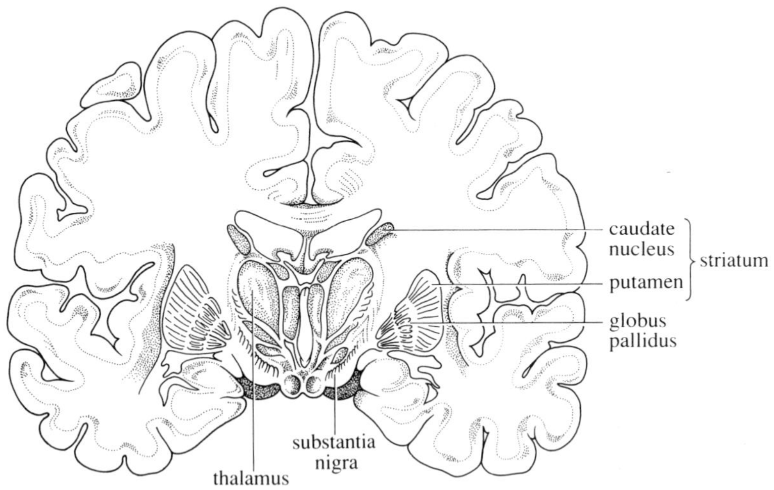

Figure 4.4 A coronal section through the human brain showing the position of the basal ganglia and the thalamus.

In fact there is not just one circuit within the basal ganglia but a number of parallel circuits, five of which have been identified to date. All have the same route or pathway that physically connect the various parts of the basal ganglia to the cortex and thalamus as shown in Figure 4.5.

□ Recall the role of the thalamus (Book 2, Section 8.8.3).

■ It is the 'receiving station' for sensory information and for outputs from the cortex.

Each of the five loops has a different function. One controls movement of the body, another is associated with movement of the eyes, others are involved in integrating motivations such as hunger, sexual arousal and fear, and yet others play a role in cognitive processes such as learning and memory. Thus the basal ganglia are extremely important in regulating cortical operations.

The loops are physically separated from one another. You studied their paths in Book 2, Section 9.4.3 and they are briefly reviewed here. One loop is shown schematically in Figure 4.5. Each circuit has a number of starting points engaging particular regions of the cortex (A, B and C in Figure 4.5). Nerve fibres from these project to specific parts of the striatum. Neurons in the striatum project to two parts, the globus pallidus and the substantia nigra. The circuit continues to the thalamus which returns the signals to the cortex. So the loops of the basal ganglia have the same start and end points. However, what happens to the signals as they pass through the basal ganglia? In order to answer this question it is necessary to look at the circuit in more detail, that is at the level of the neurons and their transmitters.

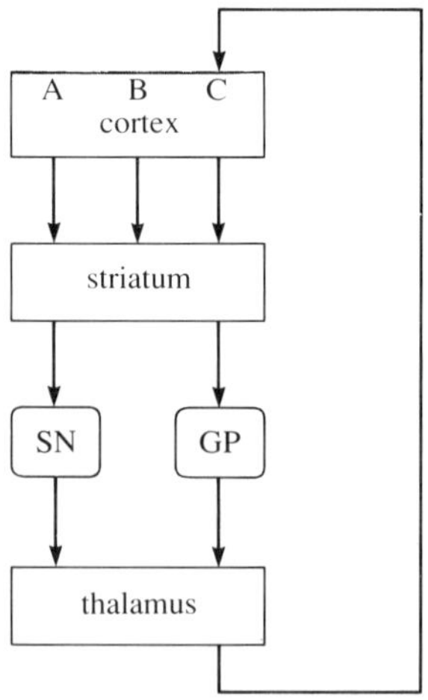

Figure 4.5 A diagram of a neuronal loop within the basal ganglia. SN: substantia nigra. GP: globus pallidus.

Transmission of information within the basal ganglia

The basal ganglia demonstrate beautifully how neurons transmit and interpret information and communicate with one another to coordinate the activities of the organism. The systems of neurons operate here as in other parts of the body, in a coordinated manner that depends both on the properties of the individual neurons in the system and on the way that the neurons are interconnected. Having stated this, however, it is important to note that the structure of the synapses and their connections in the basal ganglia do differ from the classical structure described in Book 2, Section 4.2. These structural details and their implications for study of the basal ganglia are considered in Chapter 5 of this book. This chapter concentrates on the functioning of the circuitry within the basal ganglia as a whole in order to understand what can go wrong in people with hypokinetic and hyperkinetic functional disorders.

Different groups of neurons of the basal ganglia use different neurotransmitters. Recall that the function of neurotransmitters is not only to transmit information between neurons but also to enable neurons to process this information. Transmitter binding to an excitatory receptor produces an excitatory postsynaptic potential whereas transmitter binding to an inhibitory receptor produces an inhibitory postsynaptic potential. An individual neuron may have many excitatory and inhibitory receptors activated simultaneously but the outcome will depend on the number of activated receptors of each type relative to each other. It is this process that enables neurons and the basal ganglia as a whole to act as processors of information (Book 2, Section 9.4.3).

The neurotransmitter, and the receptors to which it binds, determine not only the type of postsynaptic potential but also the magnitude of this potential. Just as the magnitude of activity (degree of depolarization) of a single neuron depends on the balance between the activity of the excitatory and inhibitory synapses, so the output of any part of the circuit of the basal ganglia (being the result of many such neuronal activities) is a result of the balance between the various inputs and outputs between its regions. Thus, for example, if an inhibitory input to a region is diminished, this in turn makes the output from that region stronger or more effective.

The overall end result of this circuit and the activity between parts of the basal ganglia is in determining the level of output from the thalamus to the cortex. A change in activity in any part of the circuit will ultimately affect the level of activity in this last part of the loop.

The motor loop

Of the five loops in the basal ganglia it is the organization of the motor loop which is important for understanding movement disorders. In fact, neurobiologists have gained an increased understanding of the functional anatomy of this loop from the study of basal ganglia disorders.

A more detailed diagram of the motor loop in a normal brain is shown in Figure 4.6 (*overleaf*). You are not expected to remember the details of the motor loop presented in this figure, but you need to understand how the circuit normally functions in order to appreciate what can go wrong. The parts of the pathway which are inhibitory are shown as solid arrows and those which are excitatory are shown as open arrows. For the time being ignore the specific transmitters and concentrate on the influence of the excitatory pathways on the inhibitory pathways and vice versa. As you can see the details are complex but what is important to note is that the strength of output from any one region depends on the level of the inputs it receives.

Examine Figure 4.6 closely. The inputs from the cortex excite the neurons of the striatum, which themselves inhibit the neurons of the globus pallidus and substantia nigra. If the inputs from the cortex to the striatum become more active this has consequences for the rest of the circuit. Since the magnitude of activity arriving at the striatum will be greater, the output from the striatum will also be stronger. Thus the inhibitory output of these cells to the globus pallidus and the substantia nigra will be stronger. At this point in the circuit, other nuclei (not shown in Figure 4.6) are involved which affect the overall output from the globus pallidus and the substantia nigra. This is such that, when the inhibitory output of the striatum is stronger, the inhibitory outputs from the globus pallidus and the substantia nigra to the thalamus are in turn stronger.

□ Referring to Figure 4.6, what would be the consequence of this to the magnitude of the excitatory output from the thalamus to the cortex?

■ The magnitude would be decreased.

This demonstrates two important points. The connections are in a circuit, converging on the thalamus, so that a change in strength at any point in the system will eventually effect the strength of output from the thalamus to the cortex. Any change in the motor loop caused by disease will ultimately affect the strength of this excitatory input to the cortex.

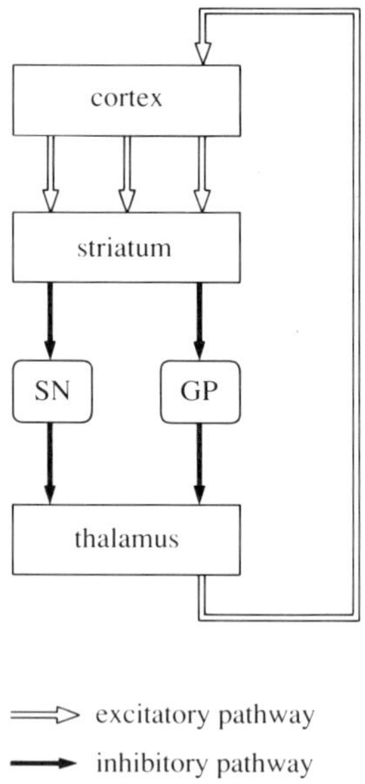

Figure 4.6 A more detailed diagram of the loop shown in Figure 4.5, showing the excitatory and inhibitory pathways in the motor loop. SN: substantia nigra. GP: globus pallidus.

Hypokinetic disorders

Now that the motor loop of the basal ganglia has been considered, this section looks at how this loop is affected in Parkinson's patients.

In order to understand what goes wrong in the basal ganglia of these people it is necessary to look at the effect of a particular neurotransmitter in the circuit. As described earlier in the course, neurochemical techniques have enabled the identification of the transmitter substances found within specific regions of the brain. You learnt in Book 2, Section 9.4.3, that the primary problem in Parkinson's disease is the death of the dopamine-producing cells in the basal ganglia. The cells of the substantia nigra produce and release the neurotransmitter dopamine (Book 2, Section 4.5.2). It is now well known that the neurochemical problem in Parkinson's disease is insufficient production of dopamine due to the death of these cells.

What is the effect of the loss of dopamine-producing cells of the substantia nigra to the rest of the circuit? How is the loss of these cells related to the clinical symptoms shown by people suffering from Parkinson's disease that were described in Section 4.3.1? As a starting point, look at the changes that occur at the cellular level of the basal ganglia in more detail. If the brain of someone who suffered from Parkinson's disease is examined microscopically and compared with that of a normal person, striking differences can be seen. Compare the two photomicrographs of the substantia nigra shown in Figure 4.7. One is from the brain of a normal person and one from someone with Parkinson's disease.

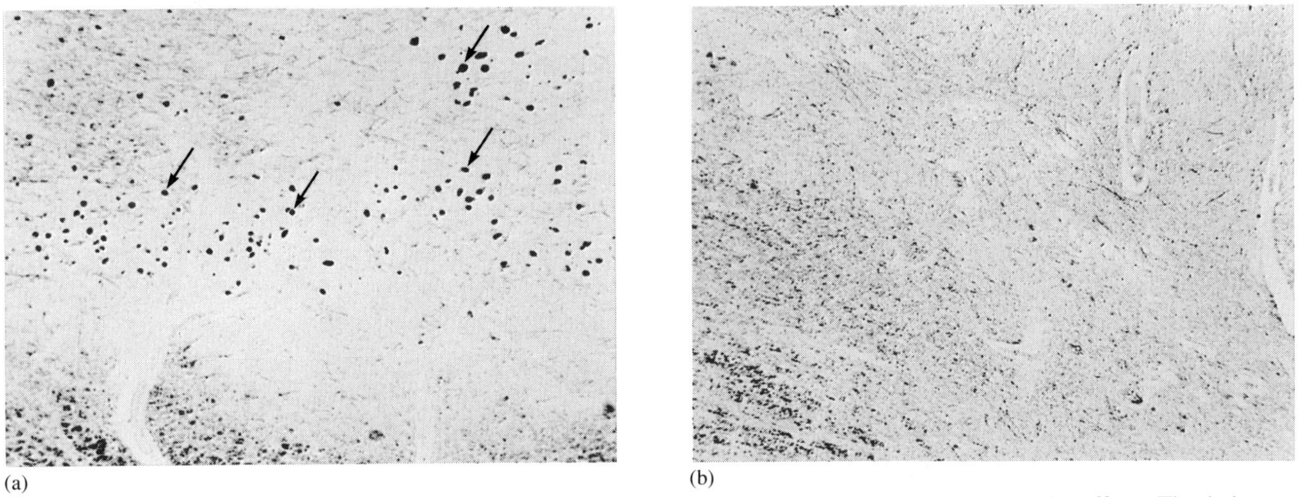

Figure 4.7 Photomicrographs of the substantia nigra of (a) a normal brain and (b) a brain from a Parkinson's sufferer. The dark pigment in (a) is melanin within dopamine neurons, some of which are arrowed. The substantia nigra of a brain with Parkinson's disease typically shows a marked reduction of pigmented neurons.

□ What is the most striking difference?

■ There are reduced numbers of neurons containing the dark pigment in the brain with Parkinson's disease compared with the normal brain.

Figure 4.8 shows the number of cells present in a Parkinson's sufferer across the whole of the substantia nigra region of the basal ganglia, compared with a normal individual of the same age. This shows that there is considerable loss of cells in the Parkinson's sufferer and that this reduction in cell number is not confined to one particular location within the substantia nigra.

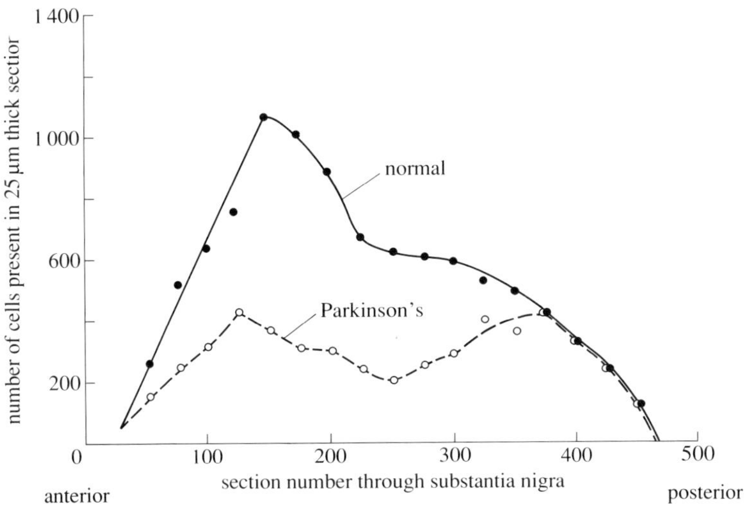

Figure 4.8 Number of cells present in a sequence of about 450 sections taken across the substantia nigra in a Parkinson's sufferer (aged 64) and in a normal individual (aged 63).

How this degenerative disorder came to be explained in terms of chemical changes was the result of a fascinating and somewhat surprising series of events.

One crucial clue was provided by a small shrub with crooked elongated roots called *Rauwolfia serpentina*. Infusions from the plant had been commonly used in India for hundreds of years as an effective mild sedative. Gandhi is recorded to have used it regularly. After the discovery in the West in the 1940s that the plant could reduce high blood pressure, various substances were extracted from the roots with a view to developing drugs. The most active extract was called reserpine. Continued medication with this extract produced dramatic side-effects, indistinguishable from Parkinson's disease. The effects of reserpine could once be seen in the expressionless or grimacing face and stooping posture of the inhabitants of chronic mental hospital wards where it was used in medication.

Arvid Carlsson and others working in Sweden in the 1950s found that reserpine also produced rigid immobility in animals other than primates, such as rats. This in turn led to a further exciting discovery. H. Ehringer and O. Hornykiewicz, also working in Sweden, found that there was a severe depletion of dopamine in the brains of affected animals. They further discovered that treatment with the dopamine precursor dopa reversed the clinical symptoms caused by reserpine.

Intrigued by these findings, Hornykiewicz and his colleagues measured the quantity of dopamine in post-mortem brain samples from people who had suffered from Parkinson's disease and people of a similar age but who had shown no symptoms of Parkinson's (controls). They discovered a large reduction of dopamine in Parkinson's sufferers, particularly in the striatum, and furthermore they found that depletion increased as the severity of the symptoms increased in these people. This work established the current working hypothesis regarding the pathology of Parkinson's disease; the clinical symptoms of slowness of movement, muscle rigidity and involuntary movements are related to the reduction of dopamine content in the striatum and other parts of the basal ganglia.

Subsequent anatomical and biochemical studies demonstrated that the pigmented neurons of the substantia nigra project widely to the striatum and that these cells are specialized to synthesize and release dopamine (Figure 4.9). Furthermore, animal studies confirmed that selective destruction of the substantia nigra led to a depletion of dopamine in the striatum and produced Parkinson-like symptoms.

Having explored the historical aspect of the findings that Parkinson's disease is associated with the loss of dopamine-producing cells of the substantia nigra, consider now the consequences of this loss to the rest of the circuit. First examine the role of dopamine-producing cells in the normal motor loop. The action of dopamine in the striatum is generally recognized as an inhibitory agent; its most probable function is to reduce the activity of the neurons in the striatum.

Thus the striatum receives two sets of inputs, one from the cortex and one from the substantia nigra.

□ Look at Figure 4.9. What is the nature of these two inputs?

■ The input from the cortex is excitatory whereas the input from the substantia nigra is inhibitory.

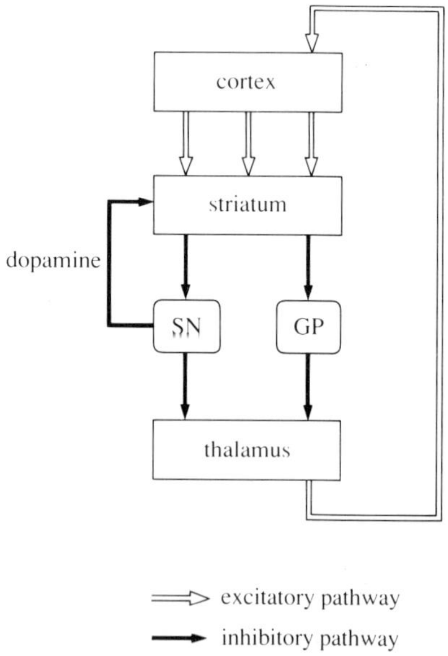

Figure 4.9 The motor loop of the basal ganglia showing the projections of the dopamine-producing cells of the substantia nigra (SN) to the striatum. GP: globus pallidus.

It is important to note that the output of the striatum reflects a balance between these two opposing inputs. You can see from Figure 4.9 that in fact there are loops within the motor loop! However what is important to note is the strength of the final output back to the cortex.

So far the role of the dopamine-producing cells in the normal circuit has been considered. Now examine what happens to the circuit in Parkinson's sufferers. This is shown diagrammatically in Figure 4.10 where the pathways with reduced activity are represented by broken arrows and those with an increased activity by double arrows.

Clearly the reduction of dopamine results in a reduction of the inhibitory input to the striatum.

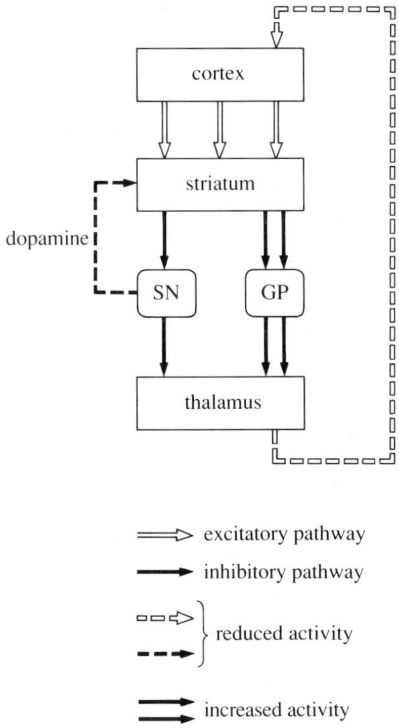

Figure 4.10 The motor loop in people suffering from Parkinson's disease. SN: substantia nigra. GP: globus pallidus.

- □ What would be the effect of this reduction on the postsynaptic cells of the striatum?
- ■ The excitatory input from the cortex would be relatively unopposed resulting in an increase in activity of striatum cells.

A further prediction might be that impairment of dopamine action, as witnessed in Parkinson's sufferers, would in turn increase the activity of the cells from the striatum that inhibit the globus pallidus (Figure 4.10). In Baltimore in 1988, Mahlon DeLong and William Miller examined monkeys suffering from drug-induced Parkinsonian symptoms. They confirmed that the globus pallidus is indeed much more inhibited than normal.

- □ From Figure 4.10 determine the effect of this increased activity in the striatum on the strength of the excitatory output from the thalamus to the cortex.
- ■ The inhibitory input to the thalamus from the globus pallidus would be increased and this would in turn reduce the activity in the pathway back to the cortex. Since this pathway is normally excitatory, the excitatory postsynaptic potential would be reduced.

This reduction in excitatory potential in the motor loop in people with Parkinson's disease is a consequence of changes in the strength of activity in the various inhibitory and excitatory inputs and outputs. Too little activity in the feedback loop to the cortex may help to explain why people with Parkinson's disease find it difficult to start moving and why their movements are slow. Thus it can be seen how a change in the level of activity at the level of neurons is translated into the clinical symptoms found at the level of the organism

Hyperkinetic disorders

Further clues to the functional anatomy of the basal ganglia have come from studies of people with hyperkinetic disorders. Patients with Huntington's disease suffer from an excess of movement with uncontrollable and rapid motor activity. Given these clinical symptoms, which contrast sharply with those of people with Parkinson's disease, it might be suspected that Huntington's sufferers have increased activity in the excitatory feedback loop to the cortex, and indeed this is found to be the case. To understand how this arises, another neurotransmitter in the basal ganglia, GABA (γ-aminobutyric acid) needs to be considered.

The initial problem in people with Huntington's disease seems to be death of neurons in the striatum which project to the globus pallidus and in particular the loss of GABA-containing neurons (see Figure 4.11). Recall that GABA is the main inhibitory neurotransmitter in vertebrates (Book 2, Section 4.5.4). The loss of these cells is the most characteristic change seen in these people at post-mortem examination although progressive degeneration spreads to other cells of the striatum and other parts of the basal ganglia. Brain imaging analysis of living subjects with Huntington's disease also shows atrophy of the striatum, which is consistent with the loss of cells.

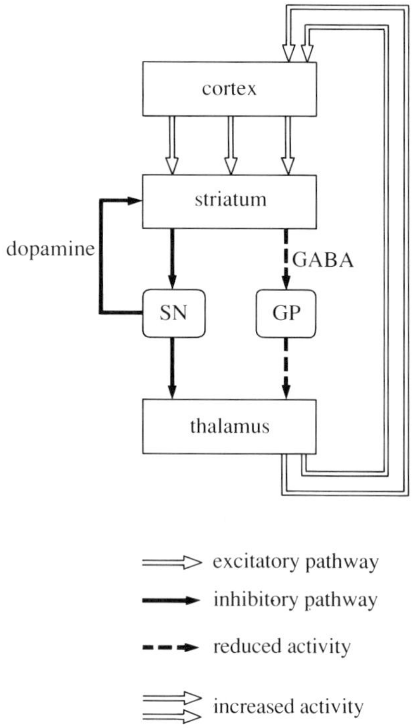

Figure 4.11 The motor loop in people suffering from Huntington's disease. SN: substantia nigra. GP: globus pallidus.

Further confirmation of the role that these cells play in Huntington's disease was made by A. R. Crossman and his co-workers in 1988 who looked at the effects of certain drugs. Recall from Book 2, Section 4.2.5 the effects of *agonist* and *antagonist* drugs on synapses which are a result of their occupying receptor sites. Different types of receptors (receptor subtypes) can be distinguished by their binding of pharmacological substances.

□ How do agonist and antagonist drugs work?

■ Agonists are substances which mimic the effect of natural transmitter molecules on the receptor whereas antagonists inhibit the action of the natural transmitter.

Crossman and his colleagues induced hyperkinetic movements in primates by infusing the drug bicuculline into the globus pallidus. This drug is a GABA receptor antagonist.

□ What input were they blocking?

■ They were blocking the GABA input from the striatum to the globus pallidus since the drug would occupy the postsynaptic membrane receptors.

So, in effect, they mimicked the pathological change in people with Huntington's disease by preventing postsynaptic activity of GABA and movement was affected as predicted.

The changes in the circuit of the basal ganglia in people with Huntington's disease are summarized in Figure 4.11.

□ From Figure 4.11 determine what would be the consequences of the death of the striatal cells, with associated reduced inhibition to the globus pallidus, for the remainder of the circuit ?

■ The changes to the rest of the circuit would be:

1 a decrease in the inhibitory output of the globus pallidus;

2 an increase in the excitatory output of the thalamus to the cortex.

Thus, hyperkinetic disorders result from reduced activity in the output from the striatum and from the globus pallidus which leads to overactivity in the pathway from the thalamus to the cortex.

This disease, as in the case of Parkinson's disease, demonstrates that a change in the amount of a particular neurotransmitter can affect the overall output from the thalamus to the cortex. The increase in activity in this feedback loop is translated into excess movement in people with Huntington's disease.

Not surprisingly, disruption can occur at different points in the circuit of the motor loop as demonstrated in the two examples here, of Parkinson's disease and Huntington's disease. The reduction in the amount of transmitter in both diseases affects the balance of excitatory and inhibitory inputs and outputs, so that the information the neurons are processing is no longer 'accurate', with the consequence that they no longer communicate effectively.

The evidence described above shows that the mechanisms underlying hypokinetic and hyperkinetic movement disorders represent two ends of the clinical spectrum. Thus, hypokinetic disorders (e.g. Parkinson's disease) are associated with excessive inhibitory output from the basal ganglia to the thalamus, and hyperkinetic disorders (e.g. Huntington's disease) with an abnormally low level of basal ganglia output to the thalamus.

Treatment

These insights as to what has gone wrong at the cellular level in these cases offers some prospects for therapy. In people with Parkinson's disease or Huntington's disease the effects of missing neurons are not restricted to the motor system. In both diseases the loss of neurons continues progressively over time and spreads to other regions. Almost certainly other loops in the basal ganglia are involved, as suggested by the wide range of symptoms that afflicted individuals may suffer.

In the case of Parkinson's disease you may recall (Book 2, Section 4.5.2) that dopamine cannot be used to alleviate symptoms since it does not pass the blood-brain barrier. However, its natural precursor dopa can enter the brain and is also absorbed when given by mouth. Dopa is believed to exert a therapeutic action by conversion into the active form of dopamine once inside the blood-brain barrier, by means of the natural route used in the body.

Unfortunately, after a number of years of treatment, many sufferers notice that the benefits of dopa treatment begin to wane. Each dose produces benefit for a shorter period of time. Determining the cause of this 'wearing-off' effect is an important aim of present research. One striking feature of Parkinson's sufferers is that the symptoms are relentlessly progressive. The most favoured explanation for the wearing-off effect of dopa treatment is that, as the pathological process continues to destroy dopamine neurons, so the storage capacity dwindles. The consequence of this is that relief of symptoms is brief because it is due only to newly-synthesized transmitter. If this hypothesis is correct then one way ahead is to develop long-acting dopamine agonists that mimic the effect of the natural transmitter. Another way forward is to look to alternative therapies.

With this aim, considerable progress in neural transplants has been made during the 1980's. Embryonic substantia nigra tissue can be successfully transplanted into the brains of sufferers but this involves major surgery and access to suitable donor material. Such transplanted tissue is removed from dead or aborted embryos. Some embryonic cells, if removed soon after abortion, continue to function after being transplanted into the brain of another individual. Whether the transplanted cells will continue to function in the long term remains an unanswered question. This topic is discussed further in Chapter 5. In addition, this technique begs the question of the ethics of using embryonic tissue, a point that is returned to in Chapter 8.

Unlike Parkinson's disease, where a means has been found to replace the deficient neurotransmitter dopamine by its natural precursor dopa, no mechanism is available to replace depleted GABA in people with Huntington's disease. Therefore, indirect treatments have been developed.

Pharmacological studies show that the loss of cells in the striatum of people with Huntington's disease results in a relative excess of dopamine in the remaining cells in the striatum. This excess has been confirmed in monkeys given high doses of dopamine agonists with the result that hyperkinetic movements are induced. So, an abnormal increase in dopamine has the opposite effect on the motor activity of the body from the reduction of dopamine as shown by Parkinson's sufferers.

□ What symptoms might humans with Parkinson's disease develop if treated with excessive amounts of dopa?

■ They might develop the hyperkinetic movements characteristic of Huntington's sufferers.

This is indeed sometimes observed. Estimating the correct dose of dopa is not always easy and accidental high doses have been given. Presumably in these people, not only has the deficiency of dopamine been made good, but there has been an 'overshoot' in quantity sufficient to 'swamp' the cells in the striatum.

This knowledge has been used to develop indirect treatment for people with Huntington's disease. It is not possible to remove the relative excess of dopamine in these patients but it is possible to block its effect.

□ Can you suggest what type of drugs should be administered?

■ Dopamine receptor anatagonists would inhibit the action of the transmitter.

Such drugs administered to people with Huntington's disease suppress abnormal movements by restoring the balance of the ratio of dopamine to GABA in the striatum.

Research continues the attempt to alleviate the symptoms of these diseases. Future therapies might involve precisely targeted injections of compounds that counteract the transmitter chemical used by inhibitory neurons. Alternatively, it might be possible to induce other neurons to synthesize missing neurotransmitter and thereby restore function to damaged circuits.

4.3.3 Causes of basal ganglia movement disorders

The above studies reveal a clear link between disease of the basal ganglia and movement disorders. In people with either Parkinson's disease or Huntington's disease there is neuronal death in restricted regions of the brain. But the significant question for prevention and treatment is why do particular neurons die in certain individuals? If the cause of this were known, development of each disease might be prevented or at least the rate of its progress within an individual slowed down.

Studies of the distribution of a disease within a population, its epidemiology, can reveal something about the biological causes of a disease. For example, there might be an association between the disease and a poor diet, cramped housing or pollution from industrial waste. Epidemiology might also reveal a pattern whereby sufferers were concentrated in a particular family. The cause in such a case might be the lifestyle of the families. However, there might be another important cause.

□ What other factor might be suspected of being the cause in these circumstances?

■ There might be a genetic effect with family members carrying an allele, or alleles, that are associated with the expression of the disease.

This section takes this broader approach to studying the causes of Parkinson's disease and Huntington's disease, by examining:

genetic factors

ageing

poisons and metabolism

One important factor to bear in mind when looking for causes is that, in people with either Parkinson's disease or Huntington's disease, the clinical symptoms appear some years after the changes at the cellular level. For example, the symptoms of Parkinson's disease only begin to appear when dopamine levels fall to as low as 20% of levels found in normal individuals of a similar age. So people develop the biological features of the disease long before the clinical symptoms are shown.

Genetic factors

There are a number of ways of exploring the possible genetic transmission of susceptibility to disease (Section 1.4.9). The commonest is by looking at the distribution and frequency of the disease in individuals in family pedigrees. Some diseases such as phenylketonuria are the result of a single gene abnormality. But Parkinson's disease shows no such clear genetic association. However, the fact that nearly 1 in 5 cases of people with Parkinson's disease have an affected relative suggests a possible genetic influence. Perhaps certain genes influence the number of cells in the substantia nigra or the longevity of these cells in some way.

Another way of analysing genetic influences is to compare the prevalence of the disease in identical and non-identical twins compared with the rest of the population. The **prevalence** of a disease is measured as the number of individuals per 1 000 of the population actually showing the disease at a given point in time. Such measurements in the case of Parkinson's disease reveal a similar prevalence in monozygotic and dizygotic twins as in the rest of the population.

□ What do these observations suggest about genetic influences on the development of Parkinson's?

■ They suggest genetic influences are absent or weak.

Any relationship between the genotype and the development of Parkinson's disease is unclear. This is in marked contrast to the development of Huntington's disease which is a classic example of a dominantly inherited genetic disorder. The doctor George Huntington treated affected people whose parents and grandparents had been treated in turn by his father and grandfather, who had been doctors in the same community. Thus Huntington was able to determine the pattern of related individuals afflicted with the disease. An example of a family pedigree for this disease is shown in Figure 4.12.

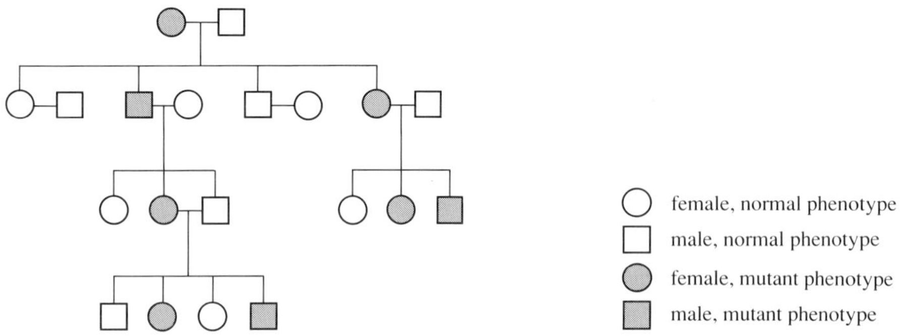

Figure 4.12 A pedigree of Huntington's disease.

□ Is one sex afflicted more frequently than the other?

■ No, both sexes appear to be affected equally.

□ Examine Figure 4.12 and determine the proportion of affected children.

■ Half the children in each generation are affected.

This pattern of inheritance with a high frequency of affected children is characteristic of the inheritance of dominant alleles.

□ Why is it particularly difficult to determine the probability of having an affected child in the case of Huntington's disease?

■ Unfortunately, sufferers are usually past their child producing years before they know they have the disease.

Having developed the symptoms, the parent suffers further in the knowledge that each of their children has a 50% chance of inheriting the disease allele.

A known genetic basis for a disease opens up alternative approaches to understanding the cause of the symptoms. If the precise gene could be identified, as in the case of phenylketonuria, and the details of its structure determined, then researchers could find the product of the normal allele. It might then be possible to treat affected people by supplying the normal product or, as in the case of phenylketonuria, by changing the diet to avoid the build-up of metabolites. This approach to understanding inherited neurological diseases is considered in more detail in Section 4.4.

Ageing

So far the effect of specific genes and/or the genotype on the development of Parkinson's and Huntingdon's diseases have been examined. Consider now the effects of the process of ageing on the development of these diseases. You have already explored this process in some detail in earlier chapters of this book where the influence of the genotype on the ageing process was considered. It is important to note that the precise function of any gene influencing the ageing process is not to determine longevity directly but to influence the process of ageing indirectly.

Huntington's disease stands apart from most other genetic diseases which become apparent at, or shortly after, birth in that symptoms are not shown until middle age. This suggests that a time factor or ageing factor is required for the symptoms to be shown. Parkinson's disease is also a disease of middle to late age.

Furthermore, it is intriguing that both Parkinson's disease and Huntington's disease have a variable age of onset of initial symptoms associated with the disorder. In the case of Huntington's disease the age of onset reaches remarkable extremes. In a few people it begins before the age of 15 and in others in their seventh decade. Why should the symptoms develop early in some people and late in others?

One possibility for explaining this difference might lie at the molecular level. It is known that many different types of mutation can occur in the same gene. This means that there can be many different alleles of the same gene even though the end result is the same, that is, the gene becomes defective. Could this difference in the age at which symptoms first develop in Huntington's patients be due to the existence of many different types of mutations within the gene associated with the disease?

The answer to this question has come from studies of a population of about 5 000 people concentrated near Lake Maracaibo in Venezuela. This population has become a major resource for exploring the genetics and expression of the disease because more than 100 living affected individuals have all inherited a defective allele that can be traced back to a common ancestor. Therefore, every individual in this population has a copy of the same defective allele with the same abnormal structure. Interestingly, the age of onset of the symptoms of the disease in this large Venezuelan pedigree varies widely.

□ What conclusion can you draw about the relationship between the age of the onset and the nature of the defective gene?

■ Since variation in the age of onset occurs even when the mutation in the gene is identical, that is all the affected individuals have the same defective allele, other genetic or environmental factors must play a role.

The existence of other genetic factors that affect the age of onset was revealed in a study by Lindsay A. Farrer and P. Michael Conneally (1985) working in the USA. It is known that the genotype of an individual influences the ageing process. For example, some people have a genotype that is more efficient at metabolizing potential poisons in the body. What Farrer and Conneally set out to determine was whether the genetic factors that influence normal ageing also affect the age that symptoms develop in Huntington's sufferers.

They compared the age of onset of the disease between parents and offspring in 214 families. How this was achieved was quite complicated but only the conclusions need concern you here. They showed that the same genetic factors that influence the normal ageing process also modify the age of onset of symptoms in people with Huntington's disease. Thus the genotype of an individual affects the expression of the particular allele associated with Huntington's disease.

These observations parallel findings from people with phenylketonuria. You may recall that these subjects have most of the symptoms alleviated by a carefully controlled intake in the quantity of phenylalanine, together with the addition of tyrosine. However, although environmental intervention renders most sufferers asymptomatic, there is extreme variability from individual to individual in the amount of phenylalanine and tyrosine needed to achieve the levels of amino acids required to prevent the associated biochemical disturbance and intellectual impairment. The variability in tolerance seems to be due to a number of factors related to the genetic background of the sufferers. These factors include the different degrees to which subjects use other pathways for metabolizing phenylalanine and the varying tolerance levels that each individual has to abnormal concentrations of accumulating breakdown products.

You may not be surprised to learn that the genotype modifies the expression of the gene associated with Huntington's disease but you may be surprised to learn that the sex of the affected parent influences the age of onset of the disease in the affected offspring. Figure 4.13 shows the percentage of subjects in the USA in 1985 with Huntington's disease according to the age at which they developed symptoms and the sex of the parent from whom they inherited the allele. The age of onset of the offspring is divided into 4 groups, juvenile (ages 3–19), early (ages 20–34), mid (ages 35–49) and late (ages 50 and above). The commonest age of onset of the disease is between 35 and 49 years.

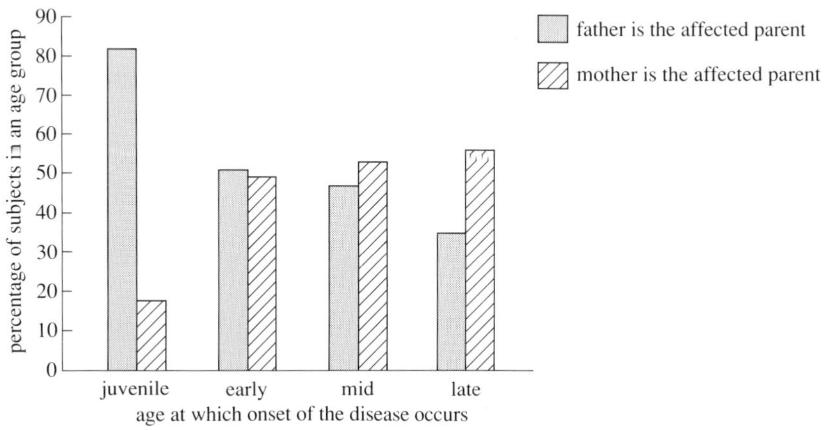

Figure 4.13 Onset of Huntington's disease at different ages and the effect of sex of affected parent on the age of onset. The number of individuals in each group was as follows: juvenile, 28; early, 172; mid, 178; late, 91.

☐ What is the striking feature about the sex of the affected parent in the juvenile age of onset group?

■ Out of a total of 28 patients with juvenile onset, 23 had affected fathers (over 82%) whereas only 5 had affected mothers.

□ Do the other age of onset groups show the same association?

■ No. Early and mid-onset patients were about evenly divided between those with an affected father and those with an affected mother. Late onset cases were more likely to have an affected mother.

These observations suggest that some factor transmitted via the mother delays the age of onset in the offspring or, alternatively, that some factor transmitted by the father causes an earlier onset. At present it is not known what this factor might be. It has been suggested that some substance in the egg might delay the age of onset in affected children. The mother also provides the environment of the womb which might affect the development of the fetus in some way.

You have seen that the symptoms in people with Huntington's disease do not develop simply as a consequence of ageing: other factors also play a part. The same is true for subjects with Parkinson's disease. Witness the presence of more active cell loss in the substantia nigra of people with Parkinson's disease compared with normal individuals of the same age. Nevertheless, a number of factors indicate that, again, ageing does contribute to the appearance and progression of the symptoms. For example, the prevalence rates vary with age, as shown in Figure 4.14.

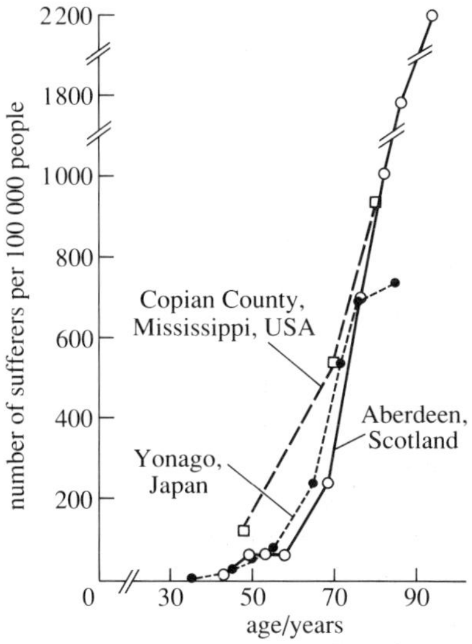

Figure 4.14 The prevalence of Parkinson's disease at different ages in three countries.

□ Examine Figure 4.14. How does the prevalence rate vary with age?

■ There is a consistent increase in the prevalence rate with increasing age.

Can any clues be gained from examining brains from individuals who have no symptoms of Parkinson's disease? Recall from Section 3.4.3 that cell loss occurs with age and that this loss varies between different regions of the brain, including the substantia nigra (Table 3.3). Figure 4.15 shows the percentage of cells remaining in the substantia nigra with increasing age.

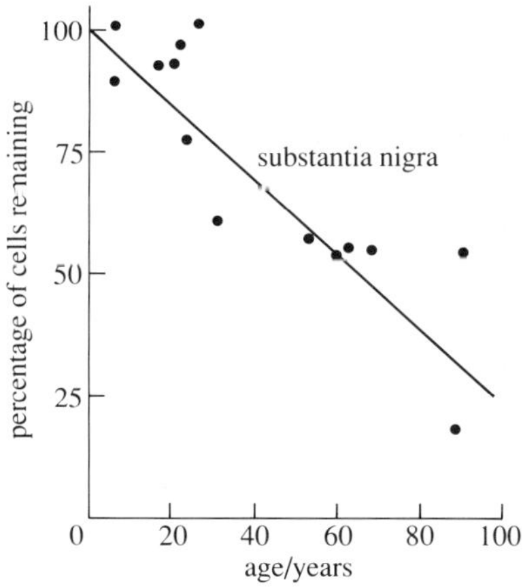

Figure 4.15 Percentage of cells remaining in the substantia nigra with increasing age.

□ What does this graph show about normal ageing?

■ Significantly, normal ageing is accompanied by attrition of cells in the CNS, including dopamine-producing cells in the substantia nigra.

Research has also shown that, accompanying the loss of cells in the substantia nigra, there is a relentless decline throughout life in levels of dopamine. This suggests that anyone living long enough might develop the disease. Perhaps Parkinson's patients merely occupy one extreme of a spectrum of age-related rates of dopamine-producing neuron degeneration. Alternatively, those individuals who happen to begin life with a diminished population of dopamine-producing neurons may be more vulnerable to this particular ageing process.

Poisons and metabolism

So far the effects of the genotype and the natural process of ageing on the development of Parkinson's and Huntington's diseases have been discussed.

Consider now the effect of environmental factors on the development of these diseases.

You saw in Section 4.2 that poisons fall into two general categories. One group contains the environmental poisons such as lead. The second group contains the chemicals or metabolites produced inside the body as products of normal cellular actions. For example, acetaldehyde, the breakdown product of alcohol is one such poison. These two categories are not clear-cut so there is overlap between them as in the case of copper.

An uneven geographical distribution of a disease implicates environmental factors, such as poisons, in its aetiology. Attempts made to correlate high prevalence rates of Parkinson's disease with environmental toxins have revealed some important pointers but as yet nothing conclusive. Prevalence rates are generally lower towards the equator; higher in rural areas than in densely populated areas; low in Japan and China; and the same in black people and white people in the USA.

The most clinching evidence to date for possible environmental poisons being involved in the development of Parkinson's disease comes from studies of neurotoxins in the laboratory. This research has given some insights into specific factors that predispose older neurons to degenerate.

Work on monkeys and rodents has revealed that a number of types of drugs affect the functioning of the basal ganglia. Here the effect of only one of these drugs, MPTP (1-methyl-4-phenyl-1,2,3,6-tetrahydropyridine!) is considered. Despite its extremely difficult name it has a number of features that make it an important tool for neurobiologists researching into the causes of Parkinson's disease. MPTP was originally produced by accident, lack of care or skill by illegal drug producers in the USA, and was inadvertently sold and consumed as heroin. The effect on its users was devastating, some developing symptoms that were virtually indistinguishable from those of Parkinson's disease. MPTP was identified as the cause and its toxic effects were confirmed in animals.

Parkinson-like symptoms can be induced in monkeys and rodents by the chemical MPTP. Animals treated with MPTP develop signs virtually identical to those found in humans with Parkinson's disease, including slowness and poverty of movement, flexed posture, muscular rigidity, and postural tremor. However, a number of other drugs can also produce similar symptoms. Recall that reserpine extracted from the plant *Rauwolfia* (described in Section 4.3.2.) produces Parkinson-type symptoms when administered to humans. However, what separates the use of MPTP from these other drugs is the fact that it produces symptoms that not only mimic those of Parkinson's disease at the clinical level, but also at the cellular level. It induces degeneration of neurons in the substantia nigra with resulting loss of dopamine in both the striatum and substantia nigra itself. It thus provides an ideal experimental model to study the development of Parkinson's disease. In addition, the effects of MPTP increase with the age of the victim, an observation that parallels Parkinson's disease.

□ Suggest two reasons why the effects of MPTP might increase with age.

■ 1 Cells may become more sensitive with age.

2 Poisons may accumulate more readily in the brain than elsewhere in the body in older victims.

There is not sufficient space to provide details of the animal experiments here but, intriguingly, neurons do not appear to become more sensitive with age to neurotoxins. It is known that toxins do accumulate in the body over time and that the usual site for such deposition is the liver. What the MPTP research suggests is that there is a change in the deposition of neurotoxins with age so that in older subjects there is a greater delivery of them to the brain rather than to other sites in the body such as the liver.

Although there is no known natural source of MPTP, it could be formed by a metabolic pathway from other substances inside the body. In fact, there is evidence to show from studies of human post-mortem brain tissue that neurotoxins do accumulate over time as part of the ageing process, particularly in people with Parkinson's disease. Some of these toxins are similar in structure to MPTP.

□ Using this information, can you suggest possible hypotheses to account for the development of Parkinson's disease?

■ One possible hypothesis is that the liver becomes damaged in some way so that neurotoxins are transported to the brain. Another possibility is that certain enzymes become defective in metabolizing potential poisons such as MPTP.

The similarities between MPTP-induced Parkinsonism and Parkinson's disease itself supports these possibilities. Thus human populations may benefit from avoidance of certain environmental toxins and/or preventative treatment. The crucial question remains as to what environmental factor or factors are responsible and how they interact with genetic factors in the individual.

These results on neurotoxins are influencing the direction of research into the cause and prevention of Parkinson's disease. The same is true for other diseases such as Huntington's disease. The time or age effect seen in sufferers before the onset of symptoms suggests that the defective allele may contribute in some way to the accumulation of toxins, to the failure to metabolize them or to secrete them.

In the case of people with Huntington's disease, recent research has focused on a particular chemical, quinolinic acid. This is a normal brain metabolite, that is, a product of normal metabolism of the body, produced during the breakdown of the amino acid tryptophan. In normal individuals this metabolite does not build up over time, but in Huntington's sufferers it accumulates.

Quinolinic acid has been further implicated as a causative agent from experimental work on animals. When quinolinic acid was injected into the brains of monkeys and rats they developed clinical symptoms which mimicked those of Huntington's sufferers. Thus in the case of Huntington's sufferers the symptoms may well develop as a consequence of the production of metabolites or poisons from a constituent of a normal diet.

□ Given this information, suggest possible roles for the allele associated with Huntington's disease.

■ Perhaps the allele in these patients alters the normal metabolism of tryptophan or possibly it alters the uptake of quinolinic acid into brain cells.

Quinolinic acid is only one of a number of known endotoxins (metabolites of the body). Other endotoxins may well accumulate as a consequence of metabolic disturbances. In the future they may also be implicated in diseases.

The outstanding problem in studying these diseases is unravelling what is the cause and what is an effect of the disease. It is important to remember when trying to establish the cause of a disease to take proper account of the complex interaction between the genotype and the internal environment and the relationship of the whole organism to the external environment. In the majority of diseases it is this *interaction* that is important as to whether a disease develops or not.

Over the next few years there will be a continued increase in knowledge of the two groups of poisons, those present in the environment and those that are produced or which accumulate in the body as a consequence of a defective metabolism. It is possible in the case of environmental poisons once identified (such as lead) for these to be reduced or removed. In the case of poisonous metabolites there is a need to increase understanding of the detailed functioning of the body.

Where a disease is associated with a single defective gene, as in the case of Huntington's disease, molecular techniques are being used to try to find the gene by separating it from all other genes on the chromosomes in a cell. This is done in the laboratory using cells from a blood sample of a sufferer. Once found, it is possible to determine the normal function of that gene. This strategy is explored in the next section.

Summary of Section 4.3

One way of learning about neurological functions is to investigate one particular region of the brain in detail through the study of disorders associated with it. The best understood example is the function of one of the circuits in the basal ganglia and the clinical symptoms of a number of crippling movement disorders.

The basal ganglia are complex structures made up of a number of regions. Neuronal circuits from the cortex connect the various parts of the basal ganglia to the thalamus and then loop back to the cortex. The level of activity from the thalamus is a balance between the excitatory and inhibitory connections of the circuit. In the case of Parkinson's disease, which is associated with insufficient production of dopamine in the substantia nigra, there is reduced activity in the feedback loop to the cortex. In contrast, in people with Huntington's disease there is increased activity in the feedback loop to the cortex. This results from the loss of GABA-containing neurons in the striatum of people with Huntington's disease. These insights into the neuropathology of the disease offer sufferers some prospects for therapy.

But the cause of excessive cell loss in particular parts of the basal ganglia in people with Parkinson's disease and Huntington's disease is unclear. The picture that is emerging for both diseases is that the genetic background and an ageing factor (or time factor), combined with the accumulation of neurotoxins, may play a part. In order to clarify this picture the effect of the drug MPTP and other potential poisons, which induce Parkinson-like symptoms, are being actively investigated as causative agents. Unlike Parkinson's disease, Huntington's disease is associated with a defective allele of a single gene. This mutant allele may contribute in some way to the accumulation of toxins.

4.4 The genetic approach to understanding neurological disease

This section concentrates on the level of genes and on the level of molecular components of individual genes. Molecular techniques can be used to diagnose a disease and to identify the precise DNA sequence of mutant alleles. However, it is not always possible to relate knowledge of changes at the molecular level to the clinical symptoms at the level of the individual because of the complexities of the intermediate levels that separate these two.

At least 3 000 disorders have been described that are associated with a single gene defect. Most of these are very rare, although there are one or two notable exceptions, such as cystic fibrosis, with an incidence in Europe of one in 2 000 live births. The basic biochemical defect is known in only a small proportion of these disorders. A few hundred have a recognized enzyme defect, and about a further hundred have a defect in a non-enzymic protein. Furthermore, there is effective treatment for only a very few of them.

There is considerable interest in identifying individual genes (and their defective alleles) that are associated in some way with the CNS. Studying the structure of these genes from affected individuals could shed some light on their pathology. An advantage of this approach is that there is no need to use brain tissue because the chromosomes are found (with the odd exception) in all the cells of the body. For molecular studies, chromosomes are usually obtained from blood samples, specifically from the white blood cells.

Where the primary cause of a disease has not been identified by means of standard biochemical techniques such as those used to define phenylketonuria, new strategies are being applied by molecular geneticists. It is known that, in mutant alleles, the sequence of bases in the DNA is changed from the normal sequence and consequently the gene product is altered.

What molecular geneticists set out to do is to try to separate the DNA sequence of a disease gene from all other DNA sequences in the cell in order to determine why the mutant allele is functionally defective. In addition, they also hope to find the product of the normal allele and thus develop an understanding of the nature of the defect. Knowledge of the underlying mechanism of a disease also facilitates the development of new therapies.

But the question is how to identify the gene itself among all the other thousands of genes in a cell. James Gusella and colleagues in 1983 working in the USA determined the approximate location of the gene associated with Huntington's disease to a region at the tip of the short arm of a particular chromosome, chromosome 4. This was done using linked genes or markers in that region (Section 1.4.9).

The way this is achieved is by identifying another gene, or marker, to which the allele for Huntington's disease is closely linked. The important point about the marker is that it must be possible to identify its presence in the DNA extracted from the sufferers' cells, even if they are too young to be certain whether they have the disease or not. Thus, even though the gene associated with Huntington's disease has not been precisely identified, its inheritance can be followed by following the known DNA marker. The more closely the marker is linked to the allele associated with Huntington's disease, the greater the chance they will be transferred together to the children. In other words, the closer the mutant allele is linked to the marker, the less likely the two alleles will be separated from each other during meiosis (the process of cell division which leads to the production of eggs and sperm).

□ Why is it important that the allele associated with Huntington's disease and the marker should not separate?

■ The DNA marker is being used as an indirect method of following the inheritance of the mutant allele.

Figure 4.16 shows two pedigrees of Huntington's disease. In each pedigree, the inheritance of a known marker is shown, but the marker is a different one for each pedigree.

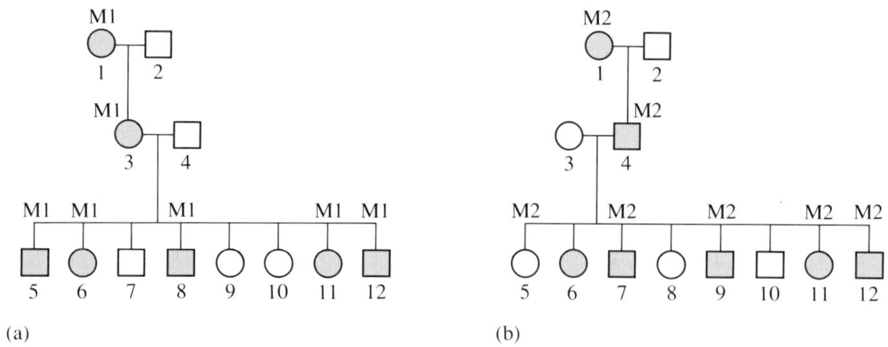

Figure 4.16 Two families with Huntington's disease. (a) Pedigree 1. (b) Pedigree 2. The pattern of inheritance of two markers M1 and M2 is also shown.

□ Study Figure 4.16. Which marker, M1 or M2, is always inherited together with the allele associated with the development of Huntington's disease?

■ M1. Every individual (1, 3, 5, 6, 8, 11 and 12 in pedigree 1 in Figure 4.16a) who develops the disease also has marker M1.

If you got this answer wrong, look at the children in the third generation of pedigree 2 (Figure 4.16b). Child 6 has Huntington's disease but no marker and child 5 is unaffected but has inherited M2. Although it is possible that the latter child might still develop the disease, child 6 of this pedigree provides clinching evidence that this marker does separate from the disease allele.

□ Which marker, M1 or M2, is most closely linked to the allele associated with Huntington's disease?

■ Almost certainly it is M1. In fact, in this particular pedigree the association between M1 and the allele associated with Huntingdon's disease is 100%.

It is important to note, however, that the pedigree represents only a small sample of individuals. It is possible that in a larger sample, separation of the Huntington's disease allele from M1 might occur.

J. F. Gusella and colleagues working with data from a large American family found a closely linked marker that separated only occasionally, in about 2–3% of all births, from the development of the Huntington's disease phenotype. Thus the presence of this marker is not absolutely reliable as a means of diagnosing the presence of the disease allele. Nevertheless, this finding of linked markers was exciting because it paved the way for pre-natal diagnosis and genetic counselling. The marker shown in pedigree 1 of Figure 4.16 can be identified in DNA extracted from any cells, even from those of a fetus. (A number of techniques are available for removing cells from an unborn fetus.) Since a fetus who has inherited this particular marker is at high risk of developing the disease, prospective parents can be offered genetic counselling and pre-natal diagnosis with the possibility of abortion of affected fetuses. (This point is taken up further in Chapter 8.)

Genetic counselling is essentially a process of communicating information about the disease, the risks of its occurrence and the psychological impact of coping with an affected offspring. The role of the counsellor is not to be directive but rather to ensure that couples have sufficient information for them to reach a decision themselves. Genetic counsellors can also offer information on the options available to couples who find the risks unacceptably high, such as contraception, adoption, artificial insemination by donor and pre-natal diagnosis.

□ Which fetuses of couples at risk of having an affected child would have to be sampled for the presence of the marker?

■ Every fetus, since only by doing the pre-natal diagnosis can the marker gene be detected and by implication the chance of the fetus being affected with Huntington's disease determined.

Pre-natal diagnosis is only worthwhile if couples consider abortion to be an option. The absence of effective therapy, however, is likely to influence any decision made by prospective parents faced with this harrowing problem.

In addition, the identification of a closely linked marker can be used for presymptomatic testing of individuals at risk.

□ Which individuals would be at risk?

■ In the case of Huntington's disease, which is associated with a dominant allele, any individual with an afflicted parent.

In a recent survey in the USA the majority of people at risk of developing Huntington's disease said they would like a predictive test (called gene screening) even though it is not 100% accurate.

□ Given that the presence of the marker is not 100% reliable as a method of diagnosis for the presence of the defective allele what are the wrong predictions that could be made when screening?

■ Occasionally a person will be diagnosed as having the marker but may not have the allele associated with Huntington's disease. In addition, a person may be diagnosed as not having the marker but may have inherited the allele associated with Huntington's disease.

Genetic screening of individuals susceptible to inherited diseases has enormous problems both for the individual and for society. Some people think it provides reassuring information for the individual because at least they know one way or the other. But it raises an extremely important question as to who controls this genetic information, an issue taken up in Chapter 8.

These findings of closely linked markers are extremely important because they open up the possibility of eventually identifying the precise DNA sequences of a mutant allele. Once the gene associated with the symptoms of Huntington's disease has been identified, pre-natal diagnosis and gene screening become much more accurate. In fact the gene associated with Huntington's disease may well be identified in the next few years.

Identification of the precise DNA sequences could in turn lead to an understanding of the basic defect and to the development of an effective treatment, as in the case of phenylketonuria. A number of genes associated with other specific diseases have now been precisely located and their DNA sequenced, such as Duchenne muscular dystrophy, cystic fibrosis and Lesch-Nyhan disease.

4.4.1 Lesch-Nyhan disease

Lesch-Nyhan disease stands out from other heritable neurological disorders because so much is known about the molecular basis of the gene involved for both normal and mutant alleles. It is therefore a useful model to examine the neurological effects of a single gene. The clinical and neurological aspects of the disease are briefly reviewed and the molecular structure and pattern of expression of the defective allele are described.

Box 4.2 A case study of Lesch-Nyhan disease

Tom is 24 months old. He suffers from mental retardation and his physical development has been much slower than that of his sisters at a comparable age. He is severely spastic (that is he has cerebral palsy) and what movements he has are uncontrolled. His vocabulary is extremely small. His most distressing feature is compulsive aggressiveness and self-mutilation of lips and fingers followed by anguished cries of pain. He was a colicky baby and first came to the attention of his doctor when his father noticed reddish spots in his nappies. These spots were identified as uric acid crystals in his urine. Tom suffers from Lesch-Nyhan syndrome.

Tom (Box 4.2) shows the typical features of this disease, namely, mental retardation, spasticity, uncontrolled movements and self-mutilation.

Lesch-Nyhan disease has a frequency of about 1 in 20 000 births. One reason for this low frequency is that it is lethal, the majority of sufferers dying in their second or third decade, usually from kidney failure. This means that afflicted individuals rarely have children and therefore do not pass the allele on to the next generation.

Individuals with Lesch-Nyhan disease have a deficiency of the enzyme hypoxanthine phosphoribosyl transferase (HPRT). HPRT is an important enzyme in the metabolism of some of the bases that form the building blocks of DNA. The enzyme regulates the production of these bases within cells. Although all cells in the human body produce HPRT, there is considerable variation in the relative levels of active enzyme between cells. Intriguingly, cells of the basal ganglia have the highest activity. Why this should be so and why neurons are so sensitive to this particular enzyme deficiency remain tantalizing questions and the subject of further research.

Molecular studies have revealed details of the complexity of the structure of the normal allele; the entire sequence of the bases has been determined. They have also revealed a range of different types of mutation associated with the disease. This means that the mutation is different in different individuals suffering from the disease. One consequence of this is to make routine identification of the defective allele in tissue samples extremely difficult. Consequently, the current method of choice for pre-natal diagnosis is an assay for HPRT enzyme activity that requires only milligrams of fetal tissue.

A lot is known about the disease at the genetic and molecular levels and at the whole organism level. However, tantalizingly little is known about the pathology of the syndrome and the associated behavioural disorder. This is a consequence partly of a lack of understanding of the intermediate levels of structure and function of the body that link the genetic level to the development of the clinical symptoms and partly of the complexity of the interactions at this intermediate level. The technique of gene therapy, where the normal allele is inserted into cells taken from the body and then these cells are reinserted into the body, does offer some hope. This technique is discussed in Chapter 8.

Techniques of molecular genetics are making a dramatic impact in neurobiology and, when combined with those of neuropathology and epidemiology, are enhancing our understanding of the development and function of the normal human nervous system.

Summary of Section 4.4

A third approach to the study of neurological diseases is a genetic one. This strategy is particularly useful where there is a known genetic basis for a disease as in the case of Huntington's disease and Lesch-Nyhan disease. Closely linked markers help geneticists localize a gene to a particular region of a chromosome. They can also be used for pre-natal diagnosis and presymptomatic testing of individuals.

Once the DNA sequence of the gene is known, however, gene screening can be done more accurately. Furthermore, it then becomes possible to determine the normal function of the gene and hopefully the nature of the disease defect.

However, results have not been quite so straightforward. In the case of Lesch-Nyhan disease for example, a lot is known about the biochemical and molecular basis of the disease. Even so, this knowledge has not yet led to the development of a therapy; nor has it provided a reliable means of pre-natal diagnosis, nor provided a clear explanation as to why neurons should be so susceptible to an imbalance in the production of DNA bases.

Nevertheless advances are being made. Through careful and detailed study, scientists, epidemiologists and doctors are increasing understanding of these diseases and this in turn increases the likelihood of developing treatments.

Summary of Chapter 4

A number of biological approaches are available to study neurological diseases in order to determine the causative factors. The primary aim in trying to understand any disease is to determine the link between the cause of the disease and the symptoms. Causative factors include genetic influences, environmental factors and an interaction between these two.

The nervous system is particularly sensitive to poisons and toxins, as shown by the effects of lead or copper in the body. However, different genotypes show varying levels of sensitivity to poisons.

Degeneration of particular neurons in the basal ganglia in people suffering from Parkinson's disease and Huntington's disease is associated with the development of specific clinical symptoms. Poisons or toxins together with an ageing factor, and, at least in the case of Huntington's disease, also a genetic factor, may be responsible for the loss of neurons in these people.

When a disease is associated with a defective allele of a single gene, as in the case of Huntington's disease and Lesch-Nyhan disease, then genetic and molecular techniques can be used to identify the gene and its normal function, and eventually the nature of the disease defect itself.

An important aspect of understanding the causative factors of a disease is the hope that it will lead to the development of a cure. The next chapter discusses the likely success of brain inplants for restitution of function of the substantia nigra in people with Parkinson's disease.

Objectives for Chapter 4

When you have completed this chapter you should be able to:

4.1 Define and use, or recognize definitions and applications of each of the terms printed in **bold** in the text.

4.2 Explain why the same concentration of a neurotoxin is associated with different responses from individual to individual. (*Questions 4.1, 4.3, 4.4 and 4.6*)

4.3 Outline the neurotoxic effects of lead in both rodents and humans. (*Question 4.2*)

4.4 Explain why some essential substances, such as copper, are neurotoxic in some individuals. (*Questions 4.3 and 4.4*)

4.5 Describe the structure and function of the basal ganglia with particular reference to the neural pathways that control movement. (*Question 4.5*)

4.6 Explain the neuropathology of the basal ganglia in people with movement disorders, such as Parkinson's disease and Huntington's disease. (*Question 4.5*)

4.7 Give a brief account of the possible causes of the loss of particular cells in people with Parkinson's disease and Huntington's disease. (*Question 4.6*)

4.8 Describe the effect of MPTP and explain why its use provides an ideal experimental model to study the development of Parkinson's disease. (*Question 4.7*)

4.9 Detect the presence of linkage of a disease phenotype to a marker from pedigree data. (*Question 4.8*)

4.10 Give a brief description of the clinical and neurological aspects of Lesch-Nyhan disease. (*Question 4.9*)

4.11 Explain why detailed molecular knowledge of gene structure and function of the mutant allele does not always reveal the nature of a neurological defect. (*Question 4.10*)

Questions for Chapter 4

Question 4.1 (*Objective 4.2*)

Some individuals in a group of male children aged 9 years, all of whom have the same amount of lead in the body, develop the symptom of lowered auditory sensitivity. Explain why not all of them develop this symptom.

Question 4.2 (*Objective 4.3*)

Why is it that the changes in behaviour of children exposed to lead do not return to normal after exposure has ceased?

Question 4.3 (*Objectives 4.2 and 4.4*)

Two children have strikingly similar phenotypes including liver disease and destruction of neurons of the brain. One is identified as having Wilson's disease, the other is genetically normal. Suggest a likely cause of the phenotype in the second child.

Question 4.4 (*Objectives 4.2 and 4.4*)

Construct a flow diagram to show how the striatum, the globus pallidus and the substantia nigra of the basal ganglia, the thalamus and the brain cortex are physically linked by a neuronal loop.

Question 4.5 (*Objectives 4.5 and 4.6*)

Post-mortem examinations of the distribution of neuropeptides in the basal ganglia of someone with a condition known as Tourette's syndrome suggested impaired function of the striatal neurons projecting to the globus pallidus. Might such a person have suffered from hypokinetic or hyperkinetic movement disorders? If so, which type and why? (You may need to refer to Figure 4.6 to answer this question.)

Question 4.6 (*Objectives 4.2 and 4.7*)
Which of the following factors are implicated in the death of particular cells in people with Parkinson's disease?

(a) Genetic background

(b) Storage of neurotoxins in the brain

(c) Efficiency of certain enzymes in metabolizing potential neurotoxins

(d) Level of neurotoxins in the diet, home or work place

Question 4.7 (*Objective 4.8*)
Which of the following processes are common both to people with Parkinson's disease and MPTP-induced Parkinsonism?

(a) Degeneration of neurons in the substantia nigra

(b) Degeneration of neurons in the thalamus

(c) Relative increase in degeneration of cells with age of victim

Question 4.8 (*Objective 4.9*)
Figure 4.17 shows a pedigree of a family with Duchenne muscular dystrophy and the inheritance of a marker (G). From the evidence provided, is G sufficiently reliable as a marker for pre-natal diagnosis of muscular dystrophy; give a reason for your answer.

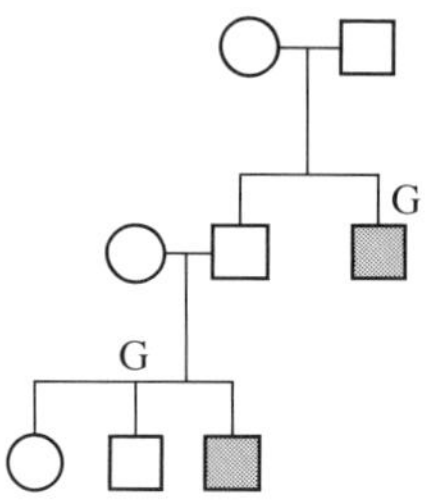

Figure 4.17 Pedigree of a family with Duchenne muscular dystrophy. The pattern of inheritance of a marker, G, is also shown.

Question 4.9 (*Objectives 4.10*)
Which one of the following statements provides evidence that a person is suffering from Lesch-Nyhan disease?

(a) The person is suffering from mental retardation.

(b) There is reduced level of activity of the enzyme HPRT in blood cells compared with controls.

(c) The person has uncontrolled movements.

Question 4.10 (*Objective 4.11*)
Explain why understanding of the pathology of Lesch-Nyhan disease is so scanty, even though a lot is known about the disease at the genetic level.

Reference

Farrar, L. A. and Conneally, P. M. (1985) A genetic model for age of onset in Huntington's disease, *American Journal of Human Genetics*, **2**, pp. 267–274.

Further reading

Oliver W. Sacks (1974) *Awakenings*, Penguin. Gives accounts of the experiences of individual sufferers of Parkinson's disease and their reactions to dopamine treatment.

CHAPTER 5
THE VULNERABLE BRAIN

5.1 Introduction

The intricate structure of the brain and the development of its complex microstructure were described in Books 2 and 4. After completion of brain development in the fetus and infant, neurons lose their capacity to replicate themselves and thus they cannot be replaced. This contrasts with many other organs of the body where cells are replaced by the normal process of cell division.

Evidence for localization of function in the human brain was presented in Book 2, Chapter 11. The sensory and motor homunculi located in the somatosensory and motor cortex of the human brain are examples of localization of function. A consequence of this localization is that restricted damage to a particular area of the brain results in specific sensory and motor deficits. For example, damage to the somatosensory cortex results in loss of sensation in restricted regions of the body whereas damage to the motor cortex results in weakness or paralysis, again restricted to localized regions of the body. Another example of localization is the left–right difference in language, memory and other cognitive functions. The evidence for asymmetry in function was derived from studies on the effects of damage to restricted parts of the brain. Where localized brain regions carry out specific and unique functions, these functions are vulnerable to even small lesions of the brain.

Although the brain is protected by the skull and is partially isolated from the metabolic products of the body by the blood–brain barrier, it can none the less sustain damage from a variety of sources. This chapter introduces the major categories of neurological disorders to which the human brain is vulnerable. It then goes on to discuss the physiological mechanisms and processes that contribute to recovery of function following brain and neuron damage, a subject that was introduced in Book 2, Section 11.3.4, and Book 4, Section 3.6.

5.2 Sources of vulnerability

This section describes the main neurological disorders of the human brain, namely vascular disorders, closed and open head injuries, epilepsy, infections, tumours and degenerative disorders. A knowledge of each of these disorders has enhanced our understanding of localization of function in the human brain. In turn, a better understanding of localization of function has led to a better diagnosis of these disorders.

5.2.1 Vascular disorders

You may recall from Section 3.5.3 that the brain is only about 2% of adult body mass but carries about 15% of the blood pumped from the heart and accounts for about 20% of the total oxygen consumption of the body. This is because the *rate* of flow of blood through the brain is greater than that through the rest of the body.

The high metabolic activity of neurons makes them susceptible to disturbances in oxygen and glucose supply. If starved of oxygen for more than about 5–10 minutes, neurons start to die and, of course, once dead they are not replaced. A cerebrovascular defect, commonly called a *stroke*, results from either bleeding from a blood vessel or from closure of a blood vessel. Blockage or rupture within a branch of the cerebral blood vessels results in failure of blood supply to regions of the brain which are supplied from the point of damage. One consequence of a stroke is a reduction in blood supply to a particular region of the brain, a phenomenon called *ischaemia*. More seriously, a stroke can result in an *infarct*, a region of dead and dying neurons. The explanation as to why neurons die so quickly after stroke has been revised over the last few years. Originally, it was thought that deprivation of oxygen leads to the immediate death of neurons. The current view is that loss of blood supply results in the accumulation of toxins that affect local neuronal activity. It is thought that NMDA receptors (Book 4, Section 5.7.3) on the cell surface become over-stimulated to the extent that affected neurons become hyperactive and 'commit suicide'. This view is supported by the fact that drugs that inhibit the activity of NMDA receptors reduce brain damage if they are administered immediately after a stroke.

The number of new cases of strokes each year in Great Britain is about 2 per 1 000 persons. About 75% occur in people aged 65 or more. (You do not need to remember such details either for stroke or for the other disorders described in this chapter; they are included only so that you will have an idea of how common the various neurological disorders are.) Approximately 35% of stroke patients die in the first three weeks and, of the survivors, about half are left disabled, usually with a paralysed arm; many are unable to walk normally.

□ What is the significance of the observation that the use of an arm is commonly affected after stroke?

■ It points to the greater vulnerability of the region of the brain that organizes the movement and/or sensory processing of this part of the body.

Not surprisingly, the symptoms of a stroke depend upon the region of the brain in which damage occurs and this in turn depends on which blood vessel is damaged. The large majority of strokes are the result of damage to, or disruption of, the arteries that carry oxygen and blood from the heart to the brain; involvement of the veins that carry away waste products and toxic substances is very rare. Strokes are not equally likely to occur in all parts of the brain because some regions of the arterial system are more susceptible to damage than others.

Figure 5.1 shows a schematic representation of the brain's blood supply. Each cerebral hemisphere is independently provided with blood through the internal carotid arteries which, in turn, divide into three main branches supplying different regions of the cortex: the anterior cerebral, middle cerebral and posterior cerebral arteries.

Figure 5.2 shows the regions of the cortex that receive their blood flow from these three branches. Each pair of diagrams in Figure 5.2 shows the distribution of the

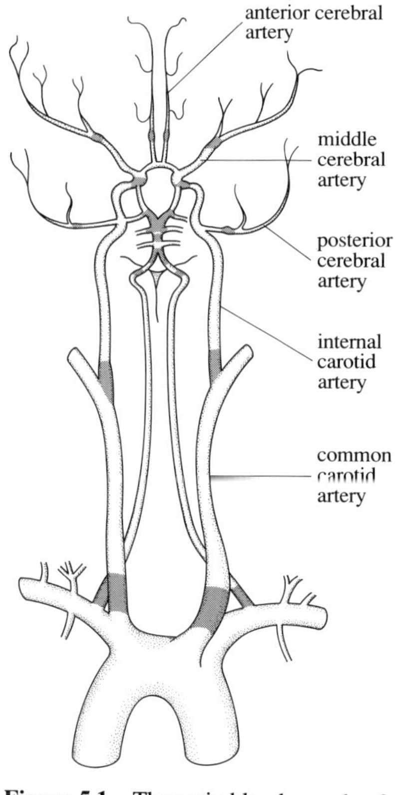

Figure 5.1 The main blood vessels of the brain: the anterior cerebral, middle cerebral and posterior cerebral arteries arise from the internal carotid artery. The shaded areas show common sites of narrowing and obstruction.

branches of one of these arteries on the lateral (upper drawing) and medial (lower drawing) surfaces of the cerebral hemisphere. From these surface arteries further branches permeate through the internal structures of the brain. The pattern of branching from these arteries is extensive, with the branches ending in fine capillaries permeating throughout the white and grey matter of the brain. The distribution of the blood supply to the deep structures of the brain from the middle, anterior and posterior arteries is shown in a coronal section in Figure 5.3. (If you need to remind yourself of the term 'coronal section', refer to Book 2, Figure 8.1.)

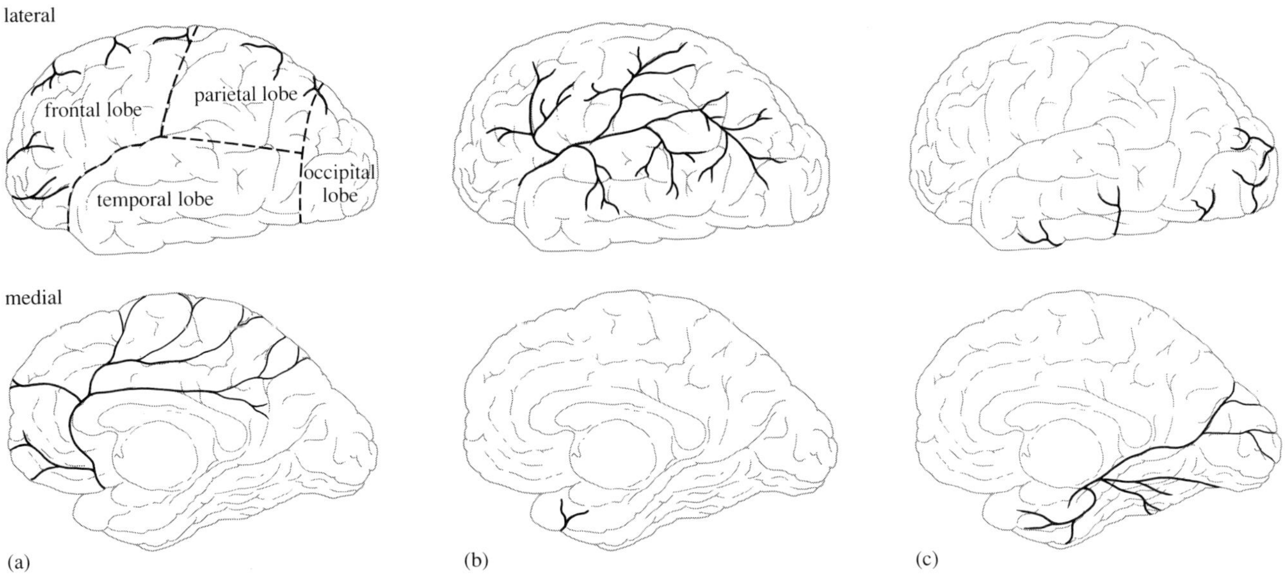

Figure 5.2 The regions of the cerebral cortex supplied by the (a) anterior, (b) middle, and (c) posterior cerebral arteries. Lateral surface of the left cerebral hemisphere (upper drawings); medial surface of the right cerebral hemisphere (lower drawings).

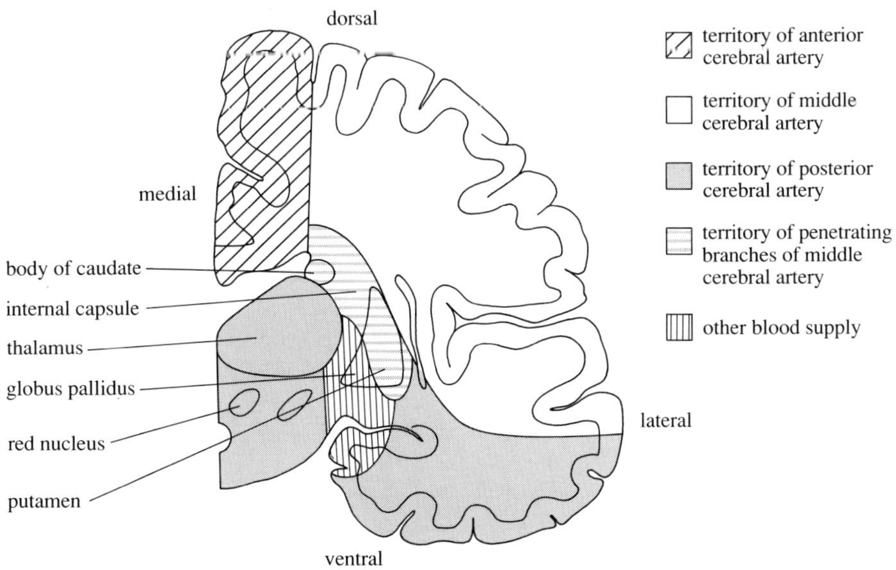

Figure 5.3 A schematic representation of a coronal section of one cerebral hemisphere showing the distribution of blood from the middle, anterior and posterior arteries to cortical and sub-cortical structures.

Because the three main branches of the internal carotid arteries supply different areas of the brain, the symptoms that result from a stroke will differ depending on which artery is affected.

- □ From your study of Figures 5.2 and 5.3 and your knowledge of the anatomical divisions of the cortex, which lobes receive their blood supply from the *middle* cerebral artery?
- ■ The middle cerebral artery supplies all four lobes of the cortex, occipital, temporal, parietal and frontal.

Because of the wide region of the cortex supplied by the middle cerebral artery, the signs of middle cerebral artery stroke depend upon which of its branches are affected. Since its territory includes somatosensory and motor cortex it is not surprising that stroke signs include muscle weakness and sensory loss.

From Figure 5.2 you can see that the branches of the middle cerebral artery are mainly found in the lateral half of the cortex and not in the medial region of the hemisphere. The 'projections' of the sensory and motor homunculi onto the somatosensory and motor cortex on either side of the central sulcus are shown in Figure 5.4.

- □ Study Figures 5.2 and 5.4. Given the distribution of the blood supply from the middle cerebral artery, (a) which parts of the body would you expect to be affected as a result of blockage of the middle cerebral artery and (b) which parts would you expect to be spared?
- ■ (a) Since the middle cerebral artery supplies the lateral part of the cortex around the central gyrus, the face and arms are likely to be damaged. (b) In contrast, the medial region of the cortex of each hemisphere around the central gyrus does not receive its blood supply from the middle cerebral artery, and so the legs will be spared.

- □ What will weakness in the left arm indicate about the location of the stroke?
- ■ The first conclusion is that the stroke occurred in the right hemisphere. The second is that the damaged tissue lies somewhere in the mid region of the motor cortex which controls the arm.

The muscles of the forehead, throat and jaw are represented in both hemispheres and are usually spared by middle cerebral artery stroke or at least show quick recovery. Following middle cerebral artery stroke in the left cerebral hemisphere, the symptoms may include aphasia.

- □ What regions of the brain do these symptoms implicate? (You may need to refer to Book 2, Section 11.2.)
- ■ Expressive aphasia is likely to result from damage to Broca's area and receptive aphasia from damage to the posterior speech region, which includes Wernicke's area (Figure 5.5).

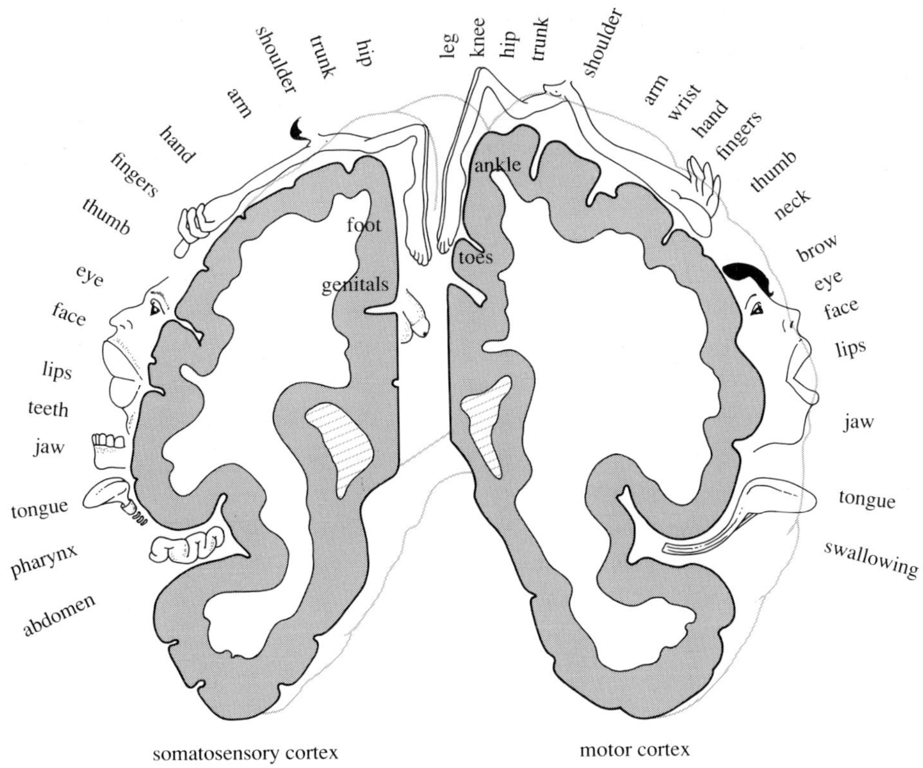

Figure 5.4 The sensory homunculus of the left side of the body, projected onto the somatosensory cortex of the right cerebral hemisphere and the motor homunculus of the right side of the body, projected onto the motor cortex of the left cerebral hemisphere .

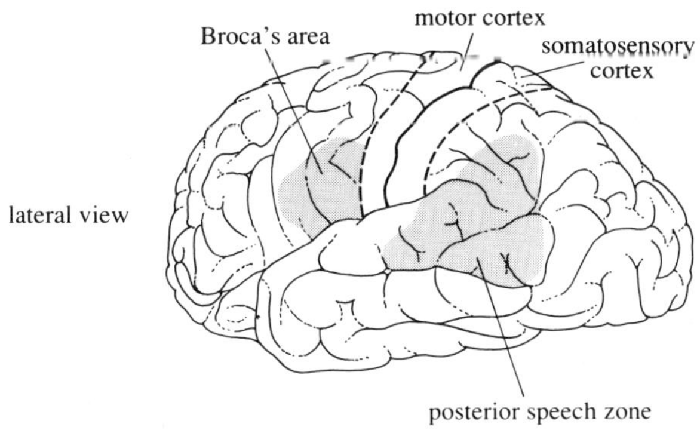

Figure 5.5 Lateral view of the brain showing the location of the speech areas (shaded).

Middle cerebral artery stroke in the region of the parietal cortex can also produce a disturbance of learned motor acts that cannot be explained by weakness or uncoordination. The disturbances may not, however, be total. For example, a sufferer might not be able to *imitate* striking a match but be able to carry out the activity if given a match to strike. Damage within the right parietal lobe can cause disturbance of spatial perception, such as copying simple diagrams, interpreting maps, finding one's way about or even putting on clothes. Damage within the parietal lobes can also result in loss of awareness of the space around the opposite side of the body and failure to use the contralateral body and limbs.

Because the middle cerebral artery also supplies blood to sub-cortical regions such as the optic radiations (the pathways from the lateral geniculate nucleus to the visual cortex; Book 3, Section 4.1, Figure 4.2), the signs of stroke may also include blindness in restricted regions of the visual field.

In contrast to the middle cerebral artery, the territory of the *anterior* cerebral artery (Figures 5.2 and 5.4) lies mainly in the medial part of each hemisphere and encroaches over the anterior surface. Where occlusion (obstruction) occurs in the branches supplying blood to the regions of the somatosensory and motor cortex that fold down into the medial surface of the hemisphere, weakness and sensory loss may result.

☐ Determine from the position of the sensory and motor homunculi shown in Figure 5.4 which regions of the body will show weakness and sensory losses.

■ The legs, particularly the distal regions.

Other symptoms of anterior cerebral artery defect may be urinary incontinence. In addition, a number of syndromes may arise because of damage to the corpus callosum which stops the flow of information between the two hemispheres and so disconnects regions on opposite sides of the brain. For example, a region of dead neurons in the corpus callosum can result in the functional disconnection of language regions in the left hemisphere from the motor region for the arm in the right hemisphere. This leaves the victim unable to carry out verbal instructions using the left arm while leaving intact the ability to carry out instructions with the right arm. Failure of the blood supply to the frontal lobes, limbic system structures and cingulate gyrus, all supplied by the anterior cerebral artery, can result in subtle behavioural disturbances.

Having discussed the territory supplied by the middle cerebral and anterior cerebral branches, this section now considers the *posterior* cerebral artery (Figure 5.2) which supplies the occipital lobe and regions of the medial temporal lobes. Failure of the blood supply to these regions results in visual field deficits, disturbances of reading and also memory loss. The posterior cerebral artery also supplies deep brain structures such as regions of the thalamus and the midbrain (Figure 5.3). Strokes in the thalamus can cause contralateral sensory loss whereas those occurring in the midbrain can produce disturbances in motor functions ranging from uncontrollable sudden movements of the limbs to paralysis.

This section has described the consequences of blockage or rupture in one of the three arterial branches supplying the brain. Because of the different distributions of the three branches and the localization of function in the human brain, failure of the blood supply will have different consequences depending on which branch is affected. The different constellations of symptoms following a stroke are used by clinicians as a basis for the initial diagnosis of the location of the stroke. An interesting point for this discussion is that some recovery can occur. About 75% of disabled stroke survivors learn to walk again, though only about 20% walk at normal speed. About 15% of those with a paralysed arm eventually regain normal arm function. People with an initial aphasia are expected to recover 25% of their memory over a period of three weeks to six months.

5.2.2 Head injuries

Although the brain is protected within the cranium it can suffer trauma from sudden blows to the head (closed head injuries) as well as from penetrating injuries that directly damage brain tissue (open head injuries). In England and Wales alone as many as 150 000 patients are admitted to hospital each year following head injury. The majority of severe cases come from road traffic accidents with about 7% being classed as severe. The transient disturbance of mental functions such as loss of consciousness and disturbances of vision and balance that can follow a blow to the head is described as concussion.

Blows to the head subject the brain to a variety of mechanical forces. The movement of the brain relative to the skull can result in several sources of damage. First, a blow to the head can result in damage to brain tissue local to the region of the blow caused by the tissue being compressed against the skull. Second, following a strong blow, the brain tissue can bounce off the skull at the point of impact so that the opposite pole of the brain is compressed against the opposite side of the skull, causing further damage. Movement and compression of the brain within the skull can also result in a twisting and shearing motion that can produce microscopic lesions of nerve fibres. This is the probable cause of the accumulating deficits seen in boxers commonly described as being ‘punch drunk’.

A further complication of closed head injuries is that neurons local to the impact may develop *oedema*, that is they retain fluid and swell up. Such neurons are less physiologically excitable. Reduction of oedema by drugs that cause the release of fluid can shorten the length of impairment.

Loss of consciousness from a closed head injury has been attributed to disruption of transmission in the fibres in the brain stem reticular formation. The duration of unconsciousness has been shown to be correlated with the extent of the damage and the longer the coma the greater the possibility of serious impairment in bodily and mental functions and of death.

Open head injuries such as those resulting from gunshot wounds produce localized lesions from both the penetrating missile and fragments of bone pushed into the brain. The opening of the skull also introduces the risk of infection. In such penetrating injuries the force transferred to the skull is very small in comparison with an automobile accident where the force of the moving body is absorbed by the skull as it hits the windscreen or ground, for example. As a result, many victims of restricted open head injuries do not lose consciousness. As you might expect, the deficits following open head injuries tend to be more specific depending on the region damaged. For example, localized gunshot wounds to the rear of the brain can result in blindness within restricted regions of the visual field.

5.2.3 Epilepsy

Epilepsy is a condition resulting from neuron disorder. An epileptic attack occurs when the neurons in regions of the cortex fire synchronously. An explanation for this activity is that inhibitory interconnections between the nerve cells become weakened and the cells' activity takes up synchronized oscillations. The synchronous activity disrupts on-going behaviour and can result in loss of consciousness and production of involuntary movements. Epileptic attacks are quite common. About one person in 20 experiences an epileptic attack in their lifetime. But only one in 200 experiences multiple attacks and is categorized as suffering from epilepsy.

Epileptic seizures fall into two categories: partial (focal) and generalized. These can be identified by their contrasting EEG records (Book 2, Section 8.8.1, Box 8.2, and Section 1.4.5 of this book). The electrical activity recorded from multiple scalp electrodes during a partial epileptic episode and a generalized epileptic event is shown in Figures 5.6a and b respectively.

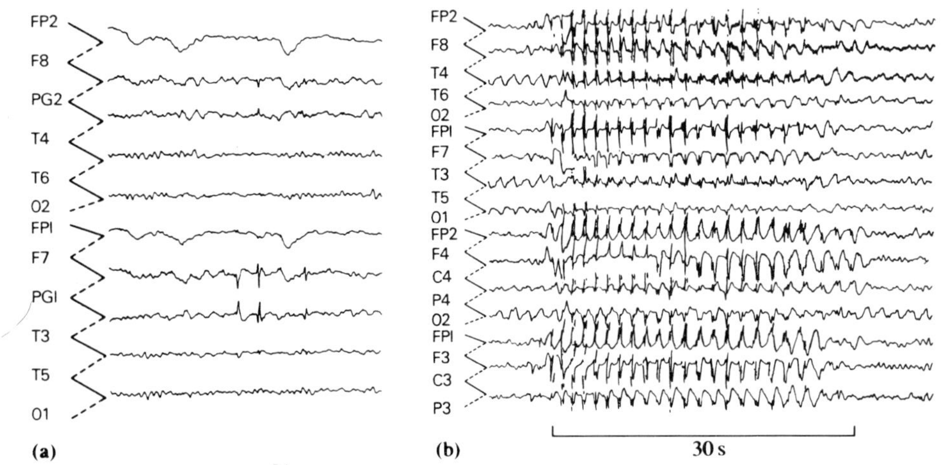

Figure 5.6 EEG electrical activity recorded from scalp electrodes during (a) a focal seizure and (b) a generalized seizure. See text for details.

The codes in the left margin of Figure 5.6 indicate the positions of the electrodes in internationally agreed standard positions. The letters 'O', 'T', 'P' and 'F' signify placement positions within the occipital, temporal, parietal and frontal lobes respectively. Notice in the record of the partial seizure that large brief spikes are seen between electrode F7 and PG1 and between PG1 and T3 over a short period of several seconds. Typically, the focus is a localized region in which a group of neurons show a synchronous rhythmic pattern of spike discharge activity. In the record of the generalized seizure, the large brief spikes occur widely throughout the brain with the seizure activity persisting for tens of seconds. This is because generalized epileptic seizures, as shown in Figure 5.6b, involve synchronous neuronal activity in large areas of the brain, with both sides of the brain being affected.

Epilepsy and localization of function

The phenomena associated with partial and generalized epileptic seizures provide another strand of information about localization of function within the brain. Studies of temporal lobe epilepsy in particular have served to complement investigations based on the deficits following lesions to the brain that destroy normal function. In 1979, David Bear, working at the National Institute of Health in the USA, reported on personality changes experienced by people suffering from temporal lobe epilepsy. Bear found that many such sufferers lose all interest in sex and this is often accompanied by an increase in social aggressiveness. These people, compared with epileptic patients with foci outside the temporal lobes, display one or more outstanding personality traits. They tend to be intensely emotional, ardently religious, extremely moralistic or lacking in humour. Bear argued that the epileptic foci act in an opposite way to the lesions that destroy normal function. His view is that the continuing discharges from localized epileptic foci act to stimulate activity in affected regions and so exaggerate their normal roles.

Significantly Bear observed differences between individuals according to which side of the head the foci occur. Individuals with foci in the right temporal lobe tend to be hyper-emotional while those with foci in the left tend to show persistent patterns of thought, such as a sense of personal destiny, moral self-scrutinizing and a penchant for philosophical explanation. Thus the picture of brain function resulting from Bear's study is of asymmetry of function of the left and right temporal lobes with the right lobe having an involvement in emotional responses and the left in the formation of ideas.

This picture is consonant with evidence described in Book 2 (Sections 11.2 and 11.3) for left–right differences in human brain functions. In most individuals, the understanding and production of speech is lateralized to the left side of the brain. Bear's findings indicate that the left temporal lobe has a wider role than just this. An interpretation of his observations is that the increase in activity generated by the epileptic foci results in modifications in the structure of the contents of speech. This implies that the left temporal lobe is involved in the underlying processes that contribute to speech, generally referred to as thinking. Thus disturbances in the balance of these processes are manifested as disturbances in thinking.

Furthermore, damage to the right temporal lobe results in disturbances in the understanding of the emotional content of speech, that is in the interpretation of the tone of the voice and the accompanying bodily posture and hand gestures. Bear's evidence complements this picture by suggesting that increased levels of activity produced by epileptic foci result in heightened levels of the emotional colouring of thought.

5.2.4 Infections

Although there is a partial separation of the bloodstream from the brain by a variety of filters known collectively as the blood–brain barrier, the central nervous system is susceptible to invasion by a wide variety of infectious agents including viruses, bacteria and parasites. Generally, infections of the nervous system spread from infections elsewhere in the body, such as the ears, nose and throat, but they can be introduced as a result of open head injuries. To the extent that infections result in the loss of neurons, they can have long-term consequences for the individual, depending on the extent and degree of localization of the loss.

5.2.5 Tumours

A brain tumour is a mass of new tissue that persists and grows independently of the surrounding structures of the brain. Brain tumours do not originate from neurons, but rather from the aberrant multiplication of glial cells or other supportive cells in the central nervous system. A tumour may develop as a localized entity or it may infiltrate surrounding tissue destroying neurons and glial cells. Because the skull has a fixed volume any increase in its contents will compress existing brain structures, resulting in disorder.

5.2.6 Degenerative disorders

Table 5.1 provides a list of the commonly occurring neurodegenerative disorders of the nervous system. Many of these have been discussed in previous chapters of this book but are included in the table for completeness. Some of these neurodegenerative disorders are associated with dementia. You should recall from Section 3.1 that the dementias involve a loss of intellect and memory and include personality changes. In a small proportion of the population, dementia is attributable to loss of brain tissue through progressive and continuing closure of small or large blood vessels within the brain, resulting in a widespread loss of nerve cells throughout the brain. This form of senile dementia is called a secondary dementia to distinguish it from a primary dementia, of which Alzheimer's disease (Section 3.7.1) is the most common.

Table 5.1 Neurodegenerative disorders

Disorder	Symptoms and possible cause
Alzheimer's disease	Loss of cholinergic neurons, development of senile plaques and neurofibrillary tangles from unknown causes
Huntington's disease	Degeneration of basal ganglia, frontal cortex and corpus callosum, due to a genetic abnormality
Pick's disease	Atrophy of frontal and temporal lobes from unknown causes
Creutzfeldt–Jakob disease	Generalized cortical atrophy caused by viruses
Korsakoff's syndrome	Atrophy of medial thalamus and mammillary bodies from chronic excessive alcohol consumption
Parkinson's disease	Loss of striatal dopamine due to degeneration of the substantia nigra from unknown causes
Multiple sclerosis	Abnormal neural activity due to loss of myelin from unknown causes
Motor neuron disease	Loss of motor neurons from unknown causes

People with Pick's disease have identical symptoms to those with Alzheimer's disease, progressive deterioration of mental faculties often accompanied by severe depression. It is only on post-mortem that the two disorders can be distinguished. The victim of Pick's disease does not show the plaques and tangles characteristic of Alzheimer's and intriguingly, the degeneration is limited to the frontal and temporal lobes of the cortex.

Creutzfeld–Jakob disease also produces dementia but is distinguished from Alzheimer's and Pick's disease by its very rapid onset. In just a few months the victim progresses through dementia, stupor, coma and death. Importantly, a transmissible virus has been discovered as the major causative factor in Creuzfeld-Jakob disease.

A striking feature of the dementias is that of neuronal loss and reduction in transmitter substance in the cortex (Section 3.4.1). The evidence for localization of cognitive functions in the cortex and the involvement of medial temporal cortex, frontal lobes and hippocampal regions in memory functions was considered in Chapter 11 of Book 2. The discussions there pointed to the close relationship between language, memory and consciousness. It is not surprising that degeneration of cortical neurons leads to the progressive loss of memory, intellect and awareness that characterize the dementias. What is not clear is whether dementia is the result of a random loss of neurons which, when it exceeds some threshold level, produces the symptoms of dementia, or, whether dementia results from the progressive degradation of specific critical cognitive systems. As described in Chapter 3, with dementia affecting some 15% of the population in the UK over 65, and with the proportion of people in that age group rapidly increasing, this is an area of research that will be given increasing attention.

Korsakoff's syndrome has been attributed to a deficiency of vitamin B_1 associated with chronic alcoholism. Post-mortem examinations of individuals who suffered from Korsakoff's syndrome reveals degeneration in the mammillary bodies of the thalamus, the medial thalamus and throughout the cortex but particularly in the frontal lobes. An individual with Korsakoff's syndrome will suffer from both retrograde and anterograde amnesia (Book 2, Section 11.4). They will also tend to make up stories rather than admit memory loss. Generally, Korsakoff's sufferers have little insight into their condition and show apathy and loss of interest. The symptoms of Korsakoff's syndrome develop very rapidly and are progressive. Massive doses of vitamin B_1 will arrest the progression of the disease but do not reverse the symptoms which are permanent and have been demonstrated to persist over 10 to 20 years.

5.2.7 Damage to neurons and nerve fibres

So far, the types of damage that the brain sustains have been categorized according to the cause and clinical symptoms shown by individuals. However, it is important to remember that, because the brain is a connected system, damage localized to one particular point may have more widespread effects. Some of these effects will be transient but others will be permanent.

The effects, both directly from the source of damage and indirectly as a consequence of the connectivity of the system, on neurons and nerve fibres in the brain include the following:

1 Functions mediated by the lost neurons are lost.

2 In time, as the axons of dead neurons die off, their target cells will be deafferented, that is, neurons connecting to the dead neurons will lose their normal synaptic inputs.

3 Nerve fibres travelling through the region will be severed or damaged, depriving their targets of their normal inputs. In turn, this can result in death of the postsynaptic neurons.

4 Neurons in regions adjacent to the damage may develop oedema and show a transient impairment of function.

5 Nerve fibres travelling through regions adjacent to the damaged regions may be traumatized by pressure or leakage of blood and their functioning transiently interrupted.

Summary of Section 5.2

The types of neurological disorders and damage of the brain were briefly reviewed. These range from damage originating externally to the brain, such as open-head injuries, to excessive degeneration and damage of neurons in specific brain regions, such as the cortex in the case of the dementias. Recovery from such damage has been mentioned following, for example, a stroke. The next two sections examine more closely the evidence for recovery of function and the mechanisms involved.

5.3 Evidence for recovery of function at the level of the individual

The preceding descriptions and discussions of the vulnerabilities and susceptibilities of the human brain to damage indicate that the brain can lose neurons and sustain damage to neurons through a wide variety of causes. The deficits that result depend upon the location of the lesions and, generally, such deficits do not show good recovery. But within this pessimistic prognosis there is ample evidence for degrees of recovery of function. This section considers the evidence for this recovery in brain-damaged patients.

Hans Teuber and his colleagues at the Massachusetts Institute of Technology (MIT) have carried out a long term study of recovery from the effects of brain damage sustained by the adult brain (Teuber, 1975). The research was based on a total of 520 servicemen wounded in World War II, or in the wars in Korea or Vietnam. Each individual had been clinically examined within a week of injury and follow-up examination occurred 20 to 30 years later. Also, because of the routine psychometric testing carried out by the US military, information such as verbal and non-verbal IQ scores was readily available. Some of the results of their investigation are shown in Figure 5.7.

The data in Figure 5.7 are broken down by the nature of the deficits and by the age of the individual at the time of the injury. The injuries were classified into four groups: motor, somatosensory, visual field and initial aphasia. The number of individuals showing an improvement in their symptoms over the 20-year period was measured and expressed as a percentage of the total number of individuals in the particular age/symptom category.

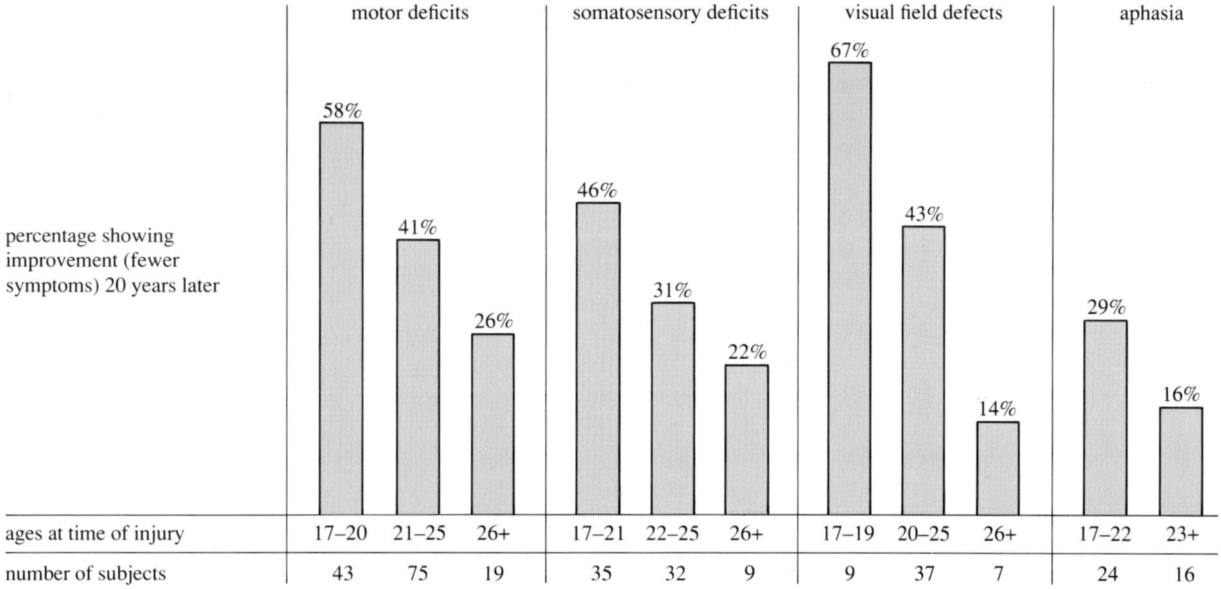

Figure 5.7 Estimated improvement between the time of the initial examination made (within one week of injury) and follow up examination (20–30 years later) for motor, somatosensory, visual field and aphasia deficits, broken down by age at time of injury.

□ Examine Figure 5.7. What is the striking feature within each injury category?

■ There is some improvement in each injury category.

□ Does the age at the time of injury affect the improvement of symptoms?

■ Within each category there is a similar trend. There is progressively less evidence for recovery with advancing age at the time of the injury.

Teuber's evidence suggests that the younger the nervous system the better the chance of recovery from brain damage.

□ Is the prognosis of recovery from brain damage uniform across the brain?

■ The data indicate that different regions of the brain have varying capabilities for recovery.

Teuber's data show that people with visual field deficits, arising from damage to the occipital lobes or sub-cortical structures such as the lateral geniculate nucleus of the thalamus, have the best chances of recovery. In contrast, damage to the speech regions, Broca's and Wernicke's areas and associated pathways, have the least capacity for recovery.

Teuber's data clearly provided evidence of improvement. However, in order to understand the processes involved at the cellular level it is important to know at what stage following brain injury most improvement occurs. This depends on a number of factors, including how localized or widespread the damage is and the type of damage sustained. The following two examples illustrate these points.

In 1979, Andrew Kertesz reviewed the progress made by people who had disorders of speech (aphasia) resulting from stroke. Using the Western Aphasia Battery, a test measuring spontaneous speech content, fluency, comprehension and repetition, he found that most recovery occurred in the first three months, with some occurring in the next six months, with little evidence for further recovery thereafter. Furthermore, recovery from the disorder is rarely absolute. Thus Kertesz's findings point to recovery of speech function after a stroke being mediated by relatively short-term processes.

In a study carried out at Montreal University, Brenda Milner reported that after surgery for treatment of epilepsy, there is a drop in intelligence test scores which returns to normal levels within twelve months. This reduction in normal functioning seems to be related to post-operative swelling around the operated region. Significantly, treatment with cortisone reduces the swelling and also lessens the decrease in intelligence test scores. Milner's evidence indicates that the take-up of fluid by neurons causing them to swell leads to impairment in their functioning. This can contribute to the loss of function seen after brain damage. Further, Milner's evidence points to a relatively long time-course for a return to normal functioning after this type of damage.

This contrasts with Kertesz's findings on recovery of damage from a stroke. In Kertesz's investigation the damage was localized to a particular area of the brain and the damage almost certainly involved some cell loss. In Milner's investigation, however, the damage was more widespread. In addition, loss of function was due to reduced efficiency of neuron function, caused for example by oedema. Given that the recovery from localized damage is relatively quick, it is surprising that recovery from generalized damage takes so long. This suggests that the type of damage sustained affects both the degree and the time course of recovery. Importantly, both Kertesz's and Milner's findings provide significant evidence for some recovery from brain damage.

5.4 Physiological repair mechanisms

Teuber's and Kertesz's data show that recovery of function does occur after brain damage. Victims of stroke and other brain injuries often show marked recovery in the months and years after the provoking incident. It is frequently reported that the symptoms in people with impairments of speech, language comprehension, paralysis or weakness of the muscles, are less severe with time. What explanations can be given to account for the apparent recovery of function after brain damage?

Spontaneous regrowth of cells, as seen in other bodily organs, is not seen in the human brain—at least not to any visible extent. The proliferation of neurons in the human brain by cell division ceases soon after birth, although the complexity of fine structure of the nervous system, indicated by the number of axon branches, continues over several years (Book 4, Chapter 3). However, Teuber's and Kertesz's findings show that there can be recovery of function even where damage occurs to the adult brain. If the brain cannot rebuild its structures by regenerating neurons, what mechanisms can be invoked to explain recovery of function after damage?

5.4.1 Reversal of oedema

One candidate for a mechanism contributing to the recovery of function seen after brain damage is the reversal of the depression of excitability of the nervous system tissue that is a side effect of damage. Local damage to motor regions of the brain can lead to contralateral paralysis that disappears with time. Also after a head injury there can be a transient period of coma, memory loss and other more subtle changes in the functioning of the brain. Milner's findings (described in Section 5.3) show that traumatic consequences of damage such as swelling of surrounding tissue and reduction in the levels of excitability in the residual nervous tissue can be expected to dissipate to a large degree over time and to contribute to a restoration of function.

5.4.2 Physiological mechanisms involving neurons

The effect of swelling of neurons is transient but the loss of neurons is permanent. However, a number of physiological mechanisms have been described that may contribute to a degree of plasticity of the damaged neural pathways. You learnt in Book 4 (Section 3.6) that plasticity of neuronal connections is not completely lost in the mature adult. The interactions between neurons that occur during development can be repeated, albeit to a limited extent, allowing the mature nervous system to adapt to injury. This section looks at mechanisms at the cellular level that might contribute in some way to recovery.

Axon sprouting

Cell bodies that are the synaptic targets of damaged neurons lose their synaptic inputs. One result of such deafferentation is that the deafferented cell releases a nerve growth factor which causes other axons and dendrites in the vicinity to start sprouting extra branches. These developing new branches extend towards the source of the nerve growth factor to make synaptic connections with the deafferented neuron.

Deafferentation of cells can have different effects at different ages, both on the degree of consequent axon sprouting and in any restructuring of dendritic trees. In 1989, Brian Kolb and Ian Whishaw, working at the University of Lethbridge in the USA, compared the effects of damage to the cortex of rats at 10 days old and when adult. They measured the extensiveness of the dendritic trees of neurons in the damaged regions of the cortex when the 10-day-old rats reached adulthood and compared each age group with control rats of the same age that had not received any damage. They found that the degree of dendritic branching and extent of the branches of the cortical neurons in the rats which received lesions when they were adult were similar to that seen in the controls. In contrast, the degree of branching and extent of the branches seen in the rats who received lesions when 10 days old was significantly greater than that in the controls (Figure 5.8). Kolb's and Whishaw's studies on the rat indicate that compensation through dendritic expansion following damage does not occur at such a high level in the adult brain as it does in the juvenile. If dendritic expansions and axon sprouting (regeneration) are related events, the implication is that in the adult rat axon sprouting also does not occur at a high level.

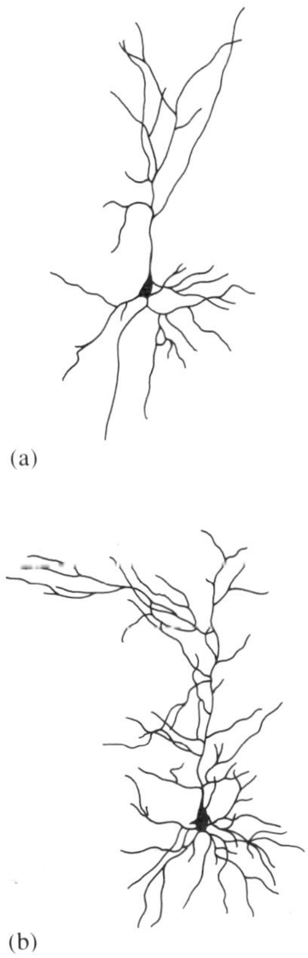

Figure 5.8 Dendritic branching of neurons from the cortex of adult rat brains of the same age. (a) Control (b) Neuron from region damaged when rat was 10 days old.

Comparisons between the dendritic trees of cortical neurons in middle-aged and elderly people (Chapter 3) reveal that elderly brains show more extensive dendritic trees than middle-aged brains. This increase in dendritic growth indicates that cortical neurons receive more synaptic inputs in elderly brains than in middle-aged brains to compensate for the progressive neuron loss that occurs during the ageing process. But it is far from clear whether the axon sprouting and dendritic enlargement in response to a progressive loss of neurons over several decades, as seen in the elderly, can occur to the same extent in response to the sudden, more concentrated tissue damage that follows a stroke or a penetrating injury to the adult brain. Most importantly, it is also not clear to what extent structural repair such as axon sprouting and dendritic enlargement can contribute to replacement of lost function, a point discussed further in Section 5.4.3.

Unmasking of silent synapses

Another factor to be considered is that the removal of the synaptic input from damaged cells can result in the unmasking of inactive 'silent' synaptic connections (Book 4, Section 3.6.2). During the development of the micro-circuitry of the brain there is active competition for synaptic space on the surfaces of neurons. During the development of CNS microstructure, active synapses may displace less active synapses to gain control of their target cell. There is also physiological evidence that some synapses are actively inhibited by others, which render them silent. Following damage, inhibited or silenced synapses can be released from inhibition and are able to activate their target neurons.

□ In what way could sprouting and unmasking of silent synapses contribute to restitution of function after damage to the brain?

■ It is difficult to see how sprouting and unmasking of silent synapses could contribute to a *direct* replacement of the functions lost through destruction of regions of nervous tissue. However, through these mechanisms, surviving pathways could be strengthened or new pathways could be created which may provide alternative routes for information.

Denervation supersensitivity

A further physiological phenomenon that may contribute to the restitution of function following brain damage is **denervation supersensitivity**. In response to removal of synaptic input, the postsynaptic membranes proliferate receptor molecules. This has been demonstrated in the dopaminergic pathways of the CNS and has the effect of making the dopaminergic system more sensitive. This is shown by the appearance of postsynaptic responses to doses of dopa that were previously ineffective. The phenomenon of denervation supersensitivity may maintain the responsiveness of pathways which have been partially deafferented.

Physiological effects of training—awakening of undamaged neurons

Studies by the German clinical physiologist J. Zihl in the early 1980s provide evidence that visual field defects such as areas of blindness resulting from damage to the primary visual cortex can be reduced with the cooperation of the individual in training. Zihl's subjects had areas of blindness in their visual field that had persisted for many years after the precipitating incident. All individuals were trained to detect a light stimulus presented briefly at the edge of the blind patch while fixating a spot of light at the centre of the visual field. The subjects were instructed to concentrate their attention on the apparent blind region while fixating the central spot.

Under these conditions of directed attention, Zihl found that the size of the blind spot became reduced and normal visual functions returned to the outer fringe of the blind region, including colour and pattern vision and binocular depth perception. After years of blindness, the visual functions reappeared after relatively short periods of training, of only 30 to 60 minutes. What is particularly interesting is the fact that years of normal visual experience are ineffective at improving vision, whereas engaging the patient's attention and motivation had a positive effect.

Zihl's explanation of the recovery of function is in terms of awakening undamaged neurons around the border of the cortical lesion whose activity had been depressed through the withdrawal of intracortical inputs following the damage. Zihl's proposal is that the depressed regions become functional again when physiological mechanisms that underlie attention increase their level of excitability. Certainly, with the years intervening between the incidence of the brain damage and the restoration of function, any plasticity attributable to induced sprouting and denervation supersensitivity would have dissipated and therefore would be unlikely to contribute to the recovery. However, the question remains as to how attention directed towards particular visual stimuli within the visual field can bring about such improvements, and what neurophysiological mechanisms might be involved.

5.4.3 Pathway restructuring after brain damage

The mechanisms of denervation supersensitivity, sprouting and unmasking of silent synapses described in Section 5.4.2 may provide a physiological basis for the restructuring and strengthening of remaining pathways rather than the construction of new pathways. This section looks at the reasons for this.

Consider two theoretical examples to show what is involved in both the strengthening of pathways and the construction of new pathways as shown in Figure 5.9 (*overleaf*). These examples use the unmasking of silent synapses as the recovery mechanism. However, the examples could equally well apply to axon sprouting.

For the sake of argument, assume that some cells in brain region A, which normally make synaptic connections with cells in region B (Figure 5.9a), have some silent synapses (axons 5 and 6). Now suppose some of these cells died leaving cells in region B deafferented, that is without their normal inputs. How is the unmasking of silent synapses likely to affect any restitution of the lost input from region A? One possibility is shown in Figure 5.9b. Suppose axons 2 and 3 of region A in Figure 5.9a die and deafferent their target cells in structure B. If axons 5 and 6 were inhibited collateral branches with silent synapses from cells 1 and 4 in structure A, these axons would now be released from inhibition. Cells in structure B that are deprived of their inputs from axons 2 and 3 may now receive input via collateral branches 5 and 6 (Figure 5.9b). This may support some restitution of the original function since the unmasked synapses originated from

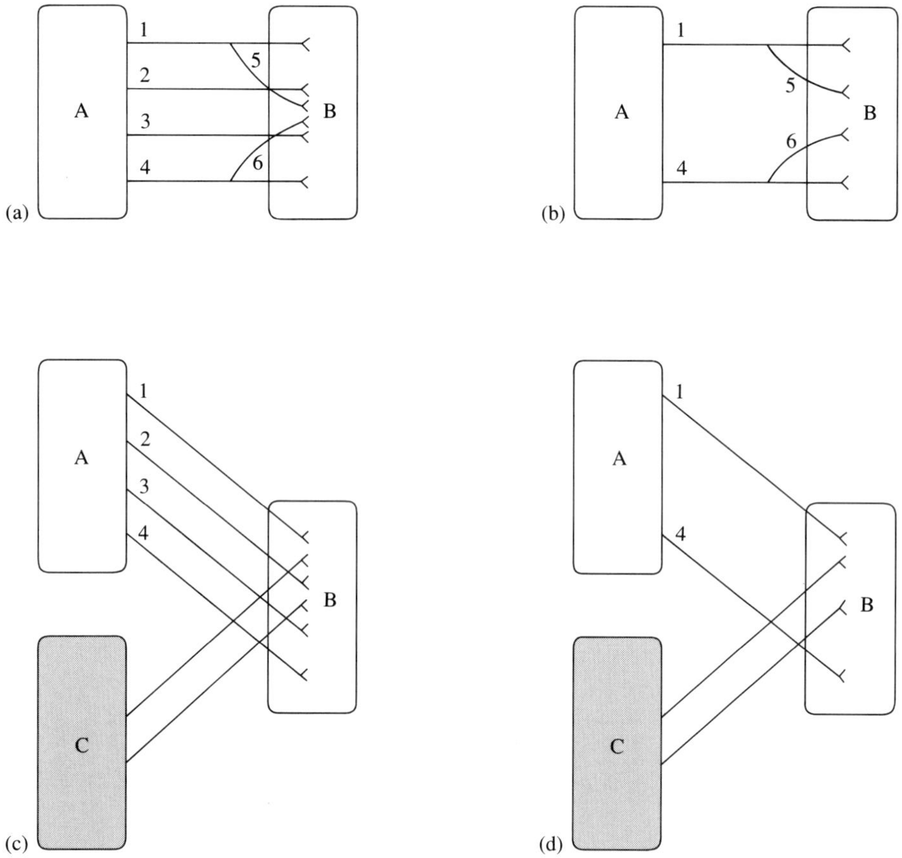

Figure 5.9 A theoretical model of restitution of input following neuron loss. See text for details.

neurons in the vicinity of those lost in region A. This would be the equivalent of strengthening a pathway.

Now suppose the unmasked synapses originate from another brain region, C, shown in Figure 5.9c and 5.9d. Here their synaptic input will not have the same informational content as the lost input. This would be equivalent to the construction of a new pathway. In this case, restitution of input may occur but it is unlikely that there will be any restitution of lost function.

Even so, this theoretical analysis of the possible effects of sprouting and unmasking of synapses suggests that, although a degree of active restructuring of remaining pathways or construction of new pathways may occur after brain damage, it is far from clear what, if any, restitution of lost function may occur in either case. The resulting problem is that specific information may be irretrievably lost.

This can best be understood by returning to an example of the visual system. Topographical organization is one of the features of the visual system. This means that each axon in visual pathways such as the optic nerve and tract running from the retina to the lateral geniculate nucleus and the optic radiations from the lateral geniculate nucleus to the visual cortex, maintain a fixed relationship with their neighbours. Because individual nerve fibres carry information about the visual events within a localized region of the retina and hence from a localized region of visual space, the effect is that any axon's receptive field will be immediately adjacent to the receptive fields of that of its neighbours. Correspondingly, the responses of specific neurons in the visual cortex represent or abstract information about events within a similarly localized region of visual space, which they in turn project to more central neurons. Localized damage to a region of the visual cortex then has the effect of producing blindness in a circumscribed region of the visual field. The result of neuron loss in the primary visual cortex is that unique information is lost.

Suppose that deafferented neurons respond by releasing a nerve growth factor that results in adjacent axons sprouting to innervate them. If these new connections formed through sprouting are physiologically viable, then the result will be that the previously deafferented neurons now have inputs which convey information that originated in a different, albeit adjacent, region of the retina from their previous inputs. The result could then be that the information about visual events that occur in one part of the retina are now conveyed to a part of the visual system that originally processed information from a different region of the retina. It is impossible to speculate about the consequences for the individual of such abnormal routeing of information!

Any possibility of restitution of function is also complicated by the fact that the afferent axons that were damaged will not regenerate. While peripheral axons do regenerate after damage, central axons in general do not (Book 4, Section 3.5). From this analysis then, damage to a topographically organized region or pathway will inevitably result in a permanent loss of information.

The results of this discussion on the contribution of physiological repair mechanisms to recovery of function following loss of central neurons are essentially pessimistic. One conclusion is that the recoveries, which are sometimes spectacular, experienced by people following strokes and other forms of brain damage are largely attributable to the dissipation of the physiological trauma (e.g. oedema) of nervous tissue surrounding the damaged region. Repair mechanisms may strengthen existing pathways or possibly even develop new pathways but it is difficult to assess whether these will make any contribution to recovery of function.

During recovery from stroke or other brain damage, a sufferer receives various therapies, such as physiotherapy to help overcome muscular weakness, or speech therapy for speech problems. As well as helping to rectify specific muscular weaknesses, such treatment also helps people to compensate for deficiencies by learning different strategies for coping with them. The strategies may involve relearning limb movements or ways to grasp objects. They may involve relearning ways of articulating speech or coping with an impaired memory. During such learning, an individual may indeed benefit from the strengthening of existing pathways and may even recruit new pathways that may have developed, but this can only be a speculative suggestion.

5.5 Restitution of function through replacement

So far in this chapter, discussion of restitution of function following brain damage has been confined to physiological repair mechanisms of the body. This section turns to repair mechanisms based on surgical intervention, developed in the 1980s, the aim of which was to reverse the progressive behavioural effects found in Parkinson's disease. This disease (Section 4.3.1) results from a progressive and selective malfunction in the dopaminergic nigro-striatal pathway. You should recall that in this disease the dopamine-producing neurons in the substantia nigra die, depriving their target neurons in the striatum of their dopaminergic inputs.

In the 1980s the treatment of Parkinson's disease assumed a great prominence for an understanding of processes of direct repair of damage. Starting with animal studies, attempts were made to replace the missing dopamine by grafting living neural tissue into the brain. At the centre of this controversial work were Anders Bjorklund and his colleagues, working at the University of Lund in Sweden, who have extended their work by collaborating with laboratories in many countries. In his early experiments Bjorklund used the laboratory rat as an animal model of Parkinson's disease, in which the basal ganglia on one side of the brain was depleted of its dopamine. This was done by injecting a toxin called 6-hydroxydopa (6-OHDA) which selectively kills dopaminergic cells and their terminals. The purpose of this model was to provide symptoms that could be quantitatively assessed, against which to test the effectiveness of the transplants.

After this treatment with 6-OHDA, which produces an asymmetrical lesion, the rats have an asymmetrical posture. Typically, if startled, they move in circles, moving in a direction away from the side of the lesion. The interpretation of the effects of the chemical lesion is that the function of one side of the basal ganglia in the initiation and maintenance of movement is reduced. As a result the intact side of the basal ganglia is dominant, producing the circling movement, just as if the wheels of one side of a car were turning faster than those on the other side. The reduction of circling behaviour was used as an index of the effectiveness of the transplants. Absence of circling behaviour was taken as evidence for the complete effectiveness of the dopaminergic grafts.

Bjorklund removed tissue from regions of the fetal rat brain destined to become the substantia nigra and grafted it into the brains of the chemically lesioned rats. In one set of experiments Bjorklund investigated the optimal age of the rat fetal neurons for grafting. The ages of the developing neurons can be described as a 7-phase sequence:

1 The cells divide and proliferate.

2 They migrate to the positions they will occupy when adult.

3 The cells aggregate to form tissues.

4 The cells differentiate to take on their adult form and structures.

5 The neurons extend their axons to find and contact their target cells.

6 The cells compete for synaptic space on their target cells.

7 The cells form and stabilize synapses.

Bjorklund found that embryonic cells in phases 2 and 3, that is before they had differentiated, provided the best results since grafts of this age could eliminate the circling behaviour. This emphasizes the success of using fetal rather than adult tissue.

The importance of the origin of the embryonic cells was confirmed by grafts of non-dopaminergic cells into the striatum. These grafts had no effect in reducing circling behaviour.

Bjorklund also investigated the most effective way of grafting the embryonic cells. He used four types of grafting method:

1 Placing the embryonic cells in the ventricles next to the chemically-lesioned striatum.

2 Placing the embryonic cells on top of the chemically-lesioned striatum.

3 Injecting a suspension of the embryonic cells directly into the chemically-lesioned striatum.

4 Placing the embryonic cells into the cortex.

The results of the four grafting methods were as follows:

Method 1 gave an average of 50% reduction of the circling behaviour.

Method 2 gave complete recovery over a 2–6 month period.

Method 3 gave complete recovery in 3–6 weeks.

Method 4 gave no improvement in circling behaviour.

Therefore the location of the grafted tissue into or on top of the striatum was crucial to success.

One possibility raised by the finding that implants into the ventricles (grafting method 1) resulted in a degree of amelioration of the turning behaviour is that the improvement was a result of the seepage of dopamine from the ventricles into the striatum. This raised the possibility that the implants into the striatum were acting simply as dopamine pumps without forming physiological connections. But, in animals which received successful grafts (method 3) into the striatum, electron-microscope studies showed that the grafted cells made abundant contacts both with neurons projecting from the striatum and with interneurons within the striatum.

□ How could you test to see if the synaptic connections were physiologically integrated?

■ One way would be to stimulate electrically other regions of the brain to see if the grafted cells responded.

In fact, electrical stimulation in regions of the brain such as the pre-frontal cortex or locus ceruleus produced physiological responses in the grafted neurons, suggesting that the grafted neurons themselves had received connections from these regions.

However, Bjorklund's results showed a further important feature. Restitution of function occurred even though there was only a small amount of synaptic replacement as in the case of method 2. Furthermore, his finding that a degree of restitution was possible even where there was no or limited significant physiological integration, as in the case of grafting method 1, also suggests that re-creation of the normal connectivity is not essential for amelioration of the Parkinson symptoms in rats.

In the above experiments the dopaminergic input to the striatum was depleted. In another series of experiments in rats, the structure of the striatum was damaged directly. Such rats showed changes in behaviour and had difficulties in carrying out tasks that involved using the forepaws to manipulate or retrieve objects such as food. These symptoms could be ameliorated by grafting embryonic striatal tissue directly into the striatum. Anatomical techniques revealed that the grafted striatal neurons became fully integrated with axons running into the graft to make synaptic contacts and out of it to make connections within other parts of the brain.

Such animal studies promised benefits to humans of embryonic tissue implantation. Bjorklund's experiments on rats suggested that restitution of function after damage through grafting fetal tissue was a possibility. In fact, there have been dozens of human Parkinson sufferers who have received grafts of human fetal tissue. At the time of writing (1992), although there are a few reports of amelioration of symptoms, the results of these experimental treatments have been difficult to evaluate. One difficulty is that the people receiving the grafts have been suffering from very advanced Parkinson's disease and, even though they show some improvements, they still remain sick and debilitated.

Other experimental human treatments for Parkinson's disease have involved transplants of a specific class of dopaminergic cells called adrenal chromafin cells taken from the patient's own adrenal gland. These cells, which have a completely different structure from neurons, are implanted into the striatum where they appear to act as physiological dopamine pumps since they cannot make physiological connections with the host neurons. This procedure has a number of practical advantages. It side-steps the severe ethical problems of obtaining and transplanting fetal tissue. It also minimizes the problems of tissue rejection because the individual's own adrenal cells do not trigger an immune response. But, the preliminary results of more than 300 human adrenal implants have shown that only about 30% of patients showed even slight improvements, many died after the operation and there is little evidence that the grafts survived for any length of time.

Section 5.5.1 goes on to evaluate grafts using human fetal tissue from regions of the brain other than the substantia nigra.

5.5.1 Dopamine replacement: a panacea or isolated instance of repair?

In contrast to the very positive results found in animal studies, implantation of dopaminergic cells into the human striatum has produced generally disappointing results. In the context of these studies of grafts of embryonic tissue from the substantia nigra, how practicable, then, are suggestions that brain grafting could be applicable to general repair of damage to CNS structures?

One factor that needs to be taken into account is that the dopaminergic nigro-striatal synapse differs substantially in its structure from the classic synapse you encountered in Chapter 4 of Book 2. Each dopaminergic neuron projecting to the striatum from the substantia nigra has an extensive postsynaptic influence throughout the striatum. Rather than terminating in discrete classical synapses, each dopaminergic neuron gives rise to an estimated 250 000 terminal varicosities. A **varicosity** is a little nodule that occurs on an axonal branch rather than at the tip (Figure 5.10). This form of synapse is peculiar to the dopaminergic system. The term varicosity means a swelling, as in varicose veins. But, unlike the varicosities seen in veins or in axons in aged brains (Section 3.4.3), the dopaminergic varicosity is not a malformation. Rather, it is a functional adaptation of the synapse. A varicosity is less structured and more 'leaky' than the classical synapse with the result that dopamine leaks into the extracellular fluid. This means that transmitter released by a varicosity has a wider sphere of influence than at a classical synapse. The result of this more diffuse pattern of connectivity is that each dopaminergic cell in the substantia nigra influences a significant proportion of the neuronal population in the striatum.

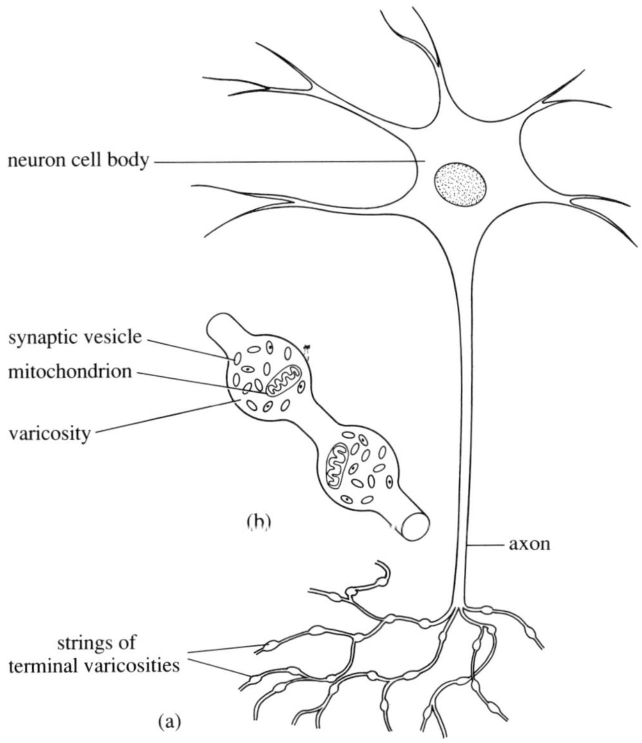

Figure 5.10 (a) A drawing of a neuron of the nigro-striatal pathway with terminal varicosities. (b) Enlarged varicosities.

A major characteristic of the nigro-striatal pathway conferred by this type of connectivity is its ability to compensate for loss of input. In the rat there are only about 7 000 dopaminergic neurons in the substantia nigra that project to the striatum, on each side of the brain. A loss of 75% of these results in considerable loss of dopamine in the striatum but yet no Parkinson-type symptoms develop. A depletion of 80–90% of dopamine is required before the symptoms appear. This finding that a very large reduction in the level of dopamine is required to produce

Parkinsonism in the rat is similar to findings in humans (Section 4.3.3), in whom Parkinson symptoms are not shown until the amount of dopamine is depleted by about 80%. The presence of the extensive connectivity also explains Bjorklund's observations that some restitution of function is possible even where only a small amount of synaptic replacement has occurred.

These characteristics of the dopaminergic system within the striatum put it into sharp contrast with other neurophysiological systems within the CNS. In the development of the visual cortex, for example, there is precise point-to-point projection from the thalamus. The microstructure of the visual cortex is built up through a period of competitive interaction for synaptic sites by the afferents from the thalamus. Experimental studies indicate that the microstructure of the visual system is fine-tuned by environmental influences. Further, its functioning as measured by visual acuity and binocular vision depends upon this fine-tuning.

The characteristics of synaptic connectivity by terminal varicosities point to a qualitatively different role for the nigro-striatal pathway. Here, the information conveyed by the nigro-striatal pathway is less specific than that carried in the visual pathways. The striatum is involved in the sub-cortical motor systems (Book 2, Section 9.4.3, and Chapter 4 of this book). In this context, it is likely that the nigro-striatal pathway is concerned not with the elaboration of motor movements themselves, but in giving movements 'permission' to occur. Put another way, the striatum is involved in selecting which behavioural plans are executed, leaving the detail of the movements to other brain structures. An interpretation of the symptoms of Parkinson's disease and of experimental depletion of dopamine in the striatum is that the influence of the nigro-striatal pathway is to provide a facilitatory role.

The important point to grasp from this analysis of the nigro-striatal projections and Parkinsonism is that it is the diffuse nature of the connections within the striatum that allows them to be reconstructed by implants of developing dopaminergic neurons. In regions of the brain showing precise topographic mapping of sensory information upon the neural tissue, such as the sensory cortex and thalamus, it is difficult to envisage how grafts of neural tissue placed in either of these regions of an adult brain could restore such intricate structure or its functions in mediating sensory processing.

The evidence that, in the rat, transplants of embryonic striatal neurons can successfully repair striatal damage is difficult to interpret. The fact that the striatum is not as structured a system as the visual system may mean it is more amenable to repair. But, given the difference in results from the implants of dopaminergic cells in rats and humans, even here the probability of positive outcomes in humans seems not straightforward.

Summary of Chapter 5

There are a number of mechanisms that contribute to the recovery of function often shown by humans after brain damage. In addition to loss of neurons and axons, damage to neural tissue can produce swelling and other transient effects. Much functional recovery can be attributed to the dissipation of these transient (but sometimes long-lasting) effects.

There are a number of physiological mechanisms, unmasking of silent synapses, awakening of inhibited neurons, denervation supersensitivity and sprouting of axons in response to release of nerve growth factor from deafferented nerve cells, which have the potential to contribute to strengthening or construction of neural circuits after brain damage resulting in cell death. In contrast to peripheral nerve axons which regenerate after damage, CNS axons do not regenerate to any extent after damage. This means that any unique information carried by damaged CNS neurons and their axons is effectively lost to structures with which they made postsynaptic connections.

In response to deafferentation, the postsynaptic neurons release nerve growth factor and cause sprouting in adjacent axons. Through this mechanism the deafferented neurons may gain new inputs, but in topographically organized pathways these axons will not have the same information content as the axons that have been lost. Deafferentation may also unmask silent synapses; but the information content of the unmasked synapses may be different from those lost.

Sprouting and unmasking, perhaps aided by denervation supersensitivity, may act to strengthen neural pathways or even construct new pathways after damage. These pathways may provide information that the brain can use either as a basis for developing new skills or to replace functions that are lost through damage. Equally, it can be argued that these pathways may simply be structural and may not contribute to replacement of lost function. Thus it is not possible to say what mechanisms at the neuron level contribute to the observed recovery at the behavioural level.

Animal studies have shown that relatively unstructured pathways such as the nigro-striatal pathway are candidates for repair of structure and replacement of lost function through implants of embryonic tissue. Despite successful trials in rats, implants in humans have given at best equivocal results. Given that any improvement of symptoms in people with Parkinson's disease following embryonic transplants is likely to be due to the diffuse nature of the connections made, they offer poor prognosis for the use of transplants for restitution of function after damage in other more structured human brain systems.

Objectives for Chapter 5

When you have completed this chapter you should be able to:

5.1 Define and use, or recognize definitions and applications of each of the terms printed in **bold** in the text. (*Question 5.5*)

5.2 Recognize the main symptoms of neurological disorders or damage to the human brain and explain how they provide information about localization of function within the brain. (*Question 5.1*)

5.3 Explain the interrelationship between localization of function in the brain and a stroke in the anterior, middle or posterior cerebral arteries. (*Question 5.2*)

5.4 Explain the neurophysiological mechanisms that contribute to the recovery of function after brain damage. (*Questions 5.3 and 5.4*)

5.5 Describe experiments of embryonic tissue implants used to repair or ameliorate deterioration or brain damage to the basal ganglia in both rats and humans. (*Questions 5.4 and 5.5*)

Questions for Chapter 5

Question 5.1 (*Objective 5.2*)

Which of the following statements about neurological disorders are wrong?

(a) A stroke can result from either occlusion of an artery or bleeding from a vein.

(b) A sufferer of epilepsy with foci in the right temporal lobe may be hyper-emotional and have difficulty in understanding the emotional content of speech.

(c) Head injuries are often accompanied by oedema, the retention of fluid in neurons.

(d) Brain tumours do not arise from neurons in the adult because neurons can no longer undergo cell division.

(e) Given the localization of cognitive functions in the cortex it is not surprising that degeneration of cortical neurons leads to loss of intellect in sufferers of dementia.

Question 5.2 (*Objective 5.3*)

For each item (a)–(d) determine whether it results from:

1 An anterior cerebral artery stroke

2 A middle cerebral artery stroke

3 A posterior cerebral artery stroke

(a) Blindness in a restricted region of the visual field

(b) Weakness in the ankles and feet in one leg

(c) Broca's aphasia

(d) A weakness in the arm on one side of the body

Question 5.3 (*Objective 5.4*)

Which of the mechanisms (a)–(f) will *not* be active in a region of an adult brain that has been damaged?

(a) Release of nerve growth factor

(b) Release of silent synapses from inhibition

(c) Denervation supersensitivity

(d) Axon sprouting

(e) Dendritic branching

(f) Increase in the number of neurons

Question 5.4 (*Objective 5.4*)
Following brain damage, people may relearn lost functions. Why are the repair mechanisms of neurons more likely to involve the strengthening of existing pathways rather than the construction of new pathways?

Question 5.5 (*Objectives 5.1 and 5.5*)
Even if grafting of fetal cells into the basal ganglia of humans becomes successful in alleviating the symptoms of Parkinson's disease, is it likely that this technique could have a wider application for the restoration of function in the brain?

Reference

Teuber, H. (1975) Recovery of function after brain injury in man, in *Outcome of severe damage to the nervous system*, Ciba Foundation Symposium no. 34, Elsevier/North Holland.

Further reading

Kolb, B. and Whishaw, I. (1990) *Fundamentals of Human Neuropsychology*, 3rd edn, W. H. Freeman and Company.

CHAPTER 6
BRAIN DAMAGE AND CONSCIOUSNESS

6.1 Introduction

The development of an understanding of the human brain, however incomplete, must rank as one of the major achievements of the last half of the 20th century. Contributions have been made from the wide range of disciplines forming the neurosciences; biochemistry, neuroanatomy, neurophysiology, neurology, psychology, and ethology. They have all added their interlocking pieces to a jigsaw-like picture of the brain and its functions. The contributions have ranged from descriptions at a molecular level of the way in which the behaviour of neurons is modified by transmitter substances, to glimpses of the way in which the strategy for controlling behaviour, which is called the mind or consciousness, depends on localized regions of the human brain.

This chapter continues discussions of brain regions that contribute to human mental activity. Cognition can be defined as the product of certain types of brain processes that lie between incoming sensory information and the execution of behaviour (Book 1, Chapter 8). This chapter will be concerned with the processes that contribute to human consciousness, particularly as shown by verbal behaviour. These processes, called *cognitive processes*, contribute to the phenomenon called mind or consciousness. Here, a simple materialist position is being adopted. This is, that the terms mind and consciousness are two descriptions of the activity of the same brain processes and systems; processes that have been shaped by evolutionary history to provide adaptive and flexible control of human behaviour.

Evidence gathered from studies of the after-effects of brain damage on verbal and non-verbal components of human cognitive processes and on the involvement of specific brain structures in memory and language functions was presented in Book 2, Chapter 11. It pointed to a picture of the human brain in which cognitive processes are not distributed diffusely within the brain. Discussions of the localization of function within the brain have both theoretical and practical significance. First, an understanding of the way in which the human mind works is important in diagnosing the nature of deficits that appear after the brain is damaged and in prescribing treatment and therapy. Second, where surgical intervention in the brain is required, a knowledge of the localization of function can help guide the surgeon both to locate the correct region and to know the potential risks of disturbing surrounding regions.

An implication of the finding that the phenomenon called mind or consciousness depends upon a range of brain processes with separate representations in the brain suggests that mental activity is itself not a *unitary* process. Rather, the implication is that mental activity is made up of a number of discrete cognitive processes or modules. It is this idea that is the focus of this chapter. The main goal will be to look at the nature of the cognitive processes themselves. Here, the sources of information are the after-effects of brain damage and particularly the fractures in cognitive processes that it can bring about.

This chapter begins with a glimpse into the controversy that led to the acceptance of the concept of localization of function in the brain (Section 6.2). It shows that science is about ideas as well as accumulating facts; ideas that have challenged religious and philosophical beliefs of the time. Do not try to remember all the details of the various positions taken or views held. Rather, you should regard the account in Section 6.2 as an interesting read and as setting the scene for the discussions that follow on the nature of cognitive processes (Section 6.3).

6.2 The emergence of localization

The descriptions of the human brain that you encountered in preceding chapters and books have encapsulated contemporary positions in neuroanatomy and neurophysiology that underpin explanations of the way in which the brain works in controlling the functioning of the body and its behaviour. The picture that is painted is of a 'late 20th-century brain', a picture that would be quite baffling to a neurophysiologist who worked in the late 19th century. Understanding of the brain has advanced in an exponential manner, with a dramatic explosion of knowledge in the last 25 years. In the 19th century and in the first 40 years or so of this century the pace of change in understanding was much more gentle and theoretical positions and perspectives that are now taken for granted took a very long time to become accepted. Sometimes debates raged over many decades. One important position which, at first sight, has some similarities with contemporary descriptions of the localization of function in the brain is the so-called school of phrenology which developed at the end of the 18th century and continued through into the first half of the 19th century.

6.2.1 Phrenology: localization of human faculties

About two hundred years ago, the phrenologists claimed to provide a complete understanding of human cognition. The human mind and behaviour, they claimed, are the result of the activity of discrete faculties localized in separate regions of the human cortex. Originally, the 'theory' of phrenology was developed in the 18th century by two Viennese physicians, Franz Josef Gall and Johan Casper Spurzheim. Gall was a respected anatomist who was the first to distinguish the functions of the grey and white matter of the brain.

Spurzheim grouped the human faculties as outlined in Table 6.1. This shows that the phrenologists' perspective is based on an analysis of human behaviour determined by 'instincts'. Essentially, this approach made the untenable assumption that any individual's behaviour was determined by fixed inherited predispositions. The phrenologists proposed that these innate predispositions were manifested through human feelings and intellectual faculties (Table 6.1) grounded in a specifiable number of so-called personality organs located in the neocortex (Table 6.2, *overleaf*). Originally, Gall proposed 26 faculties, which were later extended to the 36 shown in Table 6.2 by Spurzheim and his co-workers. (You do not need to remember the details of Tables 6.1 and 6.2.)

Table 6.1 The phrenologist's grouping of human faculties

Feelings	Intellectual faculties
1 Propensities (internal impulses to certain actions)	1 Perceptive faculties (knowledge by observation and through language
2 Sentiments (impulses prompting emotions as well as action) (a) Lower: those common to man and other animals (b) Higher: those proper to man	2 Reflective faculties (knowledge by intuition and reasoning, especially by noting comparisons)

The phrenologists' position was based on the notion that mind is based on the activity of the range of independent innate faculties listed in Table 6.2. In turn, these faculties were related uniquely to specific regions of the brain. To a large extent, but with some qualification, Gall and Spurzheim believed that any individual's personality was determined by the development of his or her faculties, which in turn, was related to the development of these underlying brain regions. Intriguingly, the map of the locations of the brain regions devised by Gall supported there being 36 faculties projected upon the skull (Figure 6.1, *overleaf*). One basic assumption of the phrenologists' view was that the topology of the skull reflected the degree of development of the underlying brain. Thus, according to Gall and Spurzheim, measurement of the regional development of the skull would provide an objective measure of individual capabilities, a claim that spawned a small industry in the UK, the Continent of Europe and the USA, of personality measurement based on identification of bumps on the head.

While some of the faculties listed in Table 6.2 are obvious from their names, others are not. If you are acquainted with French or Latin, you may guess that faculty 1, amativeness (located in the region of the cerebellum), is the propensity for passion. Located above it (Figure 6.1) faculty 2, the organ for philoprogenetiveness was described by Gall as the 'love of children'.

What was the evidence to support the views of these early phrenologists? One line of evidence used by Gall came from a single case of a soldier who lost the ability to speak after the blade of a sword had penetrated his left eye and entered the frontal lobe of the brain. Gall's localization of language on the left side of the brain (Figure 6.1, Faculty 33) is the first formal record of the recognition of the lateralization of language. More typical of Gall's method of gathering information about these faculties, however, is the now-classic case of a woman subject whom he found to have a bump on the skull at the back in the region overlying the cerebellum. From the reasons documented by Gall for attributing this region of the brain as the centre for amativeness this case study came to be known as 'Gall's passionate widow'.

In the late 18th century, when Gall was just developing his theory, the dominant explanation of the relationship between mind and brain was the view known as Cartesian dualism, originated by and named after the 17th-century French philosopher and scientist, René Descartes (Book 2, Section 11.5.3). Descartes

Table 6.2 The phrenologists' faculties.

1	Amativeness	18	Wonder
2	Philoprogenetiveness	19	Ideality
3	Concentrativeness	20	Wit
4	Adhesiveness	21	Imitation
5	Combativeness	22	Individuality
6	Destructiveness	23	Form
6a	Alimentiveness	24	Size
7	Secretiveness	25	Weight
8	Acquisitiveness	26	Colour
9	Constructiveness	27	Locality
10	Self-esteem	28	Number
11	Love of approbation	29	Order
12	Cautiouness	30	Eventuality
13	Benevolence	31	Time
14	Veneration	32	Tune
15	Conscientiousness	33	Language
16	Firmness	34	Comparison
17	Hope	35	Causality

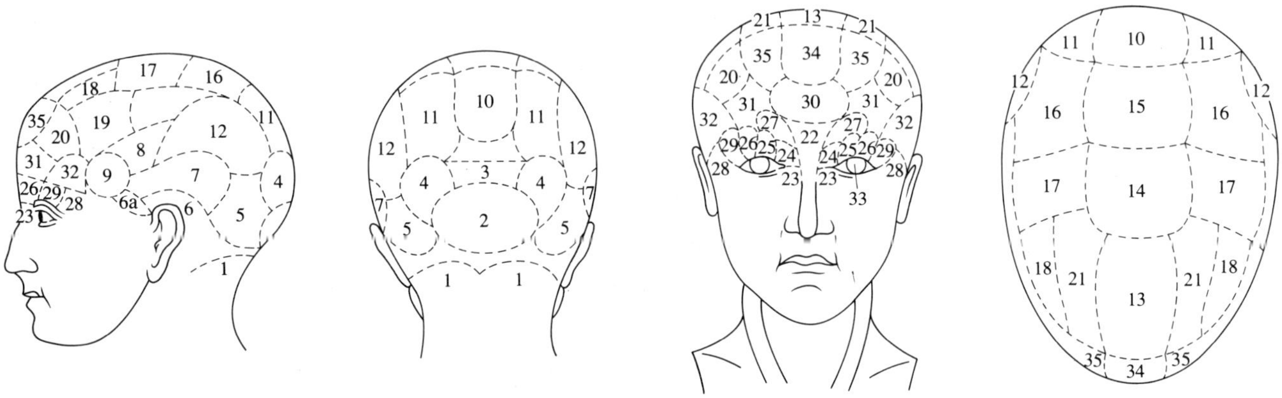

Figure 6.1 Gall's map of the localization of cognitive functions. The numbers refer to the faculties listed in Table 6.2.

distinguished between *res extensa* ('brain stuff') and *res cogitans* ('mind stuff'). *Res extensa* was the physical brain which provided the machinery for controlling the muscles and producing behaviour. Descartes proposed that both animals and humans possessed *res extensa*. Consistent with his belief, Descartes said that the difference between humans and animals lay in the fact that humans possessed an immortal soul. The manifestation of the soul was *res cogitans*, which, because of its divine nature, lay beyond human comprehension and scientific investigation.

Descartes' solution to the mind/body problem was to state that the metaphysical mind, *res cogitans*, influenced the body, *res extensa*, by acting through the pineal gland. Descartes reasoned that the pineal gland must be the crucial brain structure for this interaction because it was the only brain structure he could find that did not possess a 'double'. A single structure not showing the usual bilateral structure in which pairs of identical structures are arranged on either side of the brain, was thought to be vital for the interaction between *res cogitans* and *res extensa* because of the unitary nature of both mind and consciousness.

Gall's phrenological theory was important because it offered the first major challenge to the orthodoxy of Cartesian dualism which was the original statement of a physical relationship between mind and brain. But phrenology was rejected by the mainstream scientific community and its subsequent retreat into popular culture is aptly portrayed by the 19th century drawing shown in Figure 6.2.

Figure 6.2 Phrenology as a pastime of the fashionable.

At first sight, phrenology and its claim that specific human faculties are located in discrete regions of the human cortex seems to have a lot in common with modern 20th-century views on the relationship between mind and brain. For example, evidence for the evolution of language has been sought by examining casts of fossil pre-human and early human skulls. Also, as you will read in later sections, human cognitive processes are described in terms of 'cognitive modules' which function independently of one another.

□ What is the striking difference between the present-day understanding of the localization of functions and that of the phrenologists?

■ The major point of difference is in the nature of the functions that are localized.

In addition, the phrenologists described faculties as independent brain regions, whereas the modern understanding is that cognitive modules are independent processes whose localization does not necessarily correspond with the brain's gross anatomical divisions.

6.2.2 Non-localization views

At the beginning of the 19th century, experimental evidence emerged that appeared to give support to the contrary position that functions were not localized in the cortex. An early protagonist for this position was Pierre Flourens, a French physiologist and member of the Académie des Sciences. Flourens pioneered the use of experimental ablation (removal) in animals for the investigation of nervous system function. Flourens' subjects were mainly pigeons and chickens. Unfortunately, these are species which subsequent studies in comparative neuroanatomy have shown to have differences in brain structures from mammals. In particular, they lack the development of the cerebral cortex that typifies mammals. Flourens would not have been aware of these facts.

Flourens found that ablation of regions of the forebrain had rather general debilitating effects on the ability of his experimental animals to feed and orientate. Ablation of regions of the cerebellum produced more specific movement disorders while lesions in other brain stem regions stopped respiration and resulted in death. A lack of awareness of the differences in anatomical structure of the brains of birds and mammals compounded by the crudeness of his surgical procedures, allowed Flourens to conclude in 1824 that 'feeling, willing and perceiving are but a single and essentially unitary faculty residing in a single organ' (the cerebral hemispheres).

A further reason for the rejection of Gall's claim that faculties were localized lay in mid-19th century understanding of the functioning of the brain. Although the networks of neurons that form the brain's structure had been described, the primitive microscopy and staining procedures available in the 19th century were not sufficient for the synapse to have been identified, or its vital function to be understood. The conventional wisdom of the mid and latter 19th century was that nervous systems were composed of fused neurons, connected via their axons, so that the cytoplasm of the neurons was continuous throughout the nervous system.

□ What are the implications of a model of the brain in which there is continuity of neuronal cytoplasm throughout the entire nervous system?

■ One implication is that transmission of information along axons is bi-directional. Another is that, potentially, activity in one location of the nervous system can have an influence throughout the entire nervous system.

Because of these two features, such a nervous system would be incapable of the strict localization of function as proposed by the phrenologists. This 19th-century model of the nervous system, with its cytoplasmic continuity, then supported the belief that the brain acted in some ill-defined diffuse way. Certainly, with this concept of diffuse processing and with Flourens' experimental evidence, it is easy to see why Gall's claim for localized functions was dismissed by his scientific contemporaries.

6.2.3 Strands in the localizationist debate

Just a year after Flourens sealed the fate of the phrenologists with his conclusion that the brain acted in a non-localized way, another more fruitful strand in the development of the concept of localization of function started. Paul Broca and Carl

Wernicke (Book 2, Section 11.2) received the distinction of having cortical areas implicated in language functions named after them (1861 and 1874 respectively), but credit must also be given to others who preceded them and who also played important roles in the unravelling of the story of the localization of language. In 1825 Jean-Baptiste Bouillard in France, reported a number of clinical cases of people who had suffered from disturbances of speech. He attributed the speech disturbance to damage localized in the left frontal lobe. The next report to suggest localization of function in the human cortex came 11 years later, in 1836, when a French physician, Marc Dax, described clinical cases where language problems were associated with damage to the left hemisphere.

These verbal reports by Bouillard and Dax stimulated little interest in the scientific community and it was only Bouillard's persistence that kept the idea alive for several decades. It was Bouillard's son-in-law, Ernest Aubertin, however, who acted as the catalyst for the events that brought the idea of localization of faculties back into favour. In 1861, Aubertin spoke to the Anthropological Society of Paris about a case in which pressure on the frontal lobes caused the arrest of speech. Fortuitously, Broca attended that meeting. As fate would have it, Broca was presented just five days later with a patient called Leborgne. Leborgne suffered from paralysis of the right side of his body and, apart from the ability to mutter an oath and say the single word 'tan', was mute.

Leborgne died shortly afterwards and, on the day following his death, Broca was able to report to a meeting of the Anthropological Society that his patient's deficits were attributable to damage of the anterior region of the left cortex caused by a stroke. Leborgne's brain was preserved and housed in the museum of the Saltpetrie—a mental hospital located outside Paris. The brain (Figure 6.3) was recently subjected to modern scanning procedures. These revealed that Leborgne's brain also suffered from damage to other regions, but, fortunately, Broca had chosen the correct region to attribute the localization of language. Over the following two years, Broca accumulated seven more examples of people suffering from aphasia resulting from left hemisphere damage. This provided strong evidence for the localization of language functions in the 'third frontal convolution of the left hemisphere'.

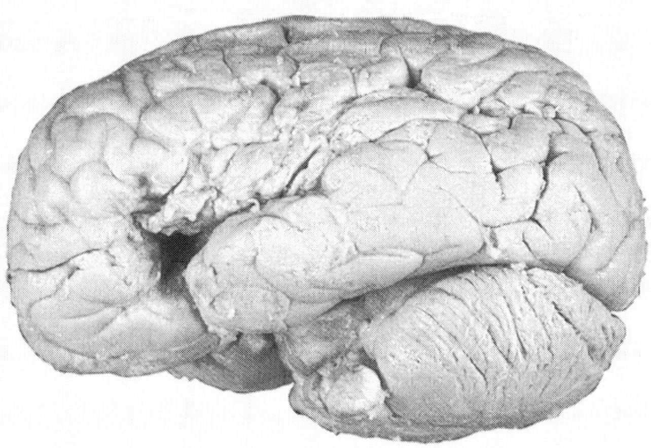

Figure 6.3 Leborgne's brain preserved after post-mortem by Broca and showing the lesion that defined Broca's area.

Broca's views were more than a challenge to the mainstream of French thought about the relationship between matter and mind; they were also a challenge to mainstream French religious opinion. In France at that time, the philosophy of Descartes, elaborated two centuries earlier, remained a dominant influence on such religious and scientific thought. Broca's identification of a region of the brain (*res extensa*) that had properties previously thought to be exclusive to the spiritual world of *res cogitans* provided a source of embarrassment to the clerical establishment. Broca's claim that speech functions were localized predominantly in the left hemisphere also went against other mid-19th-century orthodoxies: animal and human brains were thought to be both anatomically and functionally symmetrical. Another problem for Broca was the popular criticism that he was resurrecting phrenology.

Predictably, Broca came under heavy attack from both the religious and scientific communities. Surprisingly, among Broca's critics was Carl Wernicke whom, as you may recall (Book 2, Section 11.2), had described a different speech-related region in the frontal lobe where localized damage also produced impairments in language functions. But, Wernicke's criticism was not of localization of function itself, since his studies had also implicated the left rather than the right hemisphere as being involved in language. Rather, Wernicke was concerned with Broca's claim for strict localization of language to a single region of the brain.

Certainly, Broca's original identification of a region in the left temporal lobe involved in language was interpreted by some as reviving the phrenologists' concepts of discrete mental 'faculties' located in discrete regions of the cortex. Wernicke's major contribution was the suggestion that mental processes are based upon the collaboration of several localized regions of the brain linked by discrete anatomical pathways. Wernicke's ideas were taken up again in the 1950s by Norman Geschwind in the Wernicke–Geschwind model for language which promoted the analysis of deficits following brain damage in terms of the physical disconnection of brain regions (Book 2, Section 11.2.3).

In the latter part of the 19th century, accumulating evidence for localization of function and asymmetry of function in the human brain led to much more theoretically-based analyses of cognitive functions. However, there was considerable opposition to such theories because of the contemporary understanding of the brain as being composed of a network of neurons, in which each neuron was connected to every other through cytoplasmic continuity (Section 6.2.2).

One event that had a decisive influence on the successful establishment of a localizationist explanation of cerebral organization at the end of the 19th century in the UK was the Medical Reform Act passed by Parliament in 1858. This Act led to reform and re-organization of the UK hospital system. A consequence of the reforms was the establishment of specialized hospitals. One of these was the National Hospital for the Paralysed and Epileptic located in Queen's Square in Bloomsbury, London, which is today called the National Hospital for Nervous Diseases.

With the establishment of the Queen's Square hospital, neurological patients were for the first time located together in one institution. As the reputation of Queen's Square developed, it acted as a collecting point for interesting cases from all over the country. Not only did this development give the patients better treatment, it also exposed neurologists to a wider and more comprehensive range of neurological cases. This in turn encouraged more systematic studies and comparison of the effects of epilepsy, stroke and other varieties of brain damage.

The development of the speciality at Queen's Square had other significant effects on the profession of neurology. The specialized hospital attracted charitable donations which were used to support research by the neurologists. It was able to provide a coherent and directed training programme for the neurologists. Under the tutelage of notable physicians who were convinced localizationists, such as John Hughlings Jackson, David Ferrier and William Gowers, the doctrine of localization of function became firmly established. This gave UK neurologists a firm base from which to challenge the orthodox view of diffuse processing (Section 6.2.2), which permeated academic and medical physiology. It also gave neurologists in the UK a head start over their contemporaries on the Continent of Europe and in the USA in the development of the concept of localization.

The development of the doctrine of localization of human cognitive function was important practically as well as theoretically. It made feasible not only the diagnosis of the location in the brain of the sources of epilepsy, tumour and stroke, on the basis of patient symptoms, but also surgical intervention to arrest or reduce the consequences of brain disorders. The first operation to remove a brain tumour was carried out in 1884 by the surgeon Rickman Godlee. He located the tumour on the basis of the patient's symptoms, using the principles of localization. The actual operation was of limited success since the patient died 28 days later, but it marked the start of an era in brain surgery by forging a relationship between neurology and surgery, based on localizationist theories of brain function.

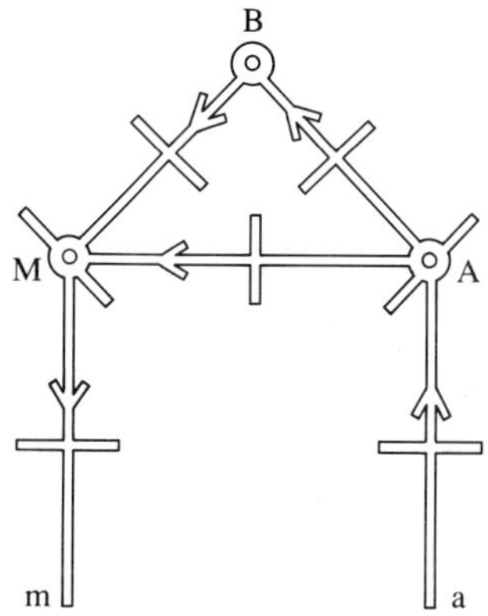

Figure 6.4 A simplified representation of Lichtheim's theory of localization of language functions. (See text for details.)

As well as promoting investigation of the mapping of function onto the brain (such as the sensory and motor homunculi), the establishment of the doctrine of localization of function led also to explanations of cognitive processes in terms of interconnected brain systems. Between 1870 and the end of the century, about 10 different models of cognitive processes were produced by reputable neurologists. The most influential of these was constructed by the German neurologist L. Lichtheim and published in 1885. A simplified version of Lichtheim's model is shown in Figure 6.4 and indicates why the period from 1860 to 1900 has been described as 'the age of the diagram makers'.

Lichtheim's model is based on a centre for auditory word-representations 'A', a centre for motor word-representations 'M' and a centre where concepts are elaborated 'B'. The points 'a' and 'm' represent auditory input and speech motor outputs. Lichtheim's model looks remarkably similar to recently developed models of cognitive structure. However, like other models of that time, Lichtheim's was vulnerable on a number of counts. First, the evidence for the functions it postulated was shaky and subjective. Second, the anatomical bases it claimed for its centres and connections was largely speculative. Third, the psychological concepts used were inadequate. It took only an onslaught on their inadequate neuro-anatomical theorizing for the entire enterprise of the diagram makers to be discredited by the vociferous critics of localization.

6.2.4 Objections to localization

The alternative to localization was called diffusionism, in which brain functions were seen as being diffusely represented throughout the cortex. Diffusionism's most vigorous champion was Charles-Edouard Brown-Séquard, a Mauritian physiologist and physician who had practised in Mauritius, Paris, London and in the USA. His influence was greatest during the 1880s when he was Professor of Medecine at the College de France.

Brown-Séquard had been a student of Hughlings Jackson, a physician at the Queen's Square hospital, and like him, believed that tumours and lesions produced seizures. Whereas Hughlings Jackson saw a cause-and-effect relationship between the site of the lesion or tumour and the symptoms, however, Brown-Séquard believed that a lesion or tumour could influence processes at a distance and could interfere with the balance of excitation or inhibition that he proposed controlled behaviour.

During the 1870s Gustav Fritz and Eduard Hitzig carried out a classic locationist study. They mapped the motor cortex by using electrical stimulation of the exposed cortex in experimental animals. Their reports of movement in different muscle groups, resulting from stimulation of different sites in the strip of cortex now known as the motor cortex, was seen as a great breakthrough for the localizationist view.

The counter arguments advanced by Brown-Séquard and other diffusionists was that, because similar results could be obtained by stimulating selected motor nerves, these results did not necessarily say anything about the organization of the cortex. Rather, they reflected the organization of the motor system based on the outflow from the segments of the spinal cord. The diffusionists thus provided alternative explanations of the phenomena cited in support of the localizationists' position, thereby pushing them into producing more compelling evidence. Sustaining the diffusionists' position were a number of key studies such as the ablation studies of Pierre Flourens (Section 6.2.2), in which he claimed that behaviour remained intact after damage to the cerebellum.

The views on cortical function that prevailed after the demise of the ideas of the diagram makers through the first 30 or 40 years of this century are illustrated in the work of the American neuropsychologist Karl Spencer Lashley, who worked at Johns Hopkins University, Maryland. Although he accepted that sensory and motor functions are localized in the strip of cortex lying around the central gyrus, and visual functions in the occipital lobe of the mammalian cortex, Lashley in 1929, proposed that the large areas surrounding these regions, called association cortex, were organized on different principles. In his experimental work, Lashley made lesions of differing sizes and locations in the association cortex of laboratory animals such as rats. He looked at the ability of laboratory rats with brain lesions to learn to negotiate mazes to receive food. His experiments showed that it was the size of the lesion rather than its location that was the factor determining the rat's ability to learn. This work convinced him that the operation of the cerebral association cortex could be described by two concepts: *equipotentiality* and *mass action*. The concept of equipotentiality proposed that any region of the cortex (outside the sensory and motor regions) had the same capability as any other. The important factor governing performance was mass action—the amount of tissue available to control a behaviour pattern.

The crucial point made by localizationists to dismiss Lashley's findings is about the nature of the information required by the rat to run one of the mazes successfully. It is argued that at each of the choice points a variety of cues will be available, including smell, vision, touch and information from the vibrissae (whiskers). Furthermore, at each point, different sensory modalities may provide the relevant cues. Disruption of capability in any sensory modality is likely, therefore, to disrupt the entire performance, because any inappropriate choice will set the animal on the wrong route. In other words, a lesion in location (a) that disrupts modality A is likely to have the same effect on performance as a lesion at location (b) that disrupts modality B, if the performance depends upon cues from both modalities. Compare this with Lashley's explanation that any region of the cortex had the same capacity as any other.

Lashley never worked with human subjects but his ideas reinforced a neurological orthodoxy that was more impressed with evidence for global deficits following brain damage than with evidence for localization of function. Because it is only in rare instances that brain damage is sharply localized and the resulting symptoms discrete, most victims of brain damage will show impairment across a range of capabilities.

6.2.5 Overview

The discussions in Section 6.2 have traced the development of the concept of localization of cognitive function within the human brain. The localization concept not only ran counter to religious orthodoxy, but also challenged philosophical orthodoxy that consciousness was unitary in nature. The demonstrations that (a) language was localized in the brain and (b) that language itself was not a unitary phenomenon, opened the way to descriptions of other cognitive processes. In this way, through the study of the after-effects of localized damage to the brain, a picture of the modular nature of cognitive processes and their relationship with consciousness has been accumulated. Section 6.3 introduces this view of consciousness and the mind, which has been developed by a branch of the neurosciences called cognitive neuropsychology.

6.3 Brain damage and the mind

The previous section described the development of the concept of localization of function in the nervous system. This debate was largely over whether specific regions of the brain mediated specific functions, for example, the sensory and motor cortices grouped around the central sulcus, and the role of Broca's and Wernicke's areas in language functions.

This section is concerned with the nature of cognitive processes. It focuses on language functions and their relationship with consciousness. The discussion in this section moves away from attempts to describe what bits of the brain do, towards describing how the brain organizes its cognitive functions, that is, a move away from studies of functional neuroanatomy to studies of cognitive neuropsychology.

What cognitive neuropsychologists set out to do is to identify the minimum number of discrete processes that make up a particular cognitive process such as writing. Their approach is based on the premise that consciousness is not a unitary process but rather is dependent upon the functioning of a number of independently functioning modules. This is discussed in Sections 6.3.2 to 6.3.4.

Before going on to consider this approach, however, Section 6.3.1 will turn to a tantalizing line of investigation that has revealed the possibility that there is more than one consciousness within the same individual.

6.3.1 Consciousness and the split-brain

Gerald Edelman's theory of consciousness as a process that depends upon a nervous system having specific structures was described in Book 2, Section 11.5.3. Edelman proposed that some primates, in particular humans, display a higher order consciousness through which they organize their perceptions of the external world and which provides the basis for communication through a symbolic language.

Consciousness is often contrasted with unconsciousness, a term used in several contexts. It is used to signify the absence of consciousness; unconsciousness may be caused by sleep, by drugs or by accident to the central nervous system. The term unconscious has a second connotation, namely forms of mental processing that occur without representation in consciousness.

Over the last decade or so, striking evidence that supports the non-unitary view of consciousness has been obtained from people who have undergone surgery on the corpus callosum for the alleviation of epilepsy. The corpus callosum (Figure 6.5) is a massive fibre tract crossing the midline of the brain and containing several million axons. In the corpus callosum, fibres run in both directions and make point-to-point connections throughout the hemispheres. Thus, any point in the cortex is connected to the mirror-image point in the opposite hemisphere. With the exception of the cortical regions that receive sensory information from or control the distal musculature (that is, the hand and foot regions of the somatosensory and motor cortex), all regions of the cortex are interconnected with their mirror-image regions.

In some people with very severe epilepsy (Section 5.2.3) the corpus callosum is surgically cut through as illustrated schematically in Figure 6.5(b). The reason for this extreme procedure is that an epileptic focus in one hemisphere, usually in a temporal lobe, bombards the mirror-image region with synchronous volleys of action potentials. As well as involving both hemispheres in epileptic seizures, the activity across the corpus callosum can result in a second epileptic focus establishing itself in the second hemisphere, thereby doubling the incidence of seizures. Cutting the corpus callosum confines the epileptic activity to within one side of the brain.

People in whom the corpus callosum has been cut are described as **split-brain** individuals because the two sides of the cortex lack the normal callosal connections. However, the term split-brain has gained popularity for another meaning; under certain circumstances these people behave as if they had two brains with separate consciousnesses.

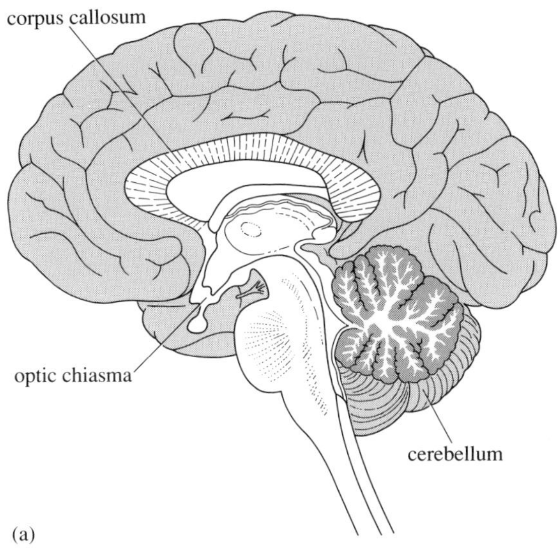

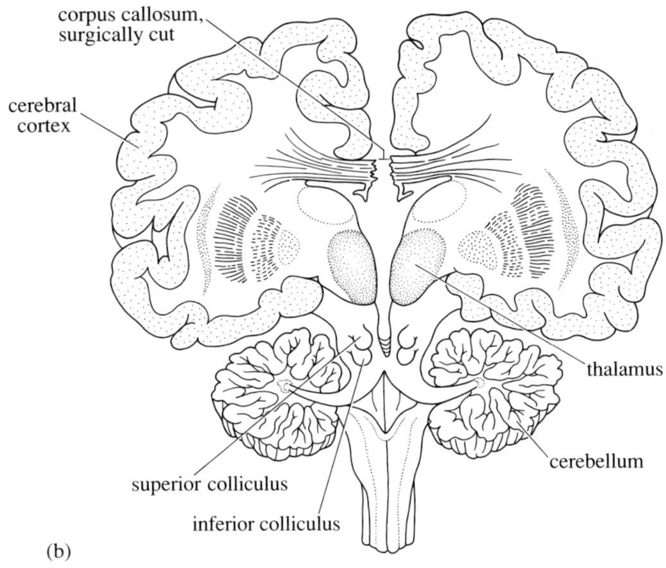

Figure 6.5 (a) Medial section and (b) coronal section of the brain showing the corpus callosum. The position of the surgical cut in split-brain people is shown in (b).

This apparent separation of consciousness, which has obvious implications for its organization, has been investigated by lateralizing incoming information, that is, confining it to one hemisphere. This relies on the fact that a visual stimulus presented in the right visual field is represented only in the left occipital lobe of the brain. Figure 6.6 shows the route that the visual information takes to the brain. Recall that the structure of the optic nerves is such that information from the left visual field reaches the right visual cortex and vice versa (Book 3, Section 4.6).

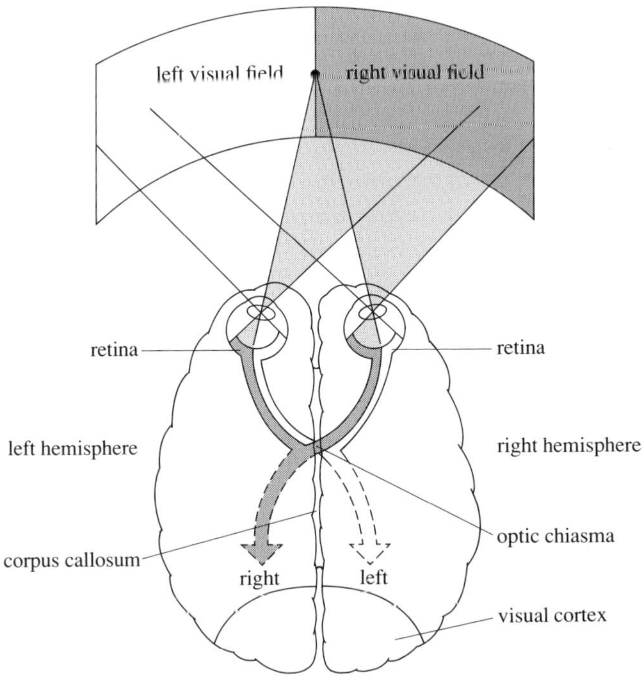

Figure 6.6 The visual projections.

Under normal conditions, eye movements ensure that information about an object that is being viewed reaches both hemispheres directly. However, when a piece of apparatus called a tachistoscope (shown schematically in Figure 6.7) is used, information can be confined to one hemisphere. The important feature of tachistoscopic presentation is that an image projected onto the screen is displayed for a period too brief for the subject to initiate an eye movement; it takes about 100 ms to move the eyes. Consider the situation under these conditions when the subject is focusing on the central spot. A visual stimulus in the left visual field, that is to the left of the subject's fixation point, is represented only in the right occipital cortex and a stimulus in the right visual field only in the left occipital cortex.

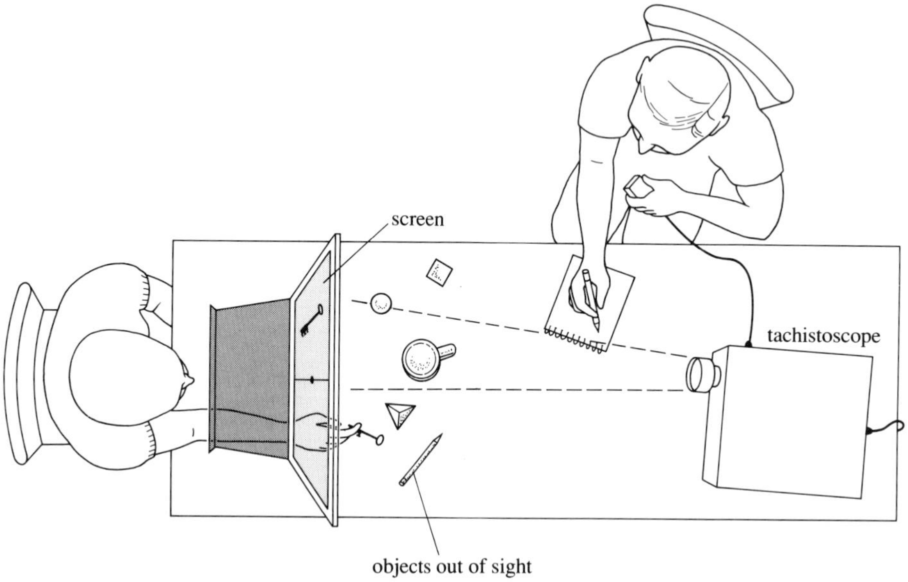

Figure 6.7 Tachistoscopic presentation.

If two projectors are used so that photographs of two different objects can be displayed simultaneously, one on each side of the fixation point, then the left and right hemispheres can be given quite different information. In the intact brain, information crosses the corpus callosum and reaches the other hemisphere. In the split-brain individual, however, the information remains lateralized. In a classic experiment, a number of objects are located out of sight under the apparatus, including objects identical with those presented on the screen. One hand is placed under the table and the subject is instructed to select the object that was shown on the screen (Figure 6.7).

Under these conditions, split-brain subjects show an intriguing phenomenon, an example of which is illustrated in Figure 6.8. When they are asked to use the right hand, they select the object presented in the right of the visual field (left hemisphere), a key for example, and they can name the object and describe why it was selected (Figure 6.8a). When asked to use the left hand, split-brain subjects also select an object shown in the left of the visual field (right hemisphere), but are unable to describe or name it (Figure 6.8b). At first sight, these findings are consistent with preceding discussions of lateralization of language to the left hemisphere.

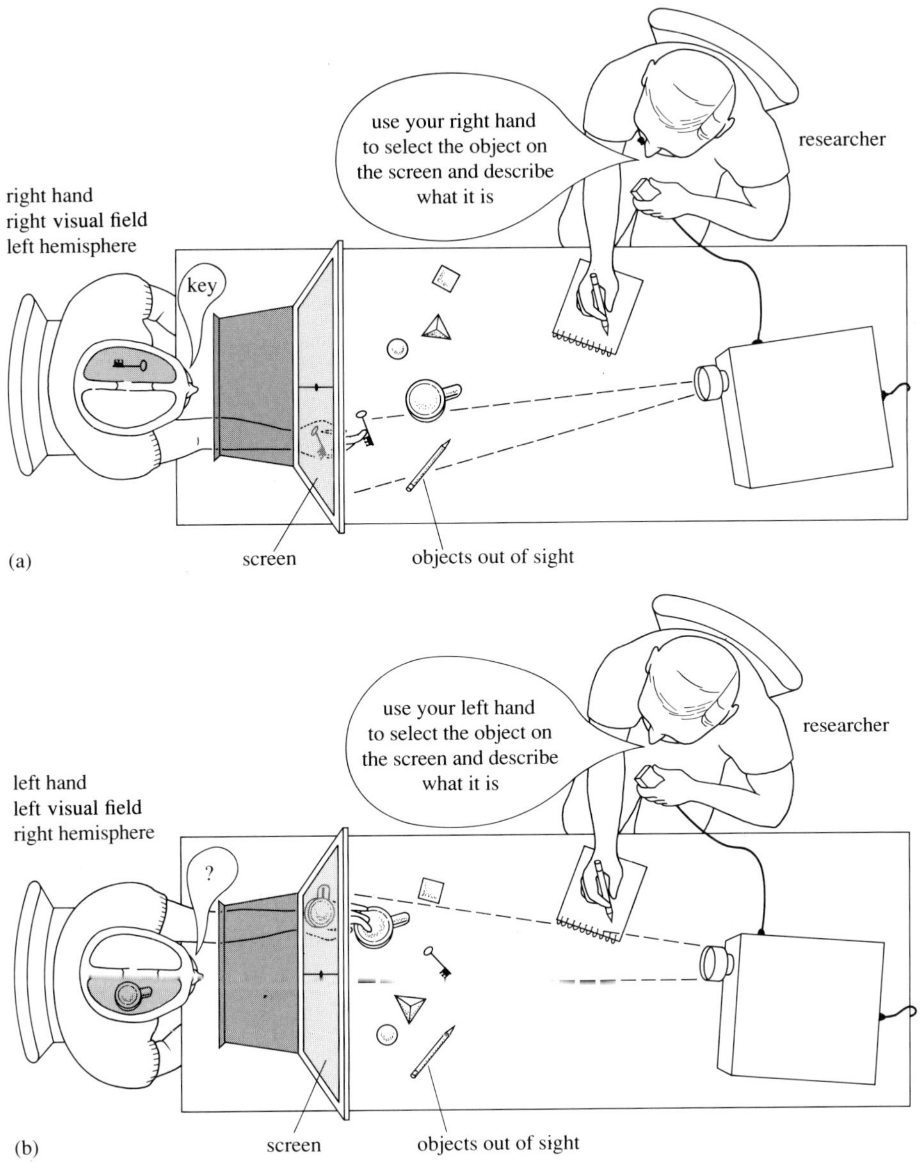

Figure 6.8 The response of split-brain individuals to visual and spoken instruction. (See text for details.)

□ How are the observations on split-brain individuals consistent with the hypothesis that language is a left-hemisphere function?

■ Split-brain subjects can name and select objects presented to the right eye (the left hemisphere). When the left hemisphere has no access to information, that is, when it is presented to the left eye (the right hemisphere), then the subject is unable to name the object.

The results also show that both the left and right hemispheres function identically in response to visual information. However, they raise the tantalizing question as to how the right hemisphere is able to carry out verbal instructions when language is confined to the left hemisphere. This suggests that *both* hemispheres have the capacity to *understand* spoken instruction.

You may be surprised to learn this. But years of split-brain research suggest that the mind is split into two functional conscious units. It is not uncommon to find that both hemispheres have the ability to comprehend language, but in the majority of subjects it is only the left hemisphere that can 'talk' and can make reports on its actions. Another important feature shown by the results is that understanding spoken instruction is a separate process from that of language production. These observations suggest that consciousness is a non-unitary process.

One individual, called P.S., studied by Michael Gazzaniga in California in the 1970s, has made the most enigmatic contribution to the question of whether independent consciousnesses exist in the separated hemispheres. In-depth studies of this patient have revealed that under certain conditions the right hemisphere *can* develop communication processes.

P.S. suffered damage to the left hemisphere during infancy and underwent an operation to cut the corpus callosum during late adolescence as a way of minimizing his epilepsy. Gazzaniga found that P.S. had developed language functions in both hemispheres. Gazzaniga was able to 'converse' with P.S.'s right hemisphere by asking P.S. to use a Scrabble set with his left hand to construct words and phrases in reply to instructions lateralized to the right hemisphere by tachistoscopic presentation. Under these conditions, Gazzaniga was able to get the right hemisphere to perform, apparently independently of the left hemisphere. The right hemisphere could not only understand and control the left hand but could also communicate, even though it had no spoken language output. Thus in this individual both hemispheres could communicate *independently* of one another.

The interpretation of the findings of Gazzaniga and others as evidence for independent consciousness in the same brain remained controversial. One critic was the British neuropsychologist, Donald MacKay, who insisted that, to make the conclusion that the right hemisphere is conscious it would have to meet several criteria. MacKay's criteria were that the right hemisphere must be able to set its own values on events, goals and response priorities. Using the Scrabble set, Gazzaniga took up MacKay's challenge and asked P.S.'s right and left hemispheres to rate on a scale from 1 to 10 each of a series of pictures of objects and scenes presented tachistoscopically. Gazzaniga claims that the differences between the ratings made by the two hemispheres indicate that P.S.'s right hemisphere made independent emotional assessment of stimuli, thus meeting MacKay's criteria for an ability to set independent values on events. This suggests the possible existence of two *independent* consciousnesses.

Gazzaniga argues that the 'I' experienced by humans is the left hemisphere language system, a consciousness that exists independently of other systems that can control behaviour. However, some of these systems meet criteria for consciousness. Here he cites the common experience of motorists who will have driven safely, sometimes for long periods, without any recollection of having done so, with the 'I' being absorbed in a 'stream of consciousness'. Gazzaniga argues further that such safe driving behaviour is not 'automatic' in the sense of being reflexive because these experiences can involve relatively demanding traffic conditions, indicating that verbal and non-verbal consciousnesses are functioning simultaneously and independently.

Gazzaniga's view is that the unitary consciousness usually experienced by humans is the result of a fragile coalition of a number of cognitive systems that are making independent judgements of the individual's interactions with the environment. Under normal conditions, through the anatomical connections of the corpus callosum, the language system produces a consensus, the results of which are observed and reported on by the language system.

6.3.2 Cognitive neuropsychology: contemporary models of mind

The analysis of split-brain subjects has thrown up fascinating insights into the non-unitary nature of consciousness and points to the relationship between the language system of the brain and the consciousness that an individual experiences. However, the implications of Gazzaniga's findings for the functioning of the normal intact brain are not clear.

Gazzaniga's interpretation of the evidence from split-brain experiments is not only challenging but also highly speculative and controversial. His findings have had little impact on the more mainstream endeavours of cognitive psychologists and cognitive neuropsychologists to describe the structures that constitute the human mind.

Systematic studies of the after-effects of brain damage often give dramatic insights into the way that cognitive functions are localized in the brain. This evidence sometimes complements and sometimes challenges the models of cognitive functions developed from studies of normal subjects. Essentially, in order to explore the cognitive processes of normal subjects, cognitive psychologists use standardized tests of cognitive performance under standard laboratory conditions.

Usually, in these studies, the results of individuals tested under an experimental condition are combined and the group scores compared with the group scores of a control population; this is a standard procedure. Although there is a history of collaboration and cross-fertilization of ideas between cognitive psychologists who work with normal subjects and those who work with subjects who have suffered brain damage, the latter are generally described as cognitive neuropsychologists. Traditionally, cognitive neuropsychologists followed similar standard procedures when working with brain damaged individuals. They tried to group individuals suffering from similar brain lesions and to compare their performance with a control population. But, during the 1980s and 1990s, the importance of the symptoms shown by *individuals* in providing critical information about the organization of cognitive processes has been recognized. Evidence from the symptoms of individuals is most important when it goes against a particular theory of the organization of cognition. This does not mean that such subjects are not tested under controlled laboratory conditions using well-defined testing procedures. But, the information is not grouped with that from people suffering from similar brain damage.

There are two major reasons for using single subjects. The first is that brain damage is seldom so localized and regular that two identical individuals can ever be found. The second is that there is also significant variability in brain structure between individuals, exacerbating the problem of what are equivalent lesions to the brain. It is argued that using group studies under such variable conditions would result in the loss of valuable differences. Certainly, studies of individuals have an excellent history in cognitive neuropsychology. You will recall the contributions to an understanding of memory made by Brenda Milner and others based on the subject H.M. (Book 2, Section 11.4.2). With time, other people with similar symptoms and underlying pathology emerged, but it was the incentive provided by Milner's studies of H.M. that stimulated the search for corroboration.

The general approach of the cognitive psychologist is to describe the minimum number of distinct units required to account for some aspect of human behaviour such as reading, writing or reasoning. The contribution of cognitive neuropsychology is to provide evidence for the way that cognitive processes fracture into parts after brain damage. Such information can be crucial in refining and developing models of cognition developed by studying the behaviour of normal subjects. Like the cognitive psychologist, the cognitive neuropsychologist is more concerned with modelling the cognitive processes themselves than with identifying their anatomical localizations and interconnections. Models of cognitive processes are the subject of the rest of this chapter.

One reason for the apparent irrelevance of the split-brain studies to the nature of the models of cognition is that, in their investigations, cognitive psychologists and neuropsychologists rely heavily on the verbal responses of their subjects. In this way, their investigations inevitably describe the characteristics of the language system itself and the resources that the language system draws upon directly. This means that their methodology does not allow the description of cognitive structures independently of the language system. Not surprisingly, it is the area of language itself, the processes that contribute to speech, reading and writing, that have produced the more interesting investigations of the processes that contribute both to human cognition and to the structure of the human mind. The development of such models (Figures 6.9 to 6.11) will be described below. The important point here is to follow the arguments as to how the processes of speech, reading and writing fracture into a number of modules, rather than to remember the details of the structure of the models.

Cognitive psychology and neuropsychology have their roots in the 1950s and 1960s. In the 1960s the impact of electronic computers and advances in information-processing systems provided a rich vein of analogies for attempts to describe the human brain. The psychologist Donald Broadbent, who headed the Applied Psychology Unit in Cambridge, described the human brain as a 'limited-capacity information processor'. In 1990, M. Eysenck and M. Keane described the information-processing approach as having at its centre the following loosely defined framework:

1 People are viewed as autonomous, intentional beings who interact with the external world.

2 The mind through which they interact with the world is a general-purpose, symbol-processing system.

3 Symbols are acted on by various processes which manipulate and transform them into other symbols which ultimately relate to things in the external world.

4 The aim of psychological research is to specify the symbolic processes and representations which underlie performance on all cognitive tasks.

5 Cognitive processes take time, so that predictions about reaction times can be made if one assumes that certain processes occur in sequence and/or have some specifiable complexity.

6 The mind is a limited-capacity processor having both structural and resource limitations.

Within this information-processing framework, the specialization of cognitive psychology and neuropsychology has been particularly successful in studies of language, perception and memory.

Figure 6.9 shows a model of the speech recognition process, proposed by Tim Shallice and Elizabeth Warrington of the National Hospital for Nervous Diseases in 1970, as a result of studies of normal and brain-damaged patients. This is a good example of work carried out within the information-processing framework laid out in the six points above.

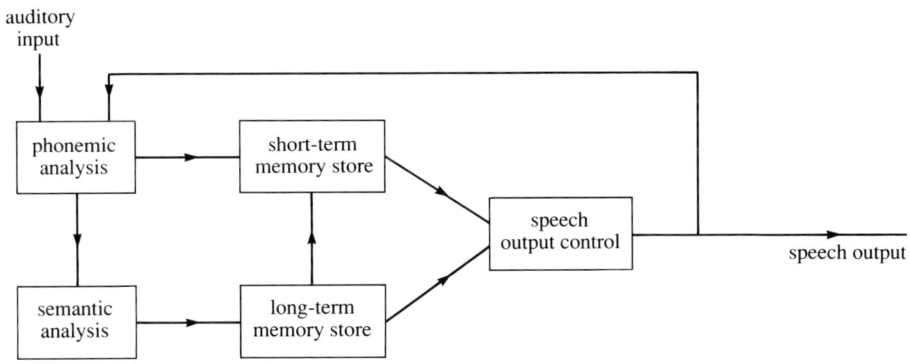

Figure 6.9 Model of the relationship between the auditory and speech recognition systems and memory store proposed by Shallice and Warrington.

Figure 6.9 identifies two stages in the processing of auditory information: the stage where the auditory components of words are recognized as phonemes (Book 3, Section 3.6), called the phonemic analysis stage. This is followed by the semantic analysis stage where the meaning of the communication is constructed. The arrowed pathways in the figure indicate the flow of information between these cognitive processes, memory stores and the production of speech.

From the 1980s, the work of Jerry Fodor of MIT (Book 2, Section 11.5.2), was influential in providing a framework which was better able to explain cognitive processes, such as the phonemic analysis and semantic analysis process shown in Figure 6.9, in terms of independent information-processing modules. (Note that these modules are not identified anatomically.) Fodor proposed that human cognitive processes depend upon the activity of numerous independent cognitive processing modules within the brain. Central to the definition of a module is that its functioning is relatively independent of other modules. A module is also domain-specific, which means that each module can process only one kind of input, such as words or faces. Another important feature of Fodor's definition of a module is that its action is mandatory, which means that its functioning is not under any form of voluntary control.

In this context, studies of normal and brain-damaged people provide evidence for the fragmentation of cognitive processes into independent modules based on the fact that some facets of performance on cognitive tasks remain unchanged, while others are disrupted or modified under specific testing conditions or after brain damage. The next two sections go on to explore examples of models of cognitive processes that have emerged from these studies.

6.3.3 Recognizing words and non-words

Figure 6.10 is an example of a contemporary model of the cognitive structures required to explain the phenomena of word recognition. The details of this model depend heavily on evidence from brain-damaged people. It shows the cognitive modules proposed by Andrew Ellis and Andrew Young in 1988 for the processes required to spell heard words in writing or in speech.

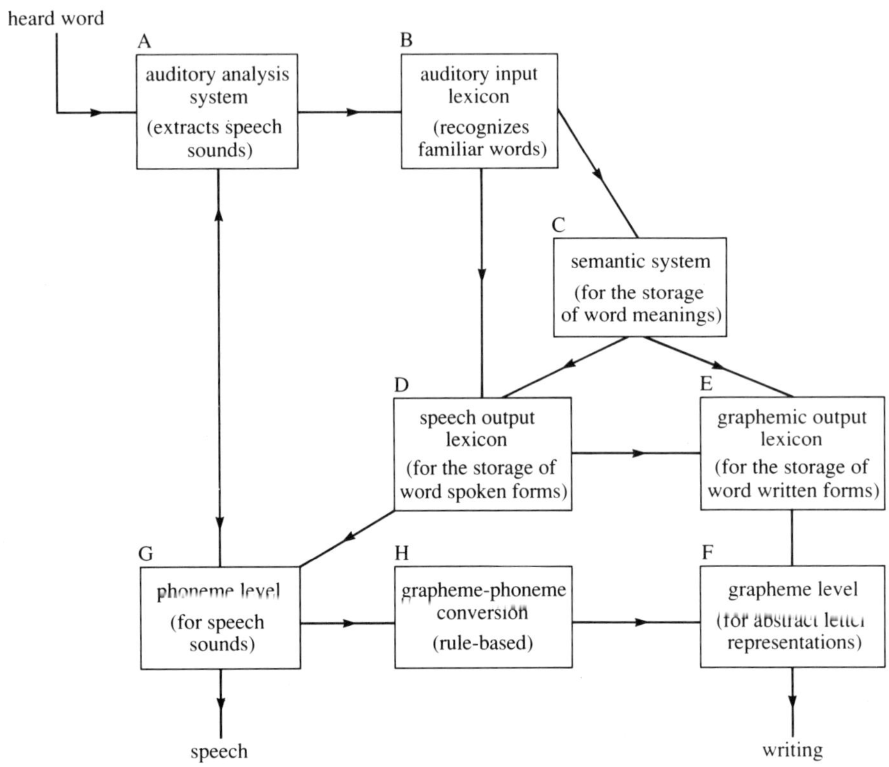

Figure 6.10 Model for the spelling of heard words.

The boxes in Figure 6.10 represent modules that Ellis and Young infer must exist independently of one another, from experimental investigations of subjects instructed to write down familiar words or 'non-words' spoken to them. Understanding of the meaning of a word occurs when the semantic system (box C) is activated. But, because non-words or unknown words can also be processed, they argue that pathways skirt around this module as well as going through it.

A strategic feature of the Ellis and Young model is the specific module for recognizing familiar words, the auditory input lexicon (box B); a lexicon is the store of words known by the subject. The process that Ellis and Young propose is that the auditory analysis system (box A) first identifies phonemes. The auditory input lexicon then compares the sequence of phonemes with the stored patterns it holds and, if a match occurs, an appropriate recognition unit is activated. This activation signifies the recognition of a known word. When a familiar word is recognized in this way it can then be processed along two major routes. Its meaning can be extracted by the semantic system (box C). Understanding a word, in the conventional sense, then depends upon activation of units in box C.

Alternatively, a familiar word can bypass the semantic system and activate the speech output lexicon (box D) directly, from where it can be processed at the phoneme level (box G). The implication of this route is that an individual can recognize a familiar word—or a sentence of familiar words, and be able to speak it internally and then write it (boxes H and F) without having understood the meaning of the word or the sentence. This is somewhat surprising but the route was found following investigations on a particular brain-damaged person. This counter-intuitive aspect of the model will be discussed in the next section.

As Figure 6.10 shows, once familiar words have been processed by either the semantic system or the speech output lexicon, there are multiple routes possible to the grapheme level. This is the stage which translates speech units into graphemes; these are representations that can be translated into the movements for writing. According to Ellis and Young, non-words or unfamiliar words are processed by another route outside the language system which involves the recognition of phonemes (box G) and are then converted into forms that can be translated into writing (boxes H and F).

The important point to understand from this model is that a process such as speech or writing has a large number of independent components or modules, each with its own function.

6.3.4 Understanding and inner speech

Ellis and Young's model can be extended to include the recognition, comprehension and naming of written words in the process of reading. Studies of a particular brain-damaged persons have revealed the fracturing of processes into independent functions.

You are familiar with the phenomenon of *inner speech*, an ability to formulate thoughts and ideas in words and to express them to yourself. A familiar example of characteristics of inner speech comes when one reads words whose pronunciation depends on a context given by the remainder of their surrounding sentence. In most cases one's inner voice generally gets the right pronunciation. For example, your inner voice will have no problem with the pronunciation of the word *tear* in the two sentences: 'Her dress had a tear in it' and ' Her eye had a tear in it'. The fact that the inner voice gets pronunciation right indicates that under normal conditions, familiar words are spoken after they have been understood, that is having been processed by the semantic system.

Ellis and Young's model can also be extended to include the cognitive modules involved in reading as shown in Figure 6.11 (*overleaf*). Note that boxes A, B, C, D and G are shared with the model for spelling of heard words in Figure 6.10. In both models, inner speech is accounted for by the arrowed line leading back from the phoneme level (box G) to the auditory analysis system (box A).

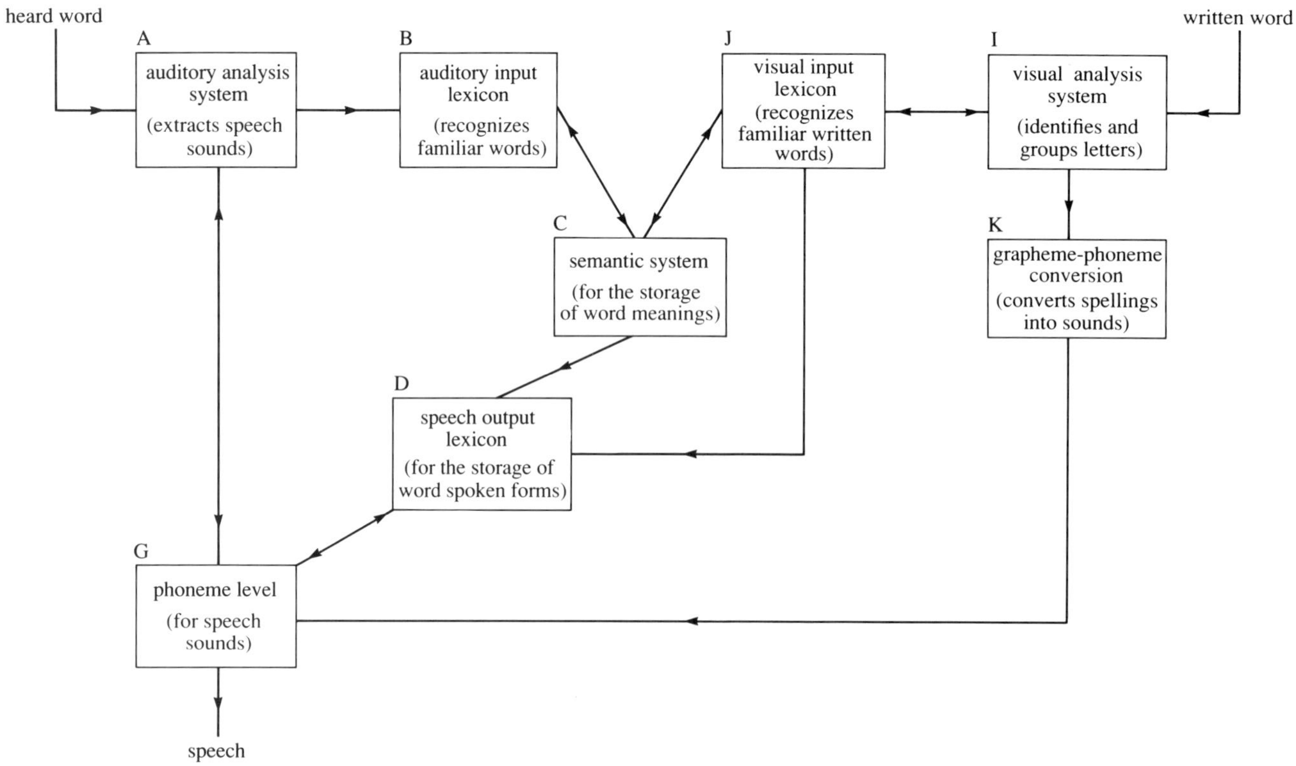

Figure 6.11 An extension of the model given in Figure 6.10 to include the recognition, comprehension and naming of written words in reading.

To name a written word or to speak it to oneself internally, the visual features of the word are processed by the visual analysis system (box I) and then by the visual input lexicon (box J) which recognizes familiar words, which are then passed to the semantic system (box C). To be vocalized internally or as speech, the word is further processed by the speech output lexicon (box D) and then passed to the phoneme level (box G) which generates the template for speech sounds.

At first sight, it is surprising that, in Ellis and Young's models, inner speech is not located more firmly with the processes proposed to underlie comprehension. But the need for the separation of the processes for representing the meaning of a familiar word and for the ability to speak it internally is shown in a dramatic example of a brain-damaged individual reported as early as 1897 by B. Bramwell. He investigated a young woman who was unable to understand speech after suffering a stroke. This woman was able to speak normally and was able to read aloud but she had great difficulty in understanding questions. For example, when asked a question like 'Do you like to come to Edinburgh?' she could repeat the question fluently but could not understand it.

- □ In terms of the model shown in Figures 6.10 and 6.11, what is the likely site of the woman's brain damage?

- ■ Because she could repeat the question fluently, it is unlikely that it was the link between the auditory analysis system (box A) and the auditory input lexicon (box B) that was disrupted. On the evidence so far, the ability to repeat words without understanding can be explained by a disruption of the link between the auditory input lexicon (box B) and the semantic system (box C).

Additionally, the young woman could write down the question. The fact that she could spell 'difficult' words like 'Edinburgh' indicates that she was not relying upon the path between auditory analysis (box A) and the phoneme level (box G) as a start of a route through boxes H and F to produce writing (Figure 6.10). If this were the case, then the writing would be expected to contain many misspellings as the words would be reconstructed from the representations of speech sounds without the help of the auditory input lexicon (box B), which identifies words as entities rather than as sequences of phonemes.

One ability the young woman discovered to help overcome her inability to comprehend speech was, on hearing a question, to write it down. Having written the question down, she found that she was then able to read it and fully understand it.

- □ With reference to Figures 6.10 and 6.11, if the woman's stroke disrupted the link between box B and box C which centres would she use for:

  (a) Repetition without understanding?

  (b) Production of heard words in writing without understanding?

  (c) Comprehension on reading a written question?

- ■ (a) The route from the auditory input lexicon (box B) through the speech output lexicon (box D) and the phoneme level module (box G).

  (b) The route from box B through boxes D, E and F.

  (c) The route through the visual analysis system (box I) and the visual input lexicon (box J) would give the semantic system (box C) access to the written words.

Bramwell's case was the first documented report of a syndrome, word-meaning-deafness, which has been reported many times since. Until fairly recently, the underlying deficit was not understood. It is only recently that models with sufficient complexity have emerged that can account for word-meaning-deafness in a coherent way.

Since then, other examples of single-case studies have had an impact on the structure of cognitive models. For example, evidence from individuals who can write reasonably well but who have no or reduced inner speech provides support for another of Ellis and Young's distinctions that are counter-intuitive and, at first sight, unnecessary. That is the distinction between the graphemic output lexicon (box E) and the speech output lexicon (box D) which, it could be argued, on grounds of parsimony, should be combined as a general lexicon of familiar or known words.

From a detailed analysis of individual brain-damaged patients by cognitive neuropsychologists, the processes of understanding and spelling of heard or written language have been found to be divisible into a number of functional units or modules. The abstract models produced attempt to show the minimum number of individual functions that are necessary to account for the observations on individuals with brain damage.

Overview and summary of Chapter 6

The theme of this chapter has been the contribution to an understanding of the human mind made by studies of the after-effects of brain damage. At the heart of such an understanding is the concept of localization of function. Over the last 250 years, explanations of human cognition based on concepts of localization developed, were rejected and then re-emerged. Starting with the observations of Paul Broca, the goal of disconnection analysis has been the identification of what different parts of the brain do. A fruitful area of work for disconnection analysis was the elaboration of the pathways that contribute to the human capability to communicate by using language. This line of investigation however, produced its most bizarre and unexpected findings in investigations of split-brain individuals. These investigations gave tantalizing glimpses of evidence for the existence of multiple consciousnesses within the same individual.

In a historical context, the consequences of lesions to the frontal and temporal lobes of the human brain, as described by Broca and Wernicke respectively, were crucial in developing the modern understanding of the organization of neural tissue outside the sensory and motor areas. The different disturbances of language functions seen after damage to Broca's and Wernicke's areas, which are not seen following damage to other brain regions, gave powerful evidence for the localization of language functions within discrete regions of the cerebral cortex.

Eventually, such evidence was vital in the rejection of equipotentiality and mass action as general explanations of the functional organization of nervous tissue. However, the acceptance of the concept of localization of function as a general explanation of brain functioning took a long time to be established. One factor that mitigated against its earlier acceptance was the legacy of the school of phrenology. A major contributory factor to the acceptance of the concept of localization was the establishment in the mid-19th century of the specialized Hospital for the Paralysed and Epileptic in London, now called the National Hospital for Nervous Diseases. This hospital brought with it a reorganization of the neurological profession in England and the establishment of a centre of practice and research that demonstrated the validity of the localizationist approach.

The later sections in the chapter described the contribution of cognitive neuropsychology to a different aspect of consciousness. The mental processes necessary for the aspects of human cognition involved in understanding spoken and written language were discussed.

Both approaches, disconnectionist and cognitive neuropsychology, have focused on language functions and their relationship with consciousness, but offer different perspectives on the relationship. The disconnectionist analysis of the split-brain subjects raised the suggestion that, although the capacity to produce language was central to an individual's experience of consciousness, other brain systems were also capable of displaying phenomena that could be described as conscious. But for the evidence for consciousness to emerge, it was necessary for split-brain subjects to be able to understand language, leaving open the possibility that aspects of language are necessary for consciousness. The apparent splitting of the mind into two functional conscious units, where one has the capacity to understand and speak and the other just to understand, is another line of evidence that goes against the individual experience that the ability to use language is part of the unitary nature of consciousness.

Further advances in an understanding of cognitive processes required a theoretical perspective that was not part of the disconnectionists' methodology. Here, the contributions of cognitive neuropsychology, drawing on concepts developed in investigations of normal subjects within the conceptual framework of cognitive psychology, have helped to build up an understanding of the organization of cognitive processes. Evidence from the deficits seen after brain damage is used to infer the nature of the processes that underlie cognition. In these analyses, the emphasis is upon the minimal number of discrete processes needed to explain the working of consciousness and their interconnections, and not in terms of their locations within the cortex or other brain regions or on the location of the pathways that provide the interconnections. On its own, the analytic procedures of cognitive psychology, with its concentration on the processes in the normal brain that underlie cognitive performance, would probably not have been able to tease out the more subtle relationships (shown in Figures 6.10 and 6.11), of the processes needed to explain reading and writing.

The anatomical locations within the brain of these cognitive processing modules are largely unknown and probably can never be known with the resolution of technology that exists today for non-invasive scanning of the human brain. Similarly, the codes that the cognitive processing modules use to communicate the messages that they transmit to each other is also unknown.

From a neurophysiological perspective, consciousness is described as a system property of a nervous system with a particular organization. Similarly, the evidence from cognitive neuropsychology is that consciousness is a system property of an organization of modular processes. The picture that emerges from contemporary studies is of a mind composed of interconnected processing units, or modules.

There is evidence that the language modules can function without there necessarily being any understanding of the information being processed. In this context, it is tempting to identify a module for understanding as the 'seat of consciousness'. But an obvious difficulty with this hypothesis is that such a module does not exist—it is what is left over after the cognitive processing modules have been described. This analysis then raises the question of whether there is a single controlling system in the brain that can be identified with consciousness or whether consciousness is really a system property of a complex organization of cognitive modules. Certainly, theoretical attempts to identify any single module or input to any module as 'the seat of consciousness' have been spectacularly unsuccessful.

In Tim Shallice's influential book on the development of cognitive neuropsychology, *From Neuropsychology to Mental Structure*, published in 1988, he concludes that

> ...it seems likely that there is no single higher level system that directly controls the operation of the lower level modules. There are a variety of higher level systems, and control is too strong a term to describe their relation to lower level systems'...'My position is to reject a common denominator [which is] the attempt to identify any particular subsystem, or better its output or some other aspect of its processing, with consciousness. Instead, I would argue that another aspect of our information-processing system holds the key. The processes of willed action, of reflection, of remembering, and of speaking about something do not involve identical subsystems or even the same control system, but it is normal for these control systems to operate in an integrated way.

Objectives for Chapter 6

When you have completed this chapter you should be able to:

6.1 Define and use, or recognize definitions and applications of each of the terms printed in **bold** in the text. (*Question 6.3*)

6.2 Distinguish between the views of diffusionists, phrenologists and localizationists on the structure and organization of the human brain. (*Questions 6.1 and 6.2*)

6.3 Explain what studies on split-brain subjects have revealed about cognitive processes. (*Questions 6.2 and 6.3*)

6.4 Describe what the models of cognitive processes developed by cognitive neurophysiologists set out to explain. (*Questions 6.2 and 6.4*)

Questions for Chapter 6

Question 6.1 (*Objective 6.2*)

For each of the statements 1–6 decide, for which, if any, of the theoretical position(s) a–c it is representative.

1 Speech is localized in the left side of the brain

2 The performance of any behaviour pattern depends upon the amount of nervous tissue available to it

3 Mental abilities depend upon the integrity of localized regions of the cortex

4 The cortex is divided into discrete and independent areas supporting innate faculties

5 Transmission of information along nerves is bi-directional

6 Local changes in the curvature of the skull correlate with the development of the underlying cortex

a Phrenologists

b Localizationists

c Diffusionists

Question 6.2 (*Objectives 6.2, 6.3 and 6.4*)
Two SD206 students are in conversation. The first says that the evidence about the structure of the mind produced by the disconnectionists and cognitive neuropsychologists essentially confirm the position of the phrenologists developed a couple of centuries ago. The second says that he cannot see any similarity.

You are given the task of arbitrating this debate. Draw up lists of points that support the positions of each student.

Question 6.3 (*Objective 6.1 and 6.3*)
What significant feature have studies of split-brain subjects revealed about cognitive processes?

Question 6.4 (*Objective 6.4*)
What do the models of cognitive neuropsychologists (Figures 6.10 and 6.11) show about the nature of cognitive processes?

Further reading

Eysenck, M. W. and Keane, M. T. (1990) *Cognitive Psychology: a student's handbook*, Lawrence Erlbaum.

Gregory, R. L. (1987) *The Oxford Companion to the Mind*, Oxford University Press.

Harrington, A. (1987) *Medicine, Mind and the Double Brain*, Princeton.

Kolb, B and Wishaw, I. Q. (1990) *Fundamentals of Human Neuropsychology*, 3rd edn, Freeman.

Shallice, T. (1988) *From Neuropsychology to Mental Structure*, Cambridge University Press.

CHAPTER 7
THE VULNERABLE MIND

7.1 The rise of mental distress

The conditions discussed hitherto in this book have been those which have traditionally been assigned to the sphere of brain disease and disorder. It is time now to consider that great range of conditions which, although clearly reflective of *mind* distress, have continued since their first identification to present problems of explanation in terms of brain function, those conditions that we defined in Chapter 1 as being diagnosed on psychiatric rather than neurological criteria.

Everyone experiences mental anguish at some time in their lives. Everyone feels, and describes themselves as feeling, 'depressed', 'anxious', even 'paranoid' or 'schizophrenic'. Sometimes these terms describing common feelings come out of everyday language (e.g. 'depressed' or 'anxious'), but they have also taken on specific clinical meanings as well. Sometimes, the reverse happens: for example, 'paranoid' and 'schizophrenic' are clinical terms that have come to assume a broader meaning. 'Psychiatric illness', 'mental illness' or 'mental disorder' are among those used by doctors, psychiatrists and other professional groups. Lay terms vary from the straightforward 'mad' or 'simple' through circumlocutions such as 'a bit touched' or 'not all there' to the street language of 'loonies' and 'nutters'. But, to return to the question raised in Section 1.2, how and when does mental *distress* become mental *disorder*? What causes the disease or disorder and what are the appropriate treatments for it?

The conditions with which this chapter is concerned have long been recognized. Accounts of people suffering from melancholia, acute anxiety, delusions, the seeing and hearing of visions, can be found in lay and religious writings throughout recorded history. As spelled out in Section 1.2.2, however, the circumstances under which such behaviour and sensations are regarded as part of the normal range of human experience and those in which they are seen as abnormal diseases or disorders, are strongly socially determined. Attempts to classify and medicalize mental distress really began in the 19th century with the development of hospitals for the insane, 'lunatic asylums' to replace and extend the old 'Bedlams'. When such institutions began to be built, or in many cases when the older Poor Law workhouses were converted to a new role, it was assumed that there were only a few hundred 'lunatics' who might require hospitalization in the entire country. But as the numbers of the asylums increased, so the estimates of the numbers of those who should be placed in them steadily rose as well, to thousands and tens of thousands; it seemed almost as if there was no limit to the numbers of the insane. Whole epidemics of hitherto rather rare conditions began to be diagnosed, notably that of hysteria, mainly in women. The fashion for this diagnosis reached its peak in the last years of the 19th century; today it is a rather rare diagnosis. Did the condition disappear, or was it given a new label, or did doctors simply cease to regard the behaviour as abnormal? You will probably have no difficulty in identifying a number of present-day equivalent fashions.

With the rise of the asylums, there also developed a group of medical specialisms concerned with the diagnosis and treatment of the mentally ill, and the beginning of attempts to distinguish and classify different types of disorder. Conditions such as 'fits' (epilepsy) and the aphasias could be separated from the melancholias and 'general paralysis of the insane,' the first becoming the province of neurology (Chapters 5 and 6), the second of psychiatry. Mental handicap, an impaired intellect, or capacity to use that intellect, became distinguished from mental distress, in which there is no problem with a person's intellect, but instead there is an incapacity to relate adequately to the external world. 'General paralysis of the insane' (GPI) was an umbrella term covering a variety of conditions. Many so diagnosed turned out to be in an advanced state of syphilis, and with the development of relatively effective treatments for syphilis in the early years of this century, the diagnosis of GPI began to disappear from the psychiatrists' lexicon. But, in the 1880s, the German psychiatrist Emil Kraepelin, observing patients admitted to asylums, concluded that certain types of insanity with an early onset and initially rather varied features all seemed to deteriorate towards a final common endpoint. He coined the term 'dementia praecox' to emphasize the progressive deterioration of mental abilities, emotional responses and integrity of the personality, which he saw as the characteristic features of the condition, and to distinguish it from idiocy or syphilis. Later, in 1911, Eugen Bleuler replaced Kraepelin's term with a new one, *schizophrenia.*

Today, textbooks offer a broad distinction between two classes of mental illness, the *psychoses* and the *neuroses*, together included within the catch-all term of psychopathology. The psychoses are the disease entities which in their full form involve insanity and include schizophrenia and manic-depressive psychosis. The psychotic patient experiences life in a disturbed way owing to delusions, hallucinations or misinterpretations. Because some clearly neurological conditions (for instance, certain types of stroke) give rise to types of behaviour that could be classified as psychotic, some psychiatrists perpetuate the Cartesian mind/brain dichotomy by making a further distinction between 'organic psychoses' and 'functional psychoses'. Schizophrenia and manic-depressive psychosis fall into the category of functional psychoses, whereas the disturbed emotions and behaviour resulting from stroke or Parkinson's disease, are organic. By contrast, the neuroses are regarded by many psychiatrists (though not by the sufferers from the conditions!) as relatively minor illnesses, in the sense that the patient experiences only such mental symptoms as anxiety or tension that are not in themselves dramatically outside the realm of normal experience. These conditions are unrelated to insanity, but are believed to arise as an exaggerated response to difficulty or stress.

7.2 Psychoses

7.2.1 Schizophrenia

Schizophrenia literally means 'split mind', (not to be confused with split-brain persons described in Section 6.3.1). A major international study of the incidence of the condition, carried out for the World Health Organization in the early 1970s, under the leadership of John Wing at the Institute of Psychiatry in London, found schizophrenia to be a disease with a world-wide distribution. Given standardized criteria, psychiatrists across the world diagnosed the condition with a surprising degree of consistency; the estimated rate at which schizophrenia could be expected to appear in the adult population during the years of highest risk (16–45) was between 0.5% and 1%. Wing and his colleagues claimed that this suggested that schizophrenia was fairly uniformly distributed across the world, which would make it unlike almost any other disease category. However, this conclusion has been disputed by Richard Warner who, using the same data, concluded in 1985 that the condition is more prevalent in advanced industrial societies than in third world rural ones. The rate of first admission of schizophrenic patients to hospital in Britain is about 15 new cases per year per 100 000 of the population. Wing himself found that both the USA and USSR (as it was then) had a rather higher incidence, but believed that this was due to the fact that in both countries rather wider diagnostic criteria were being used.

The classic picture of a schizophrenic is of a person who at the depths of their suffering feels in some fundamental way cut off from the rest of humanity. Often unable to express emotion, or to interact normally with others, or to express themselves verbally in a way that is rational to others, schizophrenics may appear blank, apathetic and dull. They may complain that their thoughts are not their own or that they are being controlled by some outside force. Dramatically ill schizophrenics appear not to be able to, nor to wish to, do anything for themselves and may take no interest in food, sexual activity, or exercise. They experience auditory hallucinations and their speech seems rambling, incoherent and disconnected to the casual listener. Here, for example, is an extract from the account of his sexual and religious beliefs given by a schizophrenic patient to his therapist.

> If you've been wronged by a woman, say, or a woman's being wronged by a man, then a person who's been wronged, if they're thinking about that person and they're hurt by them, well God will bring themselves to life in the other person without them knowing, and have a part of them, an emanation from them. And that's one of the ways that physical illness occurs, because that person will live inside another body...she can change her shape you see, that's another thing that's hard to believe for people who don't know, haven't experienced it. But Kali is the Indian Goddess of mutilation. Men who end up in hospital are very often put there by someone who, it's not a single person I hope, but it's only one woman...she has to ask God if she can hurt this person, she has to ask...the person she is inside, and she tries to make the habits of the man bad, so making him eat too much food maybe, or make him stop brushing his teeth, or making him aggressive to someone...against someone, and then she can sort of legally, legally entitle them to damage some part of his body inside you...that's what I think happens.

Some examples of writing and pictures done by schizophrenic patients are given in Figure 7.1 (*overleaf*). Some of the symptoms identified by Kraepelin as 'typifying' schizophrenia are given in Table 7.1 (*overleaf*). (There is no need to remember this list.)

Periods of acute schizophrenic withdrawal or incoherence such as the one described in the quotation are often followed by long periods of lucidity and remission. Some psychiatrists doubt whether schizophrenia is a single entity at all, or speak of core schizophrenia and a wider range of schizophrenia-like symptoms or schizoaffective disorders. Thus perhaps it is better to refer to 'the schizophrenias' rather than to the disease as a single entity. The idea of a single disease of schizophrenia may be a hangover from Kraepelin's original insistence on the unitary nature of the condition. Although it will be referred to in the singular here, as schizophrenia rather than schizophrenias, you should recognize that the diagnosis of schizophrenia in a person with a given set of symptoms can vary between doctors and cultures.

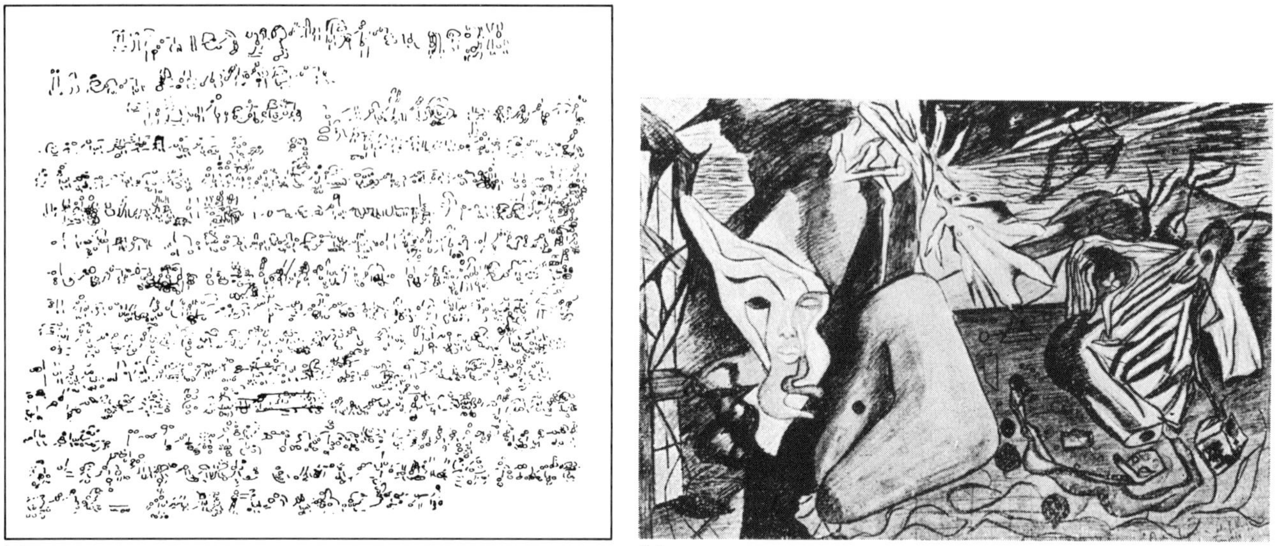

(a) (b)

Figure 7.1 (a) Writing and (b) picture produced by individuals suffering from schizophrenia.

Table 7.1 Some features of schizophrenia identified by Emil Kraepelin.

Feature	Description
Hallucinations	
Auditory	At the beginning these are usually simple noises, rustling, buzzing, ringing in the ears. Then there develops...the hearing of voices. Sometimes it is only whispering. What the voices say is, as a rule, unpleasant and disturbing. Many of the voices make remarks about the thoughts and doings of the patient. It is quite specially peculiar...that the patients' own thoughts appear to them to be spoken aloud.
Visual	Everything looks awry and wrong. People appear who are not there.
Delusions	
Paranoia	The patient notices that he is looked at in a peculiar way, laughed at, scoffed at...People spy on him...persecute him, poison the atmosphere.
Guilt	The patient has, by a sinful life, he believes, destroyed his health of body and mind.

Grandiosity	The patient is 'something better,' born to a higher place,...an inventor, a great singer, can do what he will.
Ideas of influence	Characteristic of the disease...is the feeling of one's thoughts being influenced.
Thought transference	The patient sometimes 'knows the thoughts of other people'.
Thought disorder	
Poverty of thought	There is invariably at first a loss of mental activity and therewith a certain poverty of thought.
Loose associations	The patients lose in a most striking way the faculty of logical ordering of their trains of thought....The most self-evident and familiar associations with the given ideas are absent.
Thought block	There can be a sudden blocking of their thought, producing a painful interruption in a series of ideas.
Affect (emotional expression)	
Blunting	Singular indifference...towards their former emotional relations, the extinction of affection for relatives and friends....'No grief and no joy'.
Inappropriateness	One of the most characteristic features of the disease is a frequent, causeless, sudden outburst of laughter.
Lability	Sudden oscillations of emotional equilibrium. Extraordinary violence may be developed.
Speech	
Abnormal flow	The patients become monosyllabic, sparing their words, speak hesitatingly, suddenly become mute...let all answers be laboriously pressed out of them. In states of excitement...a prodigious flow of talk may appear.
Neologisms	There may be produced...quite senseless collections of syllables, here and there some having a sound reminiscent of real words.
Autism:	Patients are more or less inaccessible,...they shut themselves away from the outside world.
Lack of drive:	The patients have lost every independent inclination for work or action.
Automatic obedience	
Waxy flexibility	The preservation of whatever positions the patient may be put in, even although they may be very uncomfortable.
Echolalia	The involuntary repetition of words said to them.
Echopraxia	The imitation of movements made in front of them.
Intellectual deterioration	The patients are distracted, inattentive, tired, dull,...their minds wander, they have no perseverance.
Deterioration of judgement	The faculty of judgement in the patient suffers.
Personality deterioration	Their thinking, feeling, and acting have lost the unity...of the psychic personality, which provides the healthy human being with the feeling of inner freedom.

7.2.2 Manic–depressive psychosis

People suffering from manic-depressive psychosis show regular oscillations in mood. In the manic, or excited, phase they are often full of grandiose (but false) ideas, such as about being possessed of great wealth or amazing secret powers, or of being a historically important person, such as Napoleon. By contrast, in the depressed phases there may be belief of impending ruin, death and destruction. Mood oscillations of this type are called *bipolar* manic-depression. In some cases however, it is only the manic or the depressed mood which occurs, and this is known as a *unipolar* condition. Just as the 'word salads' of schizophrenia can also be induced by strokes, so manic-depressive behaviour can be indicative of conditions such as brain tumours as well as of less clear-cut organic damage. The next quotation contains extracts from an interview between a patient in his 60s, suffering from manic depression, and his psychotherapist.

> I had the most acute depressions...I never had the death wish or anything like that, but I did feel very, very depressed and then I thought to myself, with hindsight again, well, with these depressions, they go—they don't stay with me—it's a bit like a ship being in the—er, what's that part of a wave—the trough of a wave is it—or which is on top? The depression, anyway, is at the bottom and sooner or later you come out of it and get up on the top of the wave again—and that's what I do. [In the depression I feel] pessimism, gloominess, extreme gloominess and inability to countenance do anything at all...lackadaisical, almost like mentally constipated...there's no hope....[At other times I feel hypermanic] a bit mad. Quite a bit mad...all the time my mind was active with other people's problems...active in a useless sort of way—purposeless active, but it made sense to me apparently...I used to imagine all sorts of things—that I was an agent for...the cluster of stars which is known as the Indian eyesight test, the Pleiades....Well I counted up to thirty of them with my naked eye, and I imagined I was an agent for them. Silly things like that. I mean its quite, unworldly...I imagined I had superhuman powers....That's about as near as I can get to it. Please don't delve too deeply into me, doctor, because it's making me a bit disturbed at the moment.

Summary of Section 7.2

The psychoses are serious illnesses which serve to separate the sufferer from the external world. They are sometimes divided into organic psychoses, resulting from forms of brain damage, and functional psychoses where it has been difficult for researchers or clinicians to detect any obvious associated brain damage. The functional psychoses include schizophrenia, the most serious, which affects some 0.5–1.0% of the adult population, and the rather rarer manic-depressive psychoses. Examples of the classification of schizophrenia, and of the speech, painting and writing of sufferers from both schizophrenia and manic-depression, were given.

7.3 Neuroses

There are many ways of subdividing the neuroses, but perhaps the simplest grouping is to consider them in two broad categories, the affective and the personality disorders. The affective disorders, which include depression and anxiety, are the most frequently diagnosed forms of mind distress. They can be very disabling, but are generally classified as 'benign illnesses' in the sense that they are not usually associated with such grave prognoses as the psychoses. Nor do the types of confusion, apparent in the examples of schizophrenic and manic-depressive thought and speech given in the quotations in the previous section, occur.

7.3.1 Affective disorders

Affect is the term used in psychology and psychiatry as a way of describing a person's responses—of emotion, attention and so forth—to events in the world around them (see Chapter 1). The affective disorders include depression, which is characterized by a variably depressive mood, but without the manic phase of a psychosis, and anxiety. The 19th-century classification of hysteria would today be grouped among the affective disorders. It is in the context of the neuroses that the issues of normality and abnormality discussed in Chapter 1 are posed most sharply, as clearly the definition of 'abnormal' depression or 'easily provoked or long lasting' anxiety that characterize such diagnoses depends on making judgments about what is normal behaviour or a normal response to a stressful situation. This involves comparing an individual's behaviour with that of people in similar situations, or a person's behaviour today with that on some previous occasion. But what are appropriate 'adaptive responses' to the world? If, for example, a man is cast into an apathetic despair about the likelihood of the world surviving a nuclear holocaust through the 1990s, or if a woman is afraid to go out of her house at night for fear of being raped or murdered, how is one to judge that these are inappropriate responses compared with those of the less sensitive majority? What the clinician or therapist attempts to do is to ask if the response is part of an overall pattern, or syndrome, that is preventing the sufferer from engaging effectively with the world within which he or she is embedded. The task then becomes one of trying to help sufferers to regain some control of their own destinies.

Consider the case of depression. When a clinician diagnoses depression, it is common to go on to attempt to distinguish between what are called *exogenous* (or reactive) and *endogenous* depressions. Exogenous depressions are supposed to be precipitated by the events in the world outside the individual, maybe a bereavement or loss of job. Conversely, endogenous depressions are said to be without obvious external precipitants and may occur cyclically at regular intervals. For example, it has long been known that suicide rates vary markedly with the seasons. It has recently become clear that, especially in northern latitudes, there is a seasonal cycle of depression in some people, who show an onset of the condition in November and offset in the spring. A subcategory of affective disorder, Seasonal Affective Disorder (SAD), has thus been recognized (and treatments such as bright light therapy offered for it).

As the great majority of the population show some seasonal mood changes, there is a real problem of deciding where the borderline is to be drawn between normal life experience and clinical condition. Perhaps everyone living in Northern Europe would benefit from winter light therapy (or skiing holidays?). But depression is also often associated with important transitions in a person's life, such as the birth of a baby (post-natal depression) or the menopause.

It is, of course, not the event in the abstract, but the context in which it occurs which seems to be important in the onset of depression—the meaning the event has for a person and the way it affects their life. The birth of a baby, with all its physiological and hormonal consequences for the mother, also affects her differently if she is living in poor housing conditions and economic insecurity than if she is living in good conditions with a secure income and support in bringing up the child. Even depressions apparently triggered by seemingly specific events do not, however, necessarily simply go away if the crisis which generated them is resolved.

It is the clinical method, described in Section 1.4.1, which is responsible for attempts to derive such classifications, and on which biological explanations of the conditions then try to build. It is important to emphasize, however, that most real-life situations are not easy to collapse into textbook categories. The likelihood is, for most sufferers, that the distinctions between different types of depression, and even between depression and anxiety, are rather unclear. The distinctions may also be unclear to the people directly concerned with treating them, at least in the first instance. In the clinical practice of most family doctors, who form the front line of prescribers of drugs to alleviate suffering, rough-and-ready diagnostic criteria, in the six minutes worth of consultation time that is, on average, likely to be available, do not allow for much of the subtlety and neatness of distinction that researchers on mental distress and disorder and its brain correlates like to make. You can feel this sense of hopelessness, and something of the complexity of the conditions which are being diagnosed as indicating disease in the words of some depressed women, as recorded by Nairne and Smith in 1984:

> In spite of having my children, the feeling that I couldn't go on battling with life any longer remained. Each time I attempted suicide I admitted defeat. All I wanted was peace of mind and death was the only way I could think of in which I might find it.
>
> Depression...well, you just feel you can't do anything, you can't do anything right and it doesn't seem to matter how hard you try you just can't seem to get to where you want to get...you try and you try and you try...you have fits of tears and just feel utter despair.
>
> I remember the first time I felt depressed. It was when I was hanging out the washing one day, just before my eldest daughter went away to college. I suddenly thought 'I'm not going to have anyone to talk to when she goes'. I realised that I hadn't been able to talk to my husband for years and that I did not have any friends I could talk to either.
>
> I suppose everybody's got to find peace of mind in their own way. Someone like me, where do I find it? I really do believe that some of us are born to go through life in the same way.

> Depression is a funny thing. One minute you will be wallowing in its depths and the next you are miles away from it.

7.3.2 Personality disorders

The other broad category included by most psychiatrists among the neuroses are grouped under the general heading of 'personality disorders', although they include a number of rather diverse conditions, from alcoholism to phobias—'irrational' fears such as agoraphobia (fear of being in busy, crowded places), claustrophobia (fear of being in enclosed spaces) and many others. Obviously such conditions vary greatly both among themselves and from the affective disorders. One example is obsessional–compulsive disorder, in which a person's thoughts are constantly interrupted by involuntary thoughts or impulses which cannot be willed away. Below is an autobiographical account of the experience of obsession, from a book by the psychologist Frederick Toates, published in 1990:

> Getting back to Odense at about 5 p.m., I went to the university to make sure that everything was in order. On leaving, I carefully checked that the laboratory equipment had all been switched off, but then went back for another confirmatory inspection. Indeed, all was well. I caught the bus to Odense town hall, where I would change to a second bus for Fruens Bøge. On arriving at the town hall, I was overcome with the feeling that something had been left switched on. There was nothing for it, but to wait for a bus to take me the 3 km or so back to university. Sure enough, all was well in the laboratory, just as a more rational bit of my brain seemed to be saying all along. Arriving by bus a second time at the town hall, the same insecurity prevailed. It was now getting late and I was torn as to what to do. A taxi was the only way to have one last look at the laboratory. The driver waited for me while I checked and then drove me to Fruens Bøge. That did finally settle the issue for that day.
>
> What was I afraid of during this checking? What pulled me to do it? I can't say, but using my conscious thoughts as evidence would suggest that it was the need to avoid a catastrophe, such as the university going up in flames. On the one level I knew all was well; on another level, I just doubted it.
>
> Did I remember switching the apparatus off? Maybe. It seemed to me that the essence was in terms of the utilization of the memory. On the one hand, intellectually, I knew that the apparatus was switched off. Had I been asked by a bookie to estimate the chances, I would have given 1 000 to 1 in favour. But, on the other hand, I couldn't use that same memory to move on to another activity. It was as if the decision mechanism would scan for problems before allowing me to start a new activity. The memory got corrupted in this decision process.
>
> The fear was that I would be responsible if a disaster were to happen, because I had left something switched on. I thought—maybe the switch I thought was down was really up. Could it be that my attention had wandered while at the crucial stage in the check? I might remember well enough touching the switch, but maybe I actually pushed the switch up

while pressing it to make 100% sure it was down. Perhaps I was deceiving myself and the memory I was then reviewing was that from yesterday's check rather than today's. One last check will do it, and so on...

Summary of Section 7.3

Neuroses are the most commonly diagnosed form of psychiatric illness, and ones where the borderlines between 'normal' and 'inappropriate' responses to the world are the least clear-cut. They include affective disorders like anxiety and depression. Also included in this category are a varying group of personality disorders. Examples of people describing their experiences of such conditions are given in the text. Attempts are often made to distinguish between 'exogenous' and 'endogenous' causes of conditions like depression, but frequently the distinction is blurred.

7.4 The epidemiology of psychopathology

The psychoses and neuroses have been described from the point of view of both clinical diagnosis and lived experience. In order to try to explain them within the framework of the brain and behavioural sciences, though, it is necessary to know something else: their incidence and epidemiology. Global figures about the numbers of people diagnosed as suffering from particular types of disorder or disease must be broken down by geographical distribution, by age, class, gender and ethnicity before their implications can be fully understood. As mentioned in Section 1.2.4, an attempt to provide an estimate of the scale of psychopathology in North America and Europe was made by Bruce Dohrenwend and his collaborators in the USA. They put together the results from, in all, 27 different studies carried out in Europe and the USA between 1950 and the late 1970s (Dohrenwend *et al.*, 1980). It was from this 'meta-analysis', as it is known, that the estimate of 21% for the total psychopathology in the adult population given in Section 1.2.4 was drawn. Table 7.2 shows the distribution of this incidence (as a percentage of the total population) according to type of diagnosis.

Table 7.2 Median and ranges of rates of psychopathology in Europe and the USA.

Type of diagnosis	Median/%	Range/%
Psychoses	1.6	0–8.3
Schizophrenia	0.6	0–2.7
Manic depressive psychoses	0.3	0–1.9
Affective disorders	9.4	0–53.5
Personality disorders	4.8	0–36.0

Dohrenwend and his colleagues went on to look at the social class and gender distribution of psychopathological diagnoses. On the basis of their meta-analysis, a person was 2.5 times more likely to be diagnosed as suffering from some type of psychopathology if they came from a lower class than if they came from a higher class social group, and, although definitions of social class vary from country to country, the ratios were remarkably similar in all countries. Note that this may mean either that the incidence of psychopathology was in reality 2.5 times higher in lower than upper classes, or that the diagnosis given by a doctor to a patient was influenced by the class position of the patient irrespective of symptoms.

Those ratios applied to all types of psychopathology. Table 7.3 shows what happens if they are broken down into different diagnoses.

Table 7.3 Prevalence of psychopathology by class.

Type of diagnosis	Average ratio low class : high class
All psychoses	2.06 : 1
Affective disorders	1 : 1
Personality disorders	1.77 : 1
All psychopathology	2.59 : 1

□ What is the most striking feature of this table?

■ Psychoses are more than twice as prevalent in lower-class people than in upper-class people, personality disorders are also higher, but there is no apparent difference between social groups in the prevalence of the affective disorders.

These studies also found that the severest disorders are, on average, twice as common in the lowest class as in the highest class. Manic–depressive psychoses are not excessively diagnosed in 'lower-class' groups and most of the difference between classes is due to the greater prevalence of schizophrenia in the lowest classes. Other studies show that schizophrenia is more commonly diagnosed in urban than rural areas. Although all such figures are open to the criticism that they are averaged from a variety of studies, they begin to indicate the questions that must be asked if we are seeking for a causal explanation of the incidence of mental distress or disorder.

Table 7.4 analyses the diagnoses according to gender. Although there are no gender differences, on average, in the diagnosis of schizophrenia, there are striking differences in manic-depressive psychoses and affective disorders, where women tend to be diagnosed nearly three times as frequently as men. By contrast, the personality disorder diagnosis is only two-thirds as common in women as men.

Table 7.4 Prevalence of psychopathology by gender.

Type of diagnosis	Average ratio women : men
Schizophrenia	1 : 1
Manic–depressive psychoses	2.96 : 1
Affective disorders	2.86 : 1
Personality disorders	0.66 : 1

Nonetheless, on the basis of the data in Tables 7.3 and 7.4, it appears that, although diagnoses such as schizophrenia show class differences but do not show gender differences, the affective disorders show gender differences but do not show class differences.

Another important type of classification that can be attempted is that by ethnic grouping. An early attempt to study this was made by a social psychologist, R. Cochrane (1971), working in Birmingham. He analysed hospital admissions in England and Wales for men and women aged over 15, broken down by country of birth. Look at Table 7.5. In doing so, bear in mind that, in 1971, none of the immigrant groups constituted more than 1% of the population and therefore, of all admissions in 1971, 91% were for people born in England and Wales. Also remember that this study was made quite a long time ago, so if a similar analysis were made today one might not get the same results.

One of the problems in assessing data is that, within some of the immigrant groups, there were big disparities in the numbers of men and women living in England and Wales—for instance, 72% of all Pakistanis living in the UK at the time were men. This reflected the pattern of immigration and immigration legislation of the period. Finally, you should note that the number of hospital admissions is, of course, only one measure of the scale of mental distress and disorder. There could be different consultation rates among the groups and many other features could account for such differences.

□ Comparing schizophrenia and affective disorders for men and women born in England and Wales, what is the most striking feature?

■ There was no sex difference in the schizophrenia admission rates, but women were nearly twice as likely as men to be admitted to hospital suffering from affective disorder.

□ Now look at the figures for schizophrenia and affective disorder for men and for women born in the Irish Republic and compare them with those for people born in England and Wales.

■ There is nearly a two-fold excess in the affective disorder rate among Irish men and a similar excess among Irish women. Women of Irish origin were three times as likely to be admitted with a schizophrenia diagnosis as were women of English or Welsh origin.

□ Now compare the figures for people born in the West Indies with those for people born in England and Wales.

■ The figures for admissions of men of West Indian birth for affective disorder were very low, lower than those for men born in England and Wales. The same is true for West-Indian-born women. The schizophrenia figures for people of West Indian origin, however, were very high, without much difference between the sexes. If you were born in the West Indies and were living in England and Wales, you were, during the 1970s, 3.5 times as likely to be admitted to a mental hospital suffering from schizophrenia as was a member of the indigenous white population.

Table 7.5 Admissions to mental hospitals in England and Wales in 1971 per 100 000 of the population over 15 (standardized for age).

Country of Birth	Schizophrenia	Manic–depressive psychoses	Affective disorders	Personality and behaviour disorders
1 *Males*				
England and Wales	87	45	48	43
Scotland	90	42	56	100
N. Ireland	96	78	121	201
Irish Republic	83	69	88	139
West Indies	290	30	19	27
India	141	31	33	36
Pakistan	158	22	36	18
Germany	99	35	41	58
Italy	71	35	22	30
Poland	189	63	42	27
USA	76	32	30	54
2 *Females*				
England and Wales	87	92	88	41
Scotland	97	99	111	67
N. Ireland	160	147	172	111
Irish Republic	254	174	165	114
West Indies	323	91	67	46
India	140	57	64	29
Pakistan	103	38	103	55
Germany	130	59	98	48
Italy	127	61	67	33
Poland	301	119	139	40
USA	133	122	78	98

No more recent comprehensive analysis of such admission figures than Cochrane's seems to exist. However, a much smaller-scale study in Nottingham by Glynn Harrison and his colleagues in 1988 also pointed to a disproportionate rate of diagnosis of schizophrenia among black people. Even more dramatically, the Nottingham study also looked at the rates of diagnosis of schizophrenia among second generation migrants from the Caribbean (that is, people either born in the UK to first-generation migrants or brought to the country before the age of 18), and found an incidence at least six times higher than that in the indigenous (white) population.

You can make similar comparisons within the other columns in Table 7.5 and relate admission rates to country of origin.

Summary of Section 7.4

Any explanation of the causes of mental distress must be able to account for the distribution of the differently diagnosed categories of psychoses and neuroses in the population, that is, their epidemiology. The global incidence of psychopathology is high, with median estimates of some one-fifth of the entire adult population in Europe or the USA. Within this global figure there are striking disparities. Thus in the UK schizophrenia is much more likely to be diagnosed in working-class than middle-class people and particularly in ethnic minority groups whose families come from the Caribbean. Neuroses such as anxiety and depression show no class differences in diagnosis but are much more commonly diagnosed in women than in men.

7.5 Explaining depression and schizophrenia

So far this chapter has given some examples of people's real life experiences and feelings of mental distress ((Section 7.1), the clinical classifications used to define and treat them (Sections 7.2 and 7.3) and the epidemiological evidence of the distribution of psychopathology (Section 7.4). The next sections can turn to the task of trying to explain the causes of such mental disorder. The remainder of this chapter concentrates on two major diagnoses, those of schizophrenia and of depression, and considers the types of explanation that have been offered. Chapter 1 grouped these explanations under three broad heads, of social, psychological (personal life history) and biological. Each will be looked at in turn.

7.5.1 Social explanations—labelling

According to the social view that was characterized in Section 1.3.1 as 'Foucaultian', schizophrenia and depression have no independent existence as attributes of a person; rather they are the ways society chooses to describe certain types of people or patterns of behaviour which are deemed unacceptable. On this basis, it could be argued that the diagnosis of schizophrenia which is given so frequently to black people, for example, is an expression of the racism of the white English host society. It would be hard to deny that elements of this type of explanation must exist as is evident from the following account of the problems faced by a Londoner, whom the authors call Calvin Johnson, who returns from a holiday in Jamaica during which:

> ...he had a strange experience: he heard a voice from God which told him to read certain verses in the Bible. With divine inspiration he...became a Rastafarian. On returning to London ...he went to his local post office with his daughter Victoria to cash a postal order—advance payment for a job he was to start later that week,.... He signed the form, but after examining it the cashier told him to wait and went behind a partition. A

quarter of an hour later he re-appeared with three policemen who rather abruptly asked him how he got the postal order. Humiliated and angry, Calvin tried to walk out of the post office with his daughter, but was pulled back. Completely losing his temper, he lashed out, hitting a policeman more by chance than intention, and was promptly arrested...While struggling, Calvin took out of his pocket and waved a 1966 shilling which he always carried with him: 'The sun was shining on it as if the lion would step out. I began to sing The Lord is My Shepherd. The police said You black bastard—you believe in God? They took me to a mad-house—Oh God, since when are we religious mad?' After some weeks in a psychiatric hospital the charge was withdrawn and Calvin went home...While remanded on the assault charge he was seen by the prison psychiatrist: 'This man belongs to Rastafarian (sic)—a mystical Jamaican cult, the members of which think they are God-like. This man has ringlet hair, a goatee straggly beard and a type of turban. He appears eccentric in his appearance and very vague in answering questions. He is an irritable character and he has got arrogant behaviour. His religious ideas are cultural. He denies any hallucinations. He is therefore not schizophrenic at the moment....' Not long after, Calvin was admitted to hospital under Section 136 of the 1959 Mental Health Act. Using this provision, the police can take anyone they feel to be in need of psychiatric attention from a public area 'to a place of safety', which usually means a psychiatric hospital....This time the trouble had started after Calvin had been smoking cannabis....Seven policemen eventually got him to the local hospital, where the doctor on duty noted: 'He was lying in the lift with two policemen on top of him. The patient was unkempt with long matted hair, talking in broken English and was difficult to follow. He frequently mentioned Christ and lions. There were ragged lacerations on his hands where he had been handcuffed. Probably a relapsed schizophrenic. Observe carefully.' (Littlewood and Lipsedge, 1982.)

□ How would the Foucaultian model explain this episode?

■ It would argue that Calvin Johnson's behaviour is not abnormal in the sense that he is ill. He is an adherent to a minority religion in an alien culture. He is also black in a white and frequently racist society. The dominant white society finds his behaviour, perhaps even his very existence, unacceptable and so labels him either as mad, to be incarcerated in a psychiatric institution, or bad, to be imprisoned.

7.5.2 'Grounded' social explanations

For the Foucaultian model, such a labelling is sufficient in itself to account for the diagnosis; the dominant social order expects black people of Caribbean origin to behave like Calvin Johnson and describes this behaviour as a disorder. Similarly, a view espoused strongly by the antipsychiatry movement of the late 1960s and early 1970s, the disproportionate labelling of working-class and poor people as schizophrenic would be a form of disempowering them, of defining their protests about the injustice of society as invalid. These arguments were given some force at the time by the evidence beginning to emerge from the Soviet Union that the regime was indeed using psychiatric diagnoses, and specifically schizophrenia, as a way of silencing political dissidents, with psychiatric hospitals serving as alternatives to prison. In the context of such arguments there are a significant number of practising psychiatrists today who would argue strongly that the time has come to abandon entirely the concept of 'schizophrenia'; and many more who would, although not wishing to abandon it entirely, worry about the implication that it is a unitary disease rather than a cluster of symptoms somewhat arbitrarily lumped together.

Despite the plausibility of such arguments, to deny the painful reality of the life experiences of schizophrenic people—and of those, family and friends, whose lives are intertwined with theirs—seems positively perverse. A more grounded (that is, one based on material factors) social explanation of Calvin Johnson's situation would accept the reality of his disturbance, albeit accepting that it was precipitated by the experience of living in a racist society. The excess diagnoses of schizophrenia among second-generation Caribbean migrants in Nottingham would be explained on the grounds that life is even more disturbing for them, as they find problems in discovering their own identity in the context of racism at school and in the streets.

How about the excess diagnoses of schizophrenia among poorer and working-class people indicated in Table 7.3? Recall Faris and Dunham's Chicago data, discussed in Section 1.4.4. Although other explanations are possible, the simplest assumption remains that there is indeed something about poverty that precipitates or exacerbates schizophrenia. A study in 1985 by the British social psychiatrist Richard Warner, based in Boulder, Colorado, looking at the fluctuations in the outcome of schizophrenia in the USA over many years, has shown a definite correlation with the boom and bust of the economic cycle, with its consequent fluctuations in unemployment.

The social fate of schizophrenics in the USA is also dramatically indicated by Table 7.6. Note that the percentage of individuals in 'Skid Row' and boarding homes who are diagnosed as schizophrenic is very large.

Table 7.6 The location of schizophrenics in the USA in the late 1970s.

Location	Number of schizophrenics	Percentage (estimated) of schizophrenics in total population	Percentage of all USA schizophrenics
'Skid Row' homeless	50 000–100 000	25	4–8
Jails and prisons	10 000–13 500	2–3	1
Boarding homes	180 000–240 000	60	15–20
Nursing homes	60 000–70 000	–	5–6
Hospitals	200 000–250 000	–	17–21
Total in non-domestic setting	500 000–673 500		42–56

How might such social explanations work in the context of the gender differences in depression? Doctors, predominantly men, might label women's behaviour as depressed more frequently than they do that of men. Alternatively, the condition of being a woman in a sexist society may induce depression. The best evidence for the latter explanation comes from research conducted by the medical sociologists George Brown and Tirrill Harris, who studied a sample of women in the inner London suburb of Camberwell in 1978. They found that at least 25% of the working-class women with children, when interviewed, appeared to be depressed to an extent that would invite a clinical diagnosis had they visited a psychiatrist. In contrast, middle-class women showed such depression at only a quarter of the working-class rate. The depressed women tended to have suffered from some serious or life-threatening event in the recent past, and were more likely to become depressed if, for instance, they had three or more children under 15 at home, or were unemployed. These are what Brown and Harris call *vulnerability factors*. Some of these factors relate to events in the woman's own childhood, such as having her mother die before she reaches the age of 11. These factors combine, they believe, to produce a condition of low self-esteem resulting, in the depressed condition, in a general sense of hopelessness. As they point out, this sense of hopelessness is not unrealistic: 'the future for many women is bleak'. In this sense, depression is an understandable response to an intolerable social condition. The concept of vulnerability factors is not confined to inner-city women; other studies have found similar factors to be predictive of the likelihood of becoming diagnosed as depressed among women living in rural conditions on a Scottish island.

7.5.3 Life history and the family

It is clear that what have been called grounded social explanations of depression and schizophrenia can only identify the circumstances which increase the likelihood of succumbing to the conditions. Not every smoker becomes a victim of lung cancer, and not every poor working-class mother in Camberwell is clinically depressed. So what distinguishes those who succumb from those who do not? The concept of vulnerability factors points to the importance of taking an individual's personal life history into account—and in doing so brings one into the terrain long occupied by psychoanalytic theories. According to this model of vulnerability factors, social and economic factors during childhood are important, the stability of the family unit and the individual's relationship with their parents during their infancy affect outcomes. Psychoanalytic theories (Section 1.3.3) claim to be able to account for the roots of a person's present distress in terms of infant and childhood experience, albeit for the most part taking the social world as given. Indeed Bleuler (Section 7.1), as much as Freud and other early psychoanalysts, stressed that schizophrenia was not the outcome of some type of physical alteration in the brain, but rather that the victims suffered from a severe disorder of the psyche having its chief source in their life histories. (Today, however, few psychoanalysts are prepared to offer to treat someone diagnosed as schizophrenic, regarding the condition as inaccessible to their techniques.)

Psychoanalysts—whether orthodox, or like the Laingian antipsychiatrists of the 1970s (Section 1.3.3), heterodox—are likely to be less interested in whether an individual's parents were unemployed at crucial periods during his or her childhood than in the individual's emotional relations with his or her mother and father. The crucial difference between orthodoxy and heterodoxy here is that orthodox psychoanalysts see the problem as that of the individual, whereas for heterodox psychoanalysts, individuals cannot be considered in isolation from their context—it is the family as a nuclear unit which is sick, to the extent that improving the condition of one of its members by psychiatric intervention will result in other family members becoming ill in their turn. For the heterodox, treatment of the schizophrenic by hospitalization or drugs is not seen as a help to the patient but as part of that person's oppression. Schizophrenia itself is seen as almost benign, a 'breakthrough' rather than a 'breakdown'. Laing claimed, for instance, that a proper analysis showed that schizophrenic speech, far from being jumbled and incoherent, revealed an inner logic and rational account of the schizophrenic's life and problems, which, granted a supportive, therapeutic community, could be resolved. Although he himself retracted many of these ideas before he died in 1989, they remain very attractive in some circles.

Even if one discounts the Laingian view of the origins of schizophrenia within the family, the consequences of living with a disturbed family member are quite disruptive, as is clear from a description by John Pringle, one of the founders of the British-based organization, The Schizophrenia Association, of living with a schizophrenic adolescent:

> One of your adolescent children begins to behave oddly. At first he just moons about. But many adolescents moon about. He or she takes to lying on the bed, presumably daydreaming. He or she becomes moody, bad-tempered, slovenly, 'difficult'. But many adolescents day-dream and are moody or difficult....To you, as a sensible parent, such behaviour is well within the limits of the normal growing-up process and no attention is paid to it. He or she will 'grow out of it', you say. But time goes on and he or she does not grow out of it. A crisis occurs when some wild display of aggression, truancy, or merely bizarre behaviour, drives you to seek expert help...your home, in an atmosphere of simulated normality, a determined show of ordinary living, is rapidly becoming preoccupied by the sick member: the ups and downs of his moods, his unpredictable vagaries of behaviour....Reactions of brothers and sisters may differ widely and their resentment may be deep if the sick one is not handled according to the ideas of each: affectionate concern; or alternatively, jealousy over the attention he is getting;....Irritation at the sufferer's 'laziness' or 'selfishness'; rejection; guilt for past teasing or bullying (shadowing the omnipresent and inescapable parental guilts); refusal to accept that he is ill at all, but only 'being himself', 'doing his own thing', that it is not he or she but 'society' which is 'all wrong'—'What he needs is to get away from those damned hospitals and live among normal people'....
>
> Opposite pressures, reflecting the ambivalencies of your situation, will be exerted on you by the fears of social embarrassment and 'stigma', both in the post-diagnosis stage and earlier....Try finding a sympathetic employer for your schizophrenic. Try getting digs where the landlady will not quickly make some excuse for getting rid of him. Try just listening in the background as he or she is floundering in front of some social security clerk who suspects he is workshy or fiddling his 'benefit'....Some members of your family may never accept the diagnosis and will quarrel

> with the parent who tries to implement it. You then have a divided and part alienated family on your hands, if you haven't one already—to say nothing of the cases where once happily married parents are themselves driven apart because one rejects or partially rejects, which the other cannot forgive....The sheer disruptive power of schizophrenia over a family is fully intelligible only to those who have been through it. But a few of the worst effects, could, I think, be avoided if the medical problem of communicating with the family were tackled at an earlier stage, and firmly. And not only with the parents, but with all the family who matter, collectively. I do not know how this is to be done, only that it ought to be done. (Pringle, 1972)

□ How does Pringle's view of the origins of schizophrenia compare with Laing's?

■ For Laing, the disorder originates as a response by the individual to the problem of family life. Pringle is primarily concerned with the problems of the rest of the family trying to respond to the schizophrenic member of the family and the disruption generated in their lives. Thus, for him, the original and primary disorder lies in the unprovoked odd behaviour of the adolescent.

□ Are these interpretations necessarily incompatible?

■ No, they are not. Both descriptions could be true, in which case the individual sufferer and their family would enter a declining spiral of relationships. Each would learn to expect certain types of behaviour and response on the part of the other.

An attempt to be more precise about the nature of the relationships between sufferers from schizophrenia and the relatives they live with has been made by a group of research psychiatrists led by Julian Leff at the Institute of Psychiatry in London in 1982. They measured what they describe as the level of 'expressed emotion' in the relatives of schizophrenics, noting particularly the number of critical comments made by relatives and the amount of hostility and over-involvement. Of course, if you found a high level of hostility and other forms of expressed emotion in terms of critical comments made by people living with a schizophrenic individual who showed the sort of behaviour described by Pringle, you could consider three possible explanations for the relationship between the person's schizophrenic behaviour and the hostility expressed by relatives.

□ Can you suggest what these might be?

■ First, the behaviour of the schizophrenic person might be so disruptive that it caused the hostility. Second, the hostility might pre-date the schizophrenic's behaviour and hence have precipitated it. Third, there might be no direct causal relationship, but a complex interaction between the behaviour of the schizophrenic person and the relatives.

The first explanation is suggested in the extract by Pringle, the second by Laing's double-bind type of analysis first discussed in Chapter 1. In fact, Leff and his colleagues claim that, by the use of statistical procedures, they have been able to rule out the possibility that the person's disturbed behaviour is directly responsible for their relatives' high levels of expressed emotion. Further, they have developed an interactive therapeutic strategy based on attempts to change the relatives' behaviour by educational programmes and group discussion, which, they claim, diminishes the rate of relapse among schizophrenic people living with relatives.

Summary of Section 7.5

This section looked at the main categories of non-biological explanation of mental disorders such as schizophrenia and depression. Schizophrenia can be seen as an example of labelling, as in the Foucaultian type of model (described in Section 1.3.1), and an example involving the arrest of a black Rastafarian was described. Or in the Laingian 'antipsychiatry' explanation, it can be regarded as a rational response to an intolerable situation. However, as an extract from the account of the parent of a schizophrenic teenager described, living with a schizophrenic member of one's family can also be immensely problematic. Social relations involve more than one person! Grounded social explanations of the excess schizophrenia diagnoses of black people living in Britain see the condition as brought about by exposure to the racism of the majority society. Similarly, grounded social explanations of depression in women identify a number of vulnerability factors based both on the childhood experience of the person and her immediate circumstances, of economic and emotional insecurity.

7.6 Biological explanations

7.6.1 Brain structure and schizophrenia

Until relatively recently, it was taken for granted that there were no gross abnormalities to be found in studying the brains of depressed or schizophrenic people; that, after all, was the implication of describing them as psychiatric rather than neurological disorders. The development of the new scanning techniques (described in Section 1.4.5) has, however, opened up a whole new era of possible study. Where once brains could only be examined directly post-mortem or by the use of the surrogate measures based on CSF, blood or urine analysis, they can now be looked at directly by PET (Book 2, Box 11.1) and MRI (Book 4, Box 5.1). Although it is still the case that such studies have found little of import in connection with neuroses, there have, from the late 1980s on, been a series of reports claiming that deviations from normality are to be found in the brains of schizophrenic persons. The data are far from conclusive, as many of the research findings seem contradictory, but they focus on two main themes. First, there are reports that the cerebral ventricles are abnormally enlarged in schizophrenia, and second, that there are unusual asymmetries between left and right hemispheres.

What such differences might mean is unclear. The ventricles (Book 2, Section 8.4.1) are sites of exchange of metabolites derived from neurons and glia and their extracellular surroundings. They and the cells which surround them seem to be vulnerable in a number of diseases and also change markedly in size in cases of neural degeneration, for example that associated with alcoholism. The issue of asymmetry is in some ways more intriguing. Although, as you know from Section

5.2 and from Book 2, Sections 11.2 and 11.3, there is marked *functional* lateralization in the brain, *structurally* right and left hemispheres are almost, though not quite, identical in size and distribution of surface landmarks such as sulci and gyri. In the 1980s however, an idea which had lain buried in the neurological literature for a century was revived, initially by the American neurologist Norman Geschwind. During an examination of a series of post-mortem human brains, Geschwind observed that there were certain small but significant left–right asymmetries in the extent of a number of cortical regions, and that these asymmetries were more marked in males than in females. Geschwind coupled the evidence of asymmetries with speculation that they played some part in cognitive differences between men and women. He claimed that there was a similar increase in asymmetry in cases of childhood autism, and that this might have something to do with the fact that the condition is more prevalent in boys than girls, the boys' 'natural' excess asymmetry making them more vulnerable. Could cerebral asymmetry help explain schizophrenia? If the observation is true, at best what it would demonstrate is a positive correlation between brain asymmetry and schizophrenia.

□ What three steps would be necessary to convert such an observed correlation into a possible causal explanation of schizophrenia?

■ 1 One would have to show that the asymmetry was necessarily related to schizophrenia, that is, that it was present in all unequivocally diagnosed instances, and that it was absent in matched but non-schizophrenic individuals.

2 It would be necessary to rule out the possibility that it was a consequence of the schizophrenia or its treatment, that is, it would need to be shown to be present prior to the onset of diagnosed schizophrenia in people who then became schizophrenic.

3 Some plausible theory would need to be developed to explain how such an asymmetry could result in the characteristic experiences and behaviours of people who suffer from the condition.

These conditions are of course counsels of perfection, and it is hard to meet them, especially when you recall the problems of the uncertainties surrounding even the initial diagnosis of schizophrenia. Indeed, granted that only a proportion of those diagnosed as schizophrenic show any specific cerebral abnormality, a characteristic strategy is to use the biological difference to lead the clinical diagnosis and argue that each specific difference is responsible for a different sub-type of schizophrenia. For instance some reports claim that there is a thickened corpus callosum in schizophrenia, whereas others find that the corpus callosum is thinner! The researchers attempt to reconcile these contradictions by arguing that one type of schizophrenia may be associated with the thickened corpus callosum, and another, with slightly different behaviour patterns, with the thinner. Perhaps the abnormalities can also account for the postulated different course of the disease in men and women. The specific structural differences that are observed in the brains of schizophrenic men (but to a much lesser extent in schizophrenic women) include enlarged ventricles and reduced size of the thalamus (Figure 7.2, *overleaf*). It is claimed that the more severe the symptoms, the greater the ventricular abnormality.

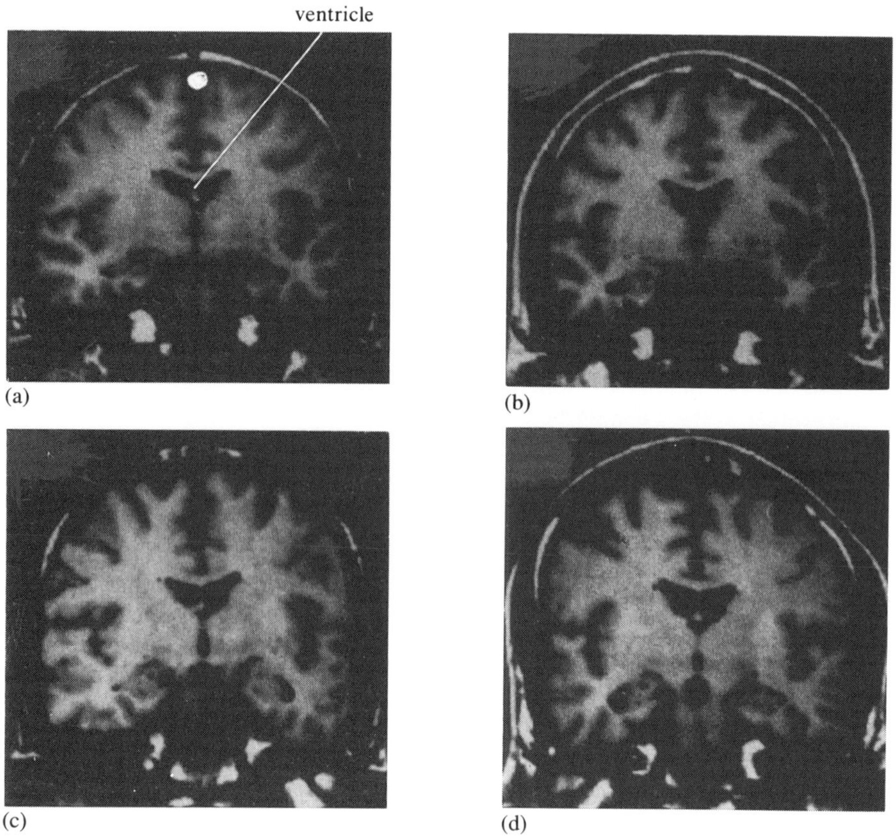

Figure 7.2 MRI coronal views from two sets of monozygotic twins; in each case, only one twin has schizophrenia. The lateral ventricles in the affected twins, (b) and (d), are larger than in the unaffected twins, (a) and (c).

There is also increasing evidence, based on EEG as well as imaging studies, that there are disorders of the medial temporal lobes, including a reduction in size, especially in the left hippocampal region. Such differences can persuasively be argued to be associated with some of the observed thought disturbances that occur in schizophrenia. Unfortunately for the clarity of such hypotheses, however, some of these differences are also found in people diagnosed as suffering not from schizophrenia but from unipolar and bipolar affective disorder (manic-depressive psychosis). Thus the specificity criterion emphasized above cannot be met.

7.6.2 Disorders of transmitter metabolism?

For a long time before the newer scanning techniques became available, the attention of researchers on both schizophrenia and depression was focused on the possible role of the neurotransmitters. This was partly for the sorts of reasons spelt out in Chapter 1, but above all because of the assumption that the mode of action of most of the drugs used effectively in the treatment or alleviations of the symptoms of both psychoses and neuroses seem to be related in some way to synaptic function. The logic of this approach is thus to argue backwards on the basis of the therapeutic effects of the drugs to the underlying biochemical system believed to be abnormal, which is perhaps not surprising, since so much research is sponsored by the pharmaceutical industry.

Apart from some early speculations about the presence of abnormal substances in the bloodstream, urine or sweat of schizophrenics (and sometimes depressives), since the 1960s most research has focused around the belief that the problem is not the presence or absence of a specific molecule, but whether there are higher or lower concentrations of particular transmitters and/or their receptors in the brain of schizophrenic people. Attention has shifted over the years between one and another class of neurotransmitters, with current fashion mostly revolving around dopamine, serotonin, GABA and the endorphins. Overproduction or underproduction of neurotransmitters in the brain in pathways concerned with analysing sensory information might result in jumbled messages in the brain, experienced as hallucinations. Altered neurotransmitter levels in pathways concerned with affect, arousal or attention might result in the characteristic symptoms of anxiety, depression or manic-depressive psychoses.

To begin with, there were just fragmentary data. In the case of schizophrenia, for example, it was noted that agents which produce hallucinations in normal subjects, drugs such as LSD, both mimic some of the experiences of schizophrenia and are chemically related to amine neurotransmitters. The introduction of dopa, initially regarded as a 'miracle drug' in the treatment of Parkinson's disease (Section 4.3.2) gave further clues, as its prolonged use resulted in sufferers experiencing disorientation, hallucination and emotional distress comparable to that of schizophrenics. Perhaps disorders of dopamine metabolism were related to schizophrenia? Post-mortem studies of tissue from people diagnosed as schizophrenic began to produce claims that there were altered levels of dopamine and other neurotransmitters, such as noradrenalin, in particular regions of the brain.

However, as always in the field of explanations of mental distress, the data are contradictory. Not all researchers have been able to confirm the altered transmitter levels, as a result of which for a time during the 1980s it was argued instead that what mattered was not so much the absolute level of any transmitter but the *ratio*, of, say dopamine to serotonin or noradrenalin. Further, there is some difficulty in reconciling the neurotransmitter hypothesis with the scanning evidence which points to the involvement of medial temporal lobe and hippocampus, as these regions are not rich in dopamine or noradrenalin and mainly seem to employ other neurotransmitters.

Neurotransmitter hypotheses are not the only ones which have been offered to account for schizophrenia; others include dietary, immunological and viral hypotheses. It is not unfair to say that, despite the wealth of hypotheses and reported positive findings, not a single molecular 'mechanism' for schizophrenia yet commands majority assent among the small army of researchers on the biochemistry of schizophrenia. It remains to be seen whether this failure is simply because research has so far been looking at the wrong molecules or biochemical processes, or whether there is a more fundamental flaw in the experimental procedures and logic being adopted.

The position is not much better in the case of the neuroses such as depression. Here again a great deal of the evidence is pharmacological. Many of the antidepressant drugs (for instance imipramine and paroxetine) have been shown to affect the serotonergic system in experimental animals, in particular binding to serotonin re-uptake sites at the synapse (Book 2, Section 4.5.5). (Benzodiazepine tranquillizers such as valium also bind to specific synaptic sites, probably related to the neurotransmitter GABA). As explained in Section 1.4.6 and Book 2, Section 5.4.1, human blood provides a source of platelets which can serve as 'surrogate neurons.' Platelets have serotonin re-uptake sites, to which drugs like imipramine also bind. Imipramine binding to the platelets can be used as a measure of the number of the re-uptake sites, and there have been a number of reports claiming that imipramine binding levels in depressed people are only about half the figure found in 'normal' controls. Individuals with a low level of such binding tend to show an increase towards control levels as their condition improves—whether as a result of taking antidepressant drugs, or as a consequence of psychotherapy, or in spontaneous remission (Figure 7.3). In this sense imipramine binding levels are 'state-markers' rather than 'trait-markers' (Section 1.4.5) for depression. However, so that you should not be left with the opinion that anything is simple and certain in this field, you should note that platelet imipramine binding levels also fluctuate with time of day and seasonally, and have also been reported to be lowered not just in clinical depression but in other conditions such as anxiety, as well as in blood taken from people who have committed suicide.

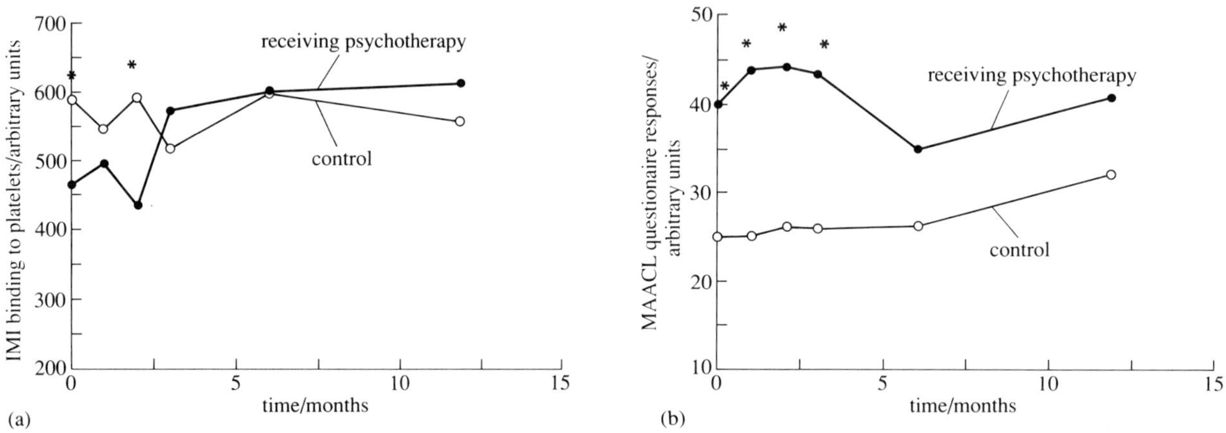

Figure 7.3 Correlations between imipramine binding to platelets and depression score in people during a year of psychotherapy. Graph (a) compares the biochemical measure of the people receiving psychotherapy (closed symbols) with a matched control group of 'normal' volunteers (open symbols). Graph (b) compares the scores of the same groups over the same period of time on a questionnaire designed to measure depression (the multiple affect adjective checklist, MAACL, shown in Figure 1.1). Note that a high score here indicates a greater degree of depression. The asterisks indicate points significantly different from the controls.

7.6.3 Changed electrical activity?

As well as changes in cerebral asymmetry and in specific neurotransmitters, there have been claims that the electrical activity of the brain, measured either as spontaneous EEG or as evoked potentials (Box 2.8.1) differs in schizophrenic patients. An example of the type of changed wave form claimed to occur is shown in Figure 7.4; however, such findings have all the ambiguities and uncertainties associated with the claims for differences in symmetry, and in addition they are subject to the criticism that no one is really very sure what the biological meaning of the observed waveforms might be. Nonetheless, there are those who believe that, if the message of the waves could be read, it would provide a real clue as to the causes of schizophrenia.

Summary of Section 7.6

Despite a multitude of physiological, neuroanatomical, biochemical and pharmacological research publications, and some widely publicized claims, it remains the case that there is no clear consensus about which neurotransmitter systems might be involved, or how, in any type of mental distress, or whether there are reliable changes occurring in any aspect of brain morphology or electrical activity. Different research groups have presented conflicting data, experiments and hypotheses, but many of these can be criticized for poor design or inadequate controls. It may simply be that there is no simple mapping of the mental state diagnosed as schizophrenia, depression or anxiety onto any single set of biochemical, physiological or anatomical markers, but rather that many different types of biological disorder can have as their 'final common pathway' the mental distress (think of the many different reasons why a car might not start on a cold wet morning!). In this sense the fact that the psychotropic drugs are generally effective would be analogous to the very general effect of a drug like aspirin in relieving pain. No one would argue that a headache and a toothache had similar biological causes simply because aspirin was effective in alleviating both sets of symptoms.

7.7 Genetic psychiatry

The biochemical studies discussed in Section 7.6 have problems both of methodology and of distinguishing between cause, consequence and correlation. Because genes do not change with the day-to-day state of the sufferer, the study of genetic differences is not fraught with the same types of problem. A genetic difference between people who suffer from a particular form of mental distress and others is always a trait and not a state marker. This in some ways makes the situation easier, in others, however, harder. For although it is easy to argue that an environmental cause can drive a person to depression or schizophrenia, and simultaneously change their brain biochemistry, by definition the genetic condition must have been present from conception, and thus must precede any external environmental cause. A genetic argument must therefore be able to offer a plausible explanation of the epidemiological data, for instance why it is that schizophrenia is more commonly diagnosed in working-class and black people. It is not unfair to say that most genetic psychologists put such epidemiological issues to one side as an additional embarrassment in an already difficult research task.

Nonetheless, it would seem at first sight a reasonable goal to try to investigate the role of genes in the causation of schizophrenia or depression, as an answer to that question might at least help in understanding a condition, if not, at present, in treating it. Certainly the belief that mental distress and disorder is 'in the genes' has at least a century of psychiatric adherence behind it. The literature of genetic psychiatry is full of emphatic assertions that disorders of intellect and will (along with alcoholism and moral weaknesses of many sorts) are fixed in the human genes. For most of this period, the search for genetic explanations has been underpinned by the false hope, held by some, that if they could be identified, a programme of positive or negative eugenics (selection for and selection against particular traits by human breeding) might help eliminate undesirable traits, whether of poverty, criminality or disease.

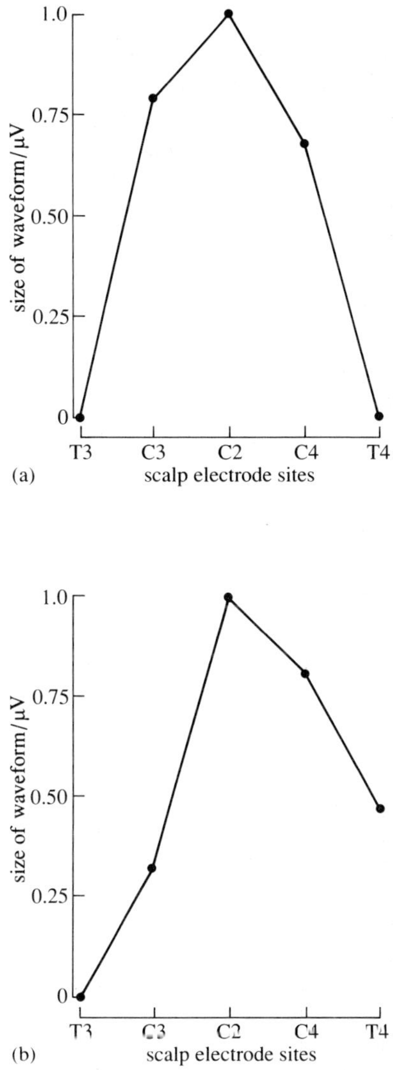

Figure 7.4 Examples of different EEG waveform (amplitude) patterns in (a) control subjects and (b) schizophrenic individuals. Controls show a left–right symmetry in such waveform patterns, whereas the amplitude in schizophrenic subjects is greater on the right side of the brain. The letters T and C refer to different scalp electrode sites.

Within the last few years however, the prospect that genetic knowledge might lead to some forms of treatment has come much closer as a result of the dramatic advances that are occurring in molecular genetics. For instance, if it were the case that a particular gene was associated with schizophrenia, and a marker for it could be found, then pre-natal diagnosis could enable prospective parents to know if a pregnancy was likely to result in a child who might become schizophrenic, and they could then decide whether to proceed with or to terminate the pregnancy. This option already exists, as you may recall from Section 4.4, for conditions such as Huntington's disease. In the longer run, identifying a specific gene might provide a clue to the biochemical processes that it helped control, and therefore to developing targetted drugs to interact with those processes. But before any of these prospects become real, how strong is the evidence that there are genes directly associated with conditions like schizophrenia and depression?

Chapter 1 discussed the available ways of 'sorting out' the effects of genes from the effects of shared environments, in particular, methods based on noting the extent of the disorder in the family of the sufferer. Methods such as kinship correlation and pedigree analysis have been available for many years; newer techniques include those of endeavouring to find a linkage (Section 1.4.9) between the condition and some marker gene, as discussed for Huntington's disease in Section 4.4.

So far as neuroses like depression are concerned, despite a variety of elaborate research programmes, neither kinship correlations nor pedigree studies reveal anything more than a slight tendency for depression to run in families. The best evidence for a genetic predisposition to mental disorder is found among the psychoses, particularly schizophrenia and manic-depressive psychosis. It is certainly well established that there is a tendency for schizophrenia to run in a family. Among parents of schizophrenics some 4% are themselves diagnosed as schizophrenic. For brothers and sisters of schizophrenics this figure is 8% and for the child of someone who is schizophrenic, 12%—well above the overall rate of incidence of schizophrenia in the general population of 0.6%. However, the Nottingham epidemiological data mentioned earlier, claiming that the incidence of schizophrenia is much higher in the children of Caribbean migrants born in Britain would be, if borne out in larger-scale surveys, a severe difficulty for the genetic argument. There is no plausible genetic mechanism to account for a kinship pattern in which the incidence of a disorder is so much greater in the offspring than their parents.

Even setting the Nottingham data to one side, without further analysis, the inheritance pattern found for schizophrenia would fit familial theories just as well as genetic ones and they would be compatible with the idea that there was some common environmental factor, whether social or chemical, that the family shared. Pedigree analyses of schizophrenia in families reveal no simple pattern of inheritance, so it is unlikely that a single gene could be responsible for the condition. Most geneticists, though not all, therefore conclude that, unlike the conditions you have met previously in this course such as phenylketonuria or Huntington's disease, predisposition to schizophrenia must be the result of the interaction of *many* different genes, which, in the context of an appropriate environment, would result in the disorder developing. Some have tried to fit this idea into the context of the brain asymmetry hypothesis described above, and have argued that it is the genes which contribute to the development of brain symmetry which are abnormal in the case of schizophrenia, an increasingly popular idea at the time of writing of this course.

Twin studies (Section 1.4.9) offer a more rigorous approach and there have been a number of such studies conducted over the past thirty years. The most substantial, made by the American biological psychiatrist Seymour Kety and his colleagues, was based on the twin registers kept for the entire population in Denmark, although the research concentrated on Copenhagen. Such studies suggest that the concordance (Section 1.4.9) for schizophrenia is higher for monozygotic twins than for dizygotic twins. Taking all the studies that have been conducted, including those from as long ago as the 1930s, the concordance for monozygotic twins ranges from 14 to 65%, but the more recent the study, the lower the concordance. In general, the older studies were less well controlled than the more recent ones. For instance, it is necessary to be sure whether the twins are really monozygotic or not. Simply asking people, which used to be all that was done, is not enough. There could obviously be good reason for a person whose twin was schizophrenic to deny that they were identical, hence the discordance among dizygotic twins would be artificially magnified. To be sure whether a twin is really monozygotic, it is necessary to do blood tests or DNA tests and not just rely on appearances. In three studies conducted in the late 1960s and early 1970s, the concordance figures are 14%, 24% and 25% for monozygotic twins, as opposed to 4%, 10% and 7% respectively for dizygotic twins. Even these newer figures, however, cannot be taken quite at face value: the monozygotic twins, for example, tend to be treated more similarly during their upbringing, and therefore to share a more similar environment, than dizygotic twins. This would seem likely to affect the results by increasing concordance in monozygotic twins.

□ How would you test this possibility?

■ You could compare the concordance of dizygotic twins with the concordance between non-twin brothers or sisters.

□ If the concordance between dizygotic twins were the result of genetic effects alone, how would you expect it to differ from the concordance between non-twin siblings?

■ There should be no difference. Genetically, dizygotic twins are no more alike than are non-twin siblings. Monozygotic twins differ from both dizygotic twins and ordinary siblings by being identical genetically.

Of course, dizygotic twins tend to be treated more similarly than ordinary siblings, so if there are family environmental factors involved, concordances between dizygotic twins should be higher than those between ordinary siblings. This indeed is the case. The same studies that find high rates of concordance for monozygotic twins also find higher rates of concordance for dizygotic twins than for ordinary siblings, a difference about as great as that between monozygotic and dizygotic twins. Thus the higher concordance found for monozygotic twins could be explained as being a result of their having been treated even more alike than the dizygotic twins.

The final type of study is based on adoption (Section 1.4.9). Again, the major studies have been those of the joint Danish–American team led by Seymour Kety. These studies are often taken as the best evidence for a genetic factor in schizophrenia. They are based on locating adults who have been diagnosed as schizophrenic and who were adopted as children. The incidence of schizophrenia in the adoptive and biological relatives of these patients is then compared with the incidence in matched non-schizophrenic individuals (controls) by assessing the diagnostic records of the relatives (where they can be traced). Of 34 schizophrenic adults who had been adopted as children, 8.7% of the biological relatives showed schizophrenia, compared with 1.9% of the biological relatives of the controls. This difference between 1.9% and 8.7% could explain the remaining evidence for the genetic basis of schizophrenia derived from this type of approach.

Critics of the genetic studies, such as the Boston-based psychologist Leo Kamin, have pointed out that even these figures are uncertain, because the actual cases of schizophrenia located among the relatives of either the schizophrenics or the controls were so few, that it was necessary to widen the definition of schizophrenia to include an entire spectrum of related personality disorders or 'psychopathology'. Double-blind procedures, in which researchers were asked to assess the people in the control groups for schizophrenia, showed that many of the controls were regarded by the researchers as schizophrenic; because they expected that some of those they were rating were likely to be schizophrenic, the researchers erred on the side of 'over-diagnosis.'

It seems therefore as if the classical methods of genetics cannot help resolve the problem of understanding the origins of schizophrenia. Uppermost in many geneticists' minds from the 1980s on, therefore, was the question of whether the new techniques, involving linkage studies, might be of more help. Linkage studies in schizophrenia and manic-depressive psychosis made headlines during the late 1980s and early 1990s.

The suggestion that there might be a genetic basis to bipolar manic-depression goes back several years, and in some ways the peculiar features of the condition, with its regular mood cycling apparently quite independent of external cues, make it seem one of the most likely of mental disorders to warrant a biological explanation. In 1987, Janice Egeland, from Miami, and her colleagues reported a study in an almost uniquely appropriate population, members of the Old Amish Order in Pennsylvania. The Old Amish are a small Christian religious group whose beliefs mean that, as well as continuing to live a life-style, and wear the clothes patterned as far as possible on those of their 17th-century puritan forebears, they also are expected to marry only within their own group. The stability of their life-style (despite the pressures of the modern America which surrounds them) means that they have good family records stretching back through several generations. Behaviour patterns which are analogous to those which would today be diagnosed as manic-depression can be traced back within Old Amish families with a better degree of certainty than in most populations, although of course such retrospective diagnoses, as we have already pointed out, have to be treated with caution. (It may be significant that the Amish have an extremely low incidence of diagnosed schizophrenia.) Figure 7.5 shows a pedigree for manic-depression in an Old Amish family.

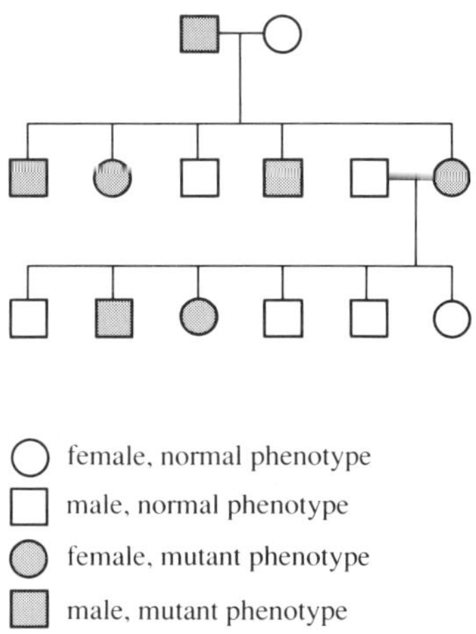

Figure 7.5 Pedigree for manic-depression in an Old Amish Order family used to calculate linkage

Egeland and her colleagues reported that in a number of individuals of this family, the manic-depression condition seemed to be linked to a particular chromosome, and specifically to the presence of a marker enzyme associated with the synthesis of amine neurotransmitters. The study generated a wave of excitement as it seemed that a real breakthrough in the genetic understanding of a mental disorder was emerging. However, it soon became clear that a similar linkage did not occur in other manic-depressive pedigrees, and, even worse, when the original Amish group was studied in more detail, and other family members included, the linkage simply disappeared. Two years later, the claim was withdrawn—it had turned out to be nothing but a statistical artefact.

A year after Egeland's claimed linkage, in 1988, Robin Sherrington and his colleagues in London reported that they had found a linkage of schizophrenia to a region of another particular chromosome in five Icelandic and two English families. Once again the excitement was intense and genetic psychiatrists began to hunt for a similar linkage in other schizophrenia pedigrees. At the time of writing (1991), no other pedigree has been shown to have a linkage of schizophrenia to this chromosome, and the Sherrington finding remains a seemingly isolated anomaly.

Summary of Section 7.7

The conclusion is that, despite extensive studies over half a century, there is still little unequivocal evidence for a marked genetic basis for schizophrenia. This is not to say that everyone has an equal genetic predisposition for schizophrenia. It may well be that, in a given environment, certain genotypes are more predisposed to schizophrenia than others. Prior to the linkage studies, pedigree, twin and adoption studies failed to disentangle conclusively genetic from environmental factors. There is however an important lesson to be drawn from the discrepancy between Sherrington and his colleagues' discovery of linkage in some pedigrees and the failure to find it in others. This is that there are many different combinations of genes that are likely to predispose a person to the condition(s) diagnosed as schizophrenia. It may be that either possession of a particular combination of genes, or exposure to a particular environment, or both produce the type of brain biochemistry which constitutes schizophrenia. Studying the rare genetic condition may help point to the biochemical mechanisms involved in the much more common environmental one.

Summary of Chapter 7

This chapter began by describing the conditions grouped within the general framework of mental distress and disorder, and tried, by the use of quotations from the words of people suffering from such conditions, to help you gain an understanding of what it felt like to be neurotic or psychotic—or to have one's experience and beliefs about the world *described* as neurotic or psychotic. Beginning with the epidemiology of the conditions, each of the three main modes of explanation of mental distress, were explored in turn, using depression and schizophrenia as model systems. The three are: caused by social and economic factors; caused by events in a person's life history (psychological factors); caused by disordered biochemistry or genes (biological factors). Most time has been devoted to the discussion of what are often regarded as the two major conflicting modes of explanation, the biological and the social, and you may feel that the chapter has been overly critical of each. The chapter has certainly not come out at the end with a tidy explanation: there is no clear-cut statement about *the* cause of any sort of psychopathology. That is at least partly because, despite the (sometimes strident) claims of those who believe in single cause explanations, there is certainly no such thing as a *unique* cause. This does not mean acceptance of a sort of simplistic middle of the road position which argues that sometimes the cause of a condition is social and sometimes it is biological. All the phenomena of mental distress, like all other 'mind' phenomena are *both* biological and social, and must in principle be able to be described in the languages of brain, of psychology and of the social context. They cannot be partitioned into *x*% of one and *y*% of the other, or reduced to being 'exclusively' the province of the biological psychiatrist or the psychoanalyst. Rather, the problem lies in the probability that 'depression' and 'schizophrenia' are not single entities. Although each term may describe similar types of behaviour, the causes are more likely to be manifold. One need not for instance assume that post-natal depression and depression following the loss of a job are precipitated by a similar set of causes or involve similar biochemical changes in the brain merely because they have the same name and may be treated with the same set of drugs. There are much more likely to be multiple states of brain biology that correspond to the phenomenon called depression. It may well be (no one knows) that certain genotypes, in the particular social and physical environment of the late 20th century in Europe or America, are especially predisposed to respond to life-threatening events by evincing depression or schizophrenia. This says nothing about how such genotypes may respond in other environments. Nor does it yet give much guidance as to how to alleviate present distress.

Objectives for Chapter 7

When you have completed this chapter, you should be able to:

7.1 Define and use, or recognize definitions and applications of each of the terms printed in **bold** in the text.

7.2 Distinguish between psychoses and neuroses. (*Question 7.1*)

7.3 Comment on the problem of interpretation of data on gender, class and ethnic distribution of psychopathology diagnoses in the UK. (*Question 7.2*)

7.4 Distinguish between three broad types of explanation of depression and schizophrenia: as caused by social and economic factors; as caused by events in a person's childhood (psychological factors); as caused by disordered biochemistry or genes (biological factors). (*Question 7.3*)

7.5 Discuss the evidence that schizophrenia is a disorder of cerebral asymmetry. (*Question 7.4*)

7.6 Summarize the evidence for a role of neurotransmitters in the causation of depression and schizophrenia. (*Question 7.5*)

7.7 Describe the principal evidence in favour of, and against, genetic theories of transmission of manic-depressive psychosis and schizophrenia. (*Question 7.6*)

Questions for Chapter 7

Question 7.1 (*Objective 7.2*)

(a) Would an otherwise fit individual with an important letter to post, who felt she could not mobilize enough energy to walk the 50 metres to the post box, be likely to be diagnosed as suffering from a neurosis or a psychosis?

(b) Would an otherwise fit individual with an important letter to post, who felt it was dangerous to go out of doors because he was likely to be followed by malevolent two-foot high Martians, be likely to be diagnosed as suffering from a neurosis or a psychosis?

(c) How sure should you be of your answers to (a) and (b)?

Question 7.2 (*Objective 7.3*)

Look at the data from the Cochrane study reproduced in Table 7.5. There were differential admission rates for Irish-born men and women, English and Welsh-born men and women, and USA-born men and women for personality disorders in England and Wales in the early 1970s. In particular, men born in Northern Ireland were admitted to hospital for personality disorders nearly five times more frequently than English or Welsh-born men and almost twice as frequently as women born in Northern Ireland. Can you conclude from this:

(a) That men have a genetic predisposition to personality disorders?

(b) That the Irish have a genetic predisposition to personality disorders?

Give reasons for your answers.

Question 7.3 (*Objective 7.4*)

(a) To what extent are personal–psychological historical explanations of mental distress and disorder compatible with theories based on social relationships?

(b) To what extent are personal–psychological historical explanations compatible with biological ones?

Question 7.4 (*Objective 7.5*)

Which of the following true statements is evidence that schizophrenia is caused by a genetic tendency towards cerebral asymmetry? Which would be arguments *against* such an explanation?

(a) The children of schizophrenic parents are more likely to be schizophrenic themselves than those from non-schizophrenic parents.

(b) PET scans reveal evidence for cerebral asymmetry in schizophrenic patients.

(c) During development, boys show a greater tendency towards cerebral asymmetry than do girls.

(d) There are suggestions that cerebral asymmetry may be associated with autism.

Question 7.5 (*Objective 7.6*)
Why do many biological psychiatrists favour the idea that there are disorders of neurotransmitter function in depression and schizophrenia?

Question 7.6 (*Objective 7.7*)
Summarize the main arguments for and against the claim that schizophrenia is a genetic disorder .

References

Cochrane, R. (1977) Mental illness in immigrants to England and Wales: an analysis of hospital admissions, *Social Psychiatry*, **12**, pp. 25–35.

Dohrenwend, B.P. *et al.* (1980) *Mental Illness in the United States: Epidemiological Estimates*, Praeger, New York.

Further reading

Brown, G. W. and Harris, T. (1978) *Social Origins of Depression: A Study of Psychiatric Disorder in Women*, Tavistock, London.

Littlewood, R. and Lipsedge, M. (1982) *Aliens and Alienists*, Penguin, Harmondsworth.

Nairne, K. and Smith, G. (1984) *Dealing with Depression*, The Women's Press, London.

Pringle, J. (1972) *Mind and Mental Health Magazine*, Spring issue.

Rose, S.P.R., Lewontin, R.C. and Kamin, L. (1984) *Not in our Genes*, Penguin, Harmondsworth.

Toates, F. (1990) *Obsessional Thoughts and Behaviour*, Thorsons, London.

Warner, R. (1985) *Recovery from Schizophrenia*, Routledge Kegan Paul, London.

CHAPTER 8
BRAIN AND BEHAVIOUR; THERAPY AND CONTROL

8.1 Introduction

The issues discussed in this book are of brain and mind function and dysfunction, of normality and abnormality, of behaviour which seems out of place, maladaptive or inappropriate—or even simply inconvenient to the person concerned or to society. In their sharpest form these issues raise questions about the extent to which the brain and behavioural sciences can contribute to human understanding of, and control over, our own selves. Such questions have in their varying forms been central to the themes of the course as a whole. They are in part philosophical, because they raise issues of the relationship between mind and brain and the nature of consciousness, which have been a rich source of debate and discussion amongst philosophers, scientists and indeed theologians for centuries. They are in part ethical, because they force one to consider the extent to which people are 'responsible' for their actions, and therefore must answer morally and legally for their acts or to what extent they are to be treated as driven by biological forces beyond conscious control, and therefore to be treated medically. They are in part political, because they are inevitably about power, the power expressed in the disproportionate labelling of socially weaker groups within the population (poorer people, women, blacks), as either mad or bad, by the dominant group (richer people, men, whites). They are also in part technological. The huge flowering of the brain and behavioural sciences in the last decades of the 20th Century has raised vistas previously undreamed of. Drugs and behavioural procedures to control and manipulate mood, to erase or improve memory and cognition, to adjust and fine-tune the mind to suit its environment, are no longer the stuff of science fiction. Today they are the day-to-day themes of laboratory research and investment decisions by pharmaceutical companies. Genetic engineering to modify or eliminate 'undesirable' phenotypes, or perhaps to propagate 'desirable' ones, and the transplantation of brain tissue to diminish the decline and fall of old age are methods which are not decades away but may well be more-or-less routine within the lifetime of this course. Not for nothing has the USA National Institutes of Health claimed the 1990s as 'The Decade of The Brain', and the Japanese government instituted an international Human Frontier Science Program with the elucidation of human memory and the biology of behaviour as its core projects.

This chapter will, therefore, look at these themes within the perspective of the course as a whole, and with the intention, not of providing definitive answers to any of the questions raised, but of providing a framework for discussion and pointing up their implications. It begins with an historical perspective on that age-old question within Western religion and philosophy, of the relationships between mind and brain. Do the brain and behavioural sciences have anything to say to resolve the question?

8.2 Mind, brain and science

8.2.1 Descartes and dualism

From the very birth of modern science in the 17th century, philosophers recognized that its methods and findings posed challenges to long-held religious views about the relationships of body to 'mind' or 'soul'. As earlier chapters have mentioned (see especially Book 2, Section 11.5.3; Book 3, Section 5.4.3, and this book, Section 6.2.1) the initial 17th century compromise was offered by the French Catholic philosopher and physicist René Descartes. For Descartes, all living organisms, with the exception of humans, were 'mere mechanism' to be understood in terms of the laws of physics and chemistry, their behaviour only a sophisticated version of clockwork automata. Non-human animals did not feel pain, for instance; their cries were analogous to the squeaking of the rusty cogs of a machine in need of oiling. Humans were different; they shared mechanism in common with other animals, but unlike them, the human mechanism was governed by a non-corporeal entity, the soul or mind, which was not subject to scientific laws but which controlled the body by way of a site of interaction within the pineal gland. Descartes believed the pineal gland to be the only brain structure which is not duplicated between left and right hemispheres (Section 6.2.1). It was therefore the only region of the brain appropriate for the task of containing and responding to the integral and indivisible nature of the soul. Thus, Descartes was able to reconcile his religious with his scientific convictions.

Descartes' view of mind-brain relationships became the basis for the philosophical position known as dualism. Dualistic thinking has persisted since Descartes' day in powerful and varying forms. It is so much a part of Western culture's natural everyday speech that it is hard to realise that it is not the only way of thinking and talking. Consider, for instance, the phrase 'I can lift my arm above my head if I choose'. This sentence seems to make perfect sense in English, yet it divides the 'I' which is making the choice from the 'arm' which is being moved as if the I was in some way non-biological and its choice was not itself an aspect of biology. However, to a scientist studying brain and behaviour it must be apparent that the I which is doing the choosing, and the act of choosing itself, cannot be cut adrift from biology. In addition, the sensation of personhood which the term I implies, and of the 'choosing' that the I does, are as much descriptions of brain processes as the lifting of the arm is a description of muscle processes. Indeed, with the increase in biological understanding that the intervening centuries have provided, it became increasingly hard to maintain the simplistic dualistic distinction that Descartes proposed. Today, only a very few at least amongst scientists would accept a version of the Cartesian position. (One of the exceptions is the Nobel prizewinning physiologist Sir John Eccles, who, as a practicing Catholic, sees dualism as important to the retention of his religious faith. Despite its central place in Western philosophical and religious history, dualism is not an inevitable way of thinking and other religious, cultural and linguistic traditions, for example the Chinese, do not make such mind/brain distinctions.

8.2.2 Reductionism and biological determinism

The development of evolutionary theory and the apparently increasing power of a physics-based science to explain the universe in the 19th century led to an increasingly materialist philosophy amongst scientists. For physiological chemists like the Dutch Moleschott, 'the brain secretes thought like the kidney secretes urine'. For the English zoologist Thomas Henry Huxley, the mind was essentially an *epiphenomenon*. This means that the real business of thinking and feeling was done by the brain, and the mind was no more relevant to this real business, no more capable of influencing its activity than was (in Huxley's archetypical 19th century metaphor), 'the whistle to the steam train'.

The 19th century was the high point of *reductionism*; the belief that everything could be explained by reducing it to its atomistic components. Thus, social phenomena were to be understood as the sum of the behaviour of individual humans. This is a position still held by some today, as exemplified by the view expressed by Margaret Thatcher when Prime Minister in the 1980s that, 'there is no such thing as society'. The behaviour of individuals was to be explained by their physiology; their physiology by their biochemistry; their biochemistry by chemistry, then physics and ultimately, therefore 'there is nothing in the world but atoms'. Such reductionism is still a widely accepted philosophical stance within biology and the brain sciences. It seems to be especially strong amongst molecular biologists. The quote in the last sentence comes from one of the discoverers of the double helix structure of DNA, James Watson. It is also shared by some philosophers; it is argued, for instance, in the well-received book *Neurophilosophy*, published in 1986 by the Californian philosopher Patricia Churchland.

Together with reductionism, the 19th century saw the growth of a viewpoint which can be described as *biological determinism*. This idea portrays human behaviour as an inevitable consequence of their biological make-up, a make-up which is given in the genes and which is itself the product of evolutionary processes. Within such a framework, both individual human differences and broad human similarities in behaviour are regarded as primarily shaped by heredity. Thus biological determinism claims that differences in intelligence between individuals, or the tendency of particular people to become schizophrenic or depressed, are largely genetic. By contrast, it sees some features of the human condition both as universal—such as a propensity towards aggression, or the divisions of labour, power and status between men and women in society. These universals, it claims, come about as the products of genetic adaptations selected during evolutionary history.

Like reductionism, biologically determinist ideas and explanations of behaviour are still widely prevalent. They were given most prominent expression by the Harvard entomologist, Edward O. Wilson in the book *Sociobiology*, published in 1975, which founded and gave its name to the subject. Wilson and many of his followers have claimed that because sociobiology lays emphasis on evolution, adaptation and selection at the level of individual genes, the 'new synthesis' it offers provides convincing genetic and deterministic arguments for the question of 'why we do what we do'—to quote the sociobiologist Robert Trivers. Since their enunciation, these comprehensive claims have been the subject of intense debate and criticism both for their scientific arguments (especially but not entirely in the context of human behaviour) and their perceived political and philosophical implications. Although all students of animal behaviour recognize the significance of genetic and adaptive factors in seeking explanations of behaviour, most would distance themselves from the sweeping claims that sociobiology's prominent protagonists have made. You can read more of the arguments in the books by Kitcher and by Rose *et al.* listed at the end of this chapter.

8.2.3 Computers and systems theory

However, although both reductionism and biological determinism have their roots in 19th century materialism, with the 20th century and especially the development of computing, the metaphors by which brain and mind were to be understood changed. No longer were there evocations of clockwork, or hydraulics or steam trains. Computers have both hardware and software. Could the mind be likened to the brain's software? Further, systems theory offered new ways of thinking about behaviour. Examples of this way of thinking can be found in Chapter 6 and in some of the chapters of Book 5. Systems theory is in many ways the reverse of reductionism. Reductionism is essentially a 'bottom-up' explanation; the behavour of an organism is 'nothing but', 'reducible to' or 'caused by' the properties of its molecules. Schizophrenia then is an abnormality of dopamine metabolism; panic disorder is an excess of lactate production. By contrast, as the systems theorists pointed out, the properties of complex and interactive systems, whether computers or neural networks, are not simply reducible to that of their components.

Reductionism cannot account for the interactions that are a property of the system as a whole rather than of its separated parts. Imagine a Martian coming to earth and trying to understand the functions of a motor car by analysing the composition of its individual components. If this were not a hard enough task in itself, the fact that the components are embedded within a system restricts the independence of any individual component. So, even if the Martian could put a car together from its parts, the functions of that car as a vehicle would be meaningless without understanding the nature of the road system and the role of individual transport in an industrial society. This constraint that the system places on its parts has so impressed some philosophers and scientists that they have argued that far from the component 'causing' the behaviour of the system, the system 'causes' the behaviour of the individual component. For example, the speed of cars is limited not by their own intrinsic mechanism but by the congestion on the road and by traffic legislation. This idea, of 'downward causation', was introduced in the 1970s by the Nobel Prizewinning neurobiologist Roger Sperry, whose work on specificity in neural development was introduced in Book 4 (Section 3.5.3) and on split-brain persons in Section 3.1.

8.2.4 Identity theory—an integrative approach

How does the approach adopted in this course relate to these very different models of the relationships of brain, mind and behaviour? To begin with, think back to the concept of *levels* (Book 2, Section 1.3).

☐ What do you understand by the term level in this context?

- A level is the term given to a particular degree of organization of the material world; thus we can speak of the molecular level, the genetic level, the cellular level, the level of the organism, and the level of the population. However, the term is also used to describe the different scientific disciplines used to study each level of organization, as physical, chemical, biochemical, physiological, psychological, behavioural and social.

The integrationist approach rejects dualism, which claims that brain and mind are completely distinct categories. It is equally unhappy, however, with a reductionism which would strive to eliminate all levels other than the most fundamental, and the Sperry solution in which upwards and downwards forms of causation operate independently, passing one another without interacting, like commuters on parallel up and down escalators. It is a materialist view, in that there is only one type of stuff in the world, not two, as dualism claims. Integrationism is a unitary view, arguing that to speak of brain and of mind is not to talk of two different things but of the same *identical* thing. Philosophers call this an *identity theory* of mind and brain.

But although the world is unitary (the philosophical term for this is *ontological unity*), it is organized on different levels and, for individuals trying to understand the world, there are many different ways of talking about it. Although the terms mind and brain refer to the same thing (just as the English word *cat* and the Italian word *gatto* refer to the same thing), for some purposes, such as most everyday speech, and also most of the concerns of psychology, mind language is appropriate; for other purposes brain language is appropriate. To say 'I am angry' is another way of saying 'My brain is in such and such a state, the neurons of certain regions of my limbic system are firing at such and such a frequency and there is an increased level of circulating adrenalin and other hormones in my blood stream'.

Depending on the purposes for which one is discussing the phenomenon, it may be better to use one language or the other, just as in some circumstances it is better to speak English and in others Italian. You can consider how such an integrationist approach might be applied to Edelman's theory of consciousness (Book 2, Section 11.5.3), the cellular model of memory (Book 4, Chapter 5) or the neurotransmitter model of schizophrenia (Section 7.6.2).

This plurality of languages to discuss a single phenomenon is known by philosophers as *epistemological diversity*. The distinction between ontology and epistemology is that the former refers to what is assumed to be the real nature of the world, whilst the latter refers to what one can know about the world, granted the strengths and weaknesses of the human mind.

However, the choice of levels is not merely about which language to use, because levels are not purely epistemological; as already said, they describe different degrees of organization of the world. Thus, although memory formation requires a complex cascade of biochemical processes (Book 4, Chapter 5), this cascade does not 'represent' the memory within the brain. The function of the cascade is to provide the biochemical housekeeping processes which produce a pattern of new physiologically active synaptic connections. It is this pattern which, most theories assume, represents the memory, and hence the proper level within the brain—the proper language, to pursue the metaphor—to speak of memory is thus that of the cellular systems, the neural networks, not that of biochemistry. Nor of course, will knowing the general principle of how memory is represented enable one to decode a *particular* pattern of connections into a statement such as 'this pattern represents my memory of my fourth birthday' or even 'this pattern represents a chick's memory of pecking at a bitter-tasting bead'. Understanding general principles will not reveal specific content.

Against this background, think back to the questions asked by Tinbergen and McFarland, with which Book 1 began. Consider how they may relate to the different levels and philosophical approaches discussed in Section 8.2. Note how each question accepts the ontological unity of the phenomenon being studied, such as drivers stopping at red lights, or birds sitting on eggs. But each answer to these questions suggests a different approach to explaining the phenomenon. These approaches differ furthermore in terms of the level at which they expect to be answered and the scientific language in which they wish this answer to be framed. In other words, they are part of the epistemological diversity of the brain and behavioural sciences.

Throughout the history of science there has been a great hope that this epistemological diversity will some day end, that there will turn out to be, as physicists currently struggle to offer, a 'grand unified theory of everything'. Such a goal may seem close, or at least thinkable for physicists; it is not close for those dealing with the vastly richer and more complex world of biology and of behaviour. Sociobiologists argued that the growth of the sciences of sociobiology and neurobiology would shortly swallow up economics, sociology, politics and psychology. Theories which attempt such a universalism are likely to do more harm than good as aids to either understanding or acting upon the world. The more modest, but nonetheless vital, assertion of ontological unity will serve better here.

Summary of Section 8.2

The history of the sciences of brain and behaviour is also the history of attempts to relate their findings to philosophical and religious ideas about the nature of the living world, and the relations of mind and brain. Theories of mind and brain include Cartesian dualism which divided the world between, on the one hand, material objects and their biology and, on the other, the immaterial soul or mind; and the 19th century emergence of materialism, reductionism and biological determinism which dismissed the mind as a mere epiphenomenon. All these positions find some support amongst today's philosophers and brain and behavioural scientists, but perhaps the viewpoint which commands most general assent is that of identity theory. This states simply that the mind is the brain, but described in a different language. The task of the neurosciences then, becomes that of learning how to translate between the languages of brain and mind. However, although identity theory asserts the ontological unity of the world, it accepts that there are many different levels of organization of matter, and translating between brain and mind in any particular case is also a task of finding the right level at which to describe the process. Although the world is unitary, there are many different questions one may ask of it, and different types of explanation are sought. These different questions represent the tasks of different sciences, and attest to the epistemological diversity with which the study of brain and behaviour must be approached. The sciences of brain and behaviour are thus very far from the physicists' claim of the goal of a 'theory of everything'.

8.3 From explanation to therapy

How do these different philosophical positions relate to the issues of mind/brain distress and disorder which have been the main topics of this book? The question is seen at its sharpest in relation to the conditions of psychopathology discussed in Chapter 7. That chapter distinguished three broad approaches to explaining mental distress and disorder, which were described as social, life-history and biological explanations. To their main protagonists, these explanations have tended to be seen as incompatible; as that chapter put it, depression may be seen as *either* caused by poor housing and economic insecurity, *or* by vulnerability factors in a person's childhood, *or* by a disorder of transmitter metabolism. Depending on how the condition is explained, a different therapeutic approach will be offered. Thus, if depression is to be explained in purely social terms, nothing short of major social change is likely to eliminate it.

8.3.1 Social therapies?

□ Recall Brown and Harris' findings about the extent and distribution of depression in Camberwell (Chapter 7). What might these findings suggest as the best way of treating depression amongst the women they interviewed?

■ Clearly, improving their housing and economic circumstances, helping them build a better relationship with their partners, and ultimately, eliminating the disparity in social and economic power between men and women which is seen as lying at the root of the greater incidence in the diagnosis of depression in women.

It is not implied here that Brown and Harris themselves explicitly advocate such a programme; they have tended to avoid drawing out the broader implications of their findings. Such large-scale solutions however, may seem unattainable short of social revolution, even though they were the staple of the radical and antipsychiatry movements of the 1960s and 1970s described briefly in Section 1.3.3. At the very least, and many therapists (perhaps especially feminist therapists) would argue that helping a woman to understand the social reasons for her depression is an integral part of providing a support network which will help her to overcome it and so regain some control over her own life.

8.3.2 Psychotherapies

By contrast, psychological life-history explanations of depression would suggest that therapy should focus on the individual. There are two broad approaches to this type of therapy. In the first the therapist endeavours to assist a person to come to terms with their own condition through an increased understanding of their own past which has led to their present state. In the second approach, the past is largely ignored and the emphasis is on using psychological methods based on learning theory to modify present behaviour. The general name for such talking therapies is *psychotherapy*. Within this broad description, there are many different and sometimes conflicting therapeutic schools. Some of the conflicts are about theory; that is, about where one should look in the individual's past or present to explain their distressed state. Some are about practice; that is, the best way of treating the condition. Some therapies involve individual one-to-one meetings between the therapist and the person being treated on a weekly or even a daily basis whereas others involve group activities.

Therapies based on psychoanalytical theory, or its multitude of successors, operate by helping the individual to recover a lost and sometimes repressed past. The aim here is to enable the person to come to terms with crises, usually crises of childhood experience, which have helped shape the current exaggerated or inappropriate, certainly maladaptive, responses to day-to-day circumstances. By recovering such experiences or, in some therapies by attempting to relive them, the person will either overcome, or at least come to terms with them. Such therapies are regarded as holistic in that they treat the whole person as an individual with both a past and a present. They do not focus on just a section of the individual such as a particular behaviour or phobia, or an assumed particular abnormal chemical.

By contrast, cognitive and behavioural therapies are not so much concerned with recovering and overcoming a lost past than with changing a current inappropriate way of thinking or behaving. Thus behaviour therapy is based firmly on the principles of conditioning (see Book 1, Chapter 6). If behaviour is controlled by contingencies of reinforcement and the therapist rewards desired behaviour, and denies reward for undesired behaviour, then according to this type of theory, the reinforced person should alter their behaviour so as to maximize the reward. Such therapies have been used in the treatment of phobias and obsessive-compulsive behaviour, though as will be seen below, in other contexts they have a less benign role. In particular, whatever the intentions of the therapist, the dividing line between denying reward and administering punishment is a fine one.

Psychotherapy arouses strong feelings; it has both its passionate advocates and its critics. Advocates point to its holistic methods and claim that, unlike biological psychiatry which treats a symptom, it endeavours to understand the causes of a person's mental distress and to help resolve those causes. Opponents claim that it is expensive, a 'middle class indulgence', at best useless for those with anything other than trivial problems, and at worst a confidence trick played out on a vulnerable group. The problem is that, unlike the case for more biological forms of treatment such as new drugs, it is extremely difficult to design an objective test of the effectiveness of psychotherapy.

□ Suppose that of 20 people coming into a particular form of psychotherapy for depression, 18 claim to feel substantial improvement over a year of treatment. How could you verify the claim that this proved the effectiveness of the therapy?

■ With great difficulty! You could test whether they had 'objectively' improved over time by administering a questionnaire-type test (such as that shown in Figure 1.1) at the beginning and end of the treatment. However, even if the improvement was statistically very marked, you could not rule out the possibility that they would have improved anyhow over the period by spontaneous remission. An alternative approach would be to compare the 20 people with a group of untreated, but otherwise comparable controls, an approach that is discussed further below.

In the case of drug trials, this is controlled by giving half the people in the trial the new drug and half either an inert substance (a placebo; see Book 3, Section 5.9.1) or a drug of proven efficacy. You can then compare the effects on the two groups of people. The interesting feature of placebos for neurotic subjects is that about 30% of all individuals show improvement with the placebo by comparison with taking no treatment at all, so the new drug has to do better than that in order to be regarded as effective. But it is very hard to think of what could be a placebo in the case of psychotherapy. Control groups in trials of psychotherapy tend therefore to use either a matched 'normal' population, or a group of individuals on a psychotherapist's waiting list who are not yet in therapy. Suggestions that subjects treated by untrained therapists should provide the control group have been dismissed as unethical and neither of the other two types of control can really properly resolve the problem of spontaneous remission.

To return to the orginal question, however, even if individuals improve dramatically under the care of the therapist, this will not prove that the theory behind the therapy is correct. The therapist may have a particularly strong personality or be an excellent healer, who will help people to improve because such a strong belief in the effectiveness of methods is held. Of course, the therapist may nonetheless be self-deluding about the explanation for the improvement.

8.3.3 Biological psychiatry

By contrast with psychotherapy, the rationale of biological psychiatry lies straightforwardly in the assumption that the key to understanding and treating mental distress lies in a disordered biochemistry or physiology. Consequently, if the biology can be adjusted by appropriate intervention, then so will the mind. Treatments offered may be pharmacological or physiological. The most extreme cases have involved psychosurgery, which is a surgical intervention in the brain.

As you have seen (Book 2, Section 4.5.8 and this book Section 7.6.2), most of the effective psychotropic drugs are known to work by interacting with specific transmitter-receptor systems. They may serve as receptor agonists or antagonists, or they may affect the re-uptake sites for the neurotransmitter, or one of the enzyme systems involved in its synthesis or degradation.

Remember that, just because the drugs act on particular transmitter systems, it does *not* follow that it is these systems whose disorder is the cause of the distress. This issue has been discussed in earlier chapters. The essential point is that the drugs may be treating and alleviating the symptoms of a condition rather than its cause. Further, all drugs have multiple effects on many different biochemical systems; putting a chemical spanner into the works of the brain will result in a chain of events. The actual disordered biochemical process underlying the psychiatric condition may be remote from the immediate and obvious site of action of the drug.

An alternative to pharmaceutical intervention is the use of physiological methods, notably that of electroconvulsive therapy or ECT. In the 19th century it was thought that madness could be cured by administration of a vigorous shock, for example by immersing the patient in cold water. Modern methods are in some way an extension of that belief. Thus from the 1920s onwards a variety of techniques to induce prolonged unconsciousness were tried, including insulin therapy, which deprives the brain of glucose and hence produces coma. During the 1930s these treatments were largely superseded by ECT, which orginated with Ladislas Von Meduna in Budapest and Ugo Cerletti in Milan. They believed, wrongly, that schizophrenia and epilepsy were mutually exclusive conditions, so that, if they could generate the convulsions that mimicked epilepsy they might be able to cure the schizophrenia. Early techniques involved inducing seizures with drugs like camphor or metrazol. Later, they turned to the use of electrodes applied to the skull. These produced wild convulsions, including the thrashing of limbs. To control this and to diminish the danger of fractures, ECT is now combined with the use of anaesthetics or muscle relaxants.

In fact it turned out that ECT has no beneficial effect in schizophrenia. However, by the 1950s it was being used in the treatment of depression, and here it *was* claimed to be of benefit. It has remained as a fairly frequently used form of treatment, despite a continuing backdrop of concern about its efficacy and humaneness. ECT causes a massive transient depolarization of nerve cells, and a sort of electrical storm of activity passes across the brain. As you will recall from Book 4 (Section 5.3.2), recent memory depends on the brain's electrical activity, so the application of ECT causes amnesia for events immediately preceding the shock. In addition, there is a random increase in the death of neurons and other rather diffuse effects, including (according to some ingenious tests devised by the San Diego-based neuropsychologist Larry Squire), a degree of loss of older well-established memories. If there *is* a therapeutic effect of ECT, and opinions are divided, then the reasons within neurophysiology are obscure.

Nonetheless, many psychiatrists are convinced that ECT is of benefit in treating certain sorts of depression, particularly postnatal and menopausal depression, which seem extremely resistant to other forms of interventive therapy. Many sufferers speak strongly in its favour, one well-known example here being the comedian Spike Milligan. Others dread it and may see the offer of ECT in a psychiatric hospital as a form of threat—how they will be treated if they don't 'behave' and improve their condition. (This may be regarded as a form of negative behaviour therapy, therefore!) The evidence for the long-term therapeutic effects of ECT is in fact not strong. For example, in 1981 two reports appeared. One described a double-blind clinical trial in which people suffering from depression were given ECT. They showed significant improvement over a matched control group. The other described a similar trial, but in this case the patients were followed up for six months after treatment. This second report found that, initially, there was a small but significant improvement in the rate of remission for people who received ECT compared with the control group, but by the end of the six-month period both groups had improved so greatly that there was no longer any difference between them. Thus no long term benefits could be attributed to ECT, but it did seem to speed up recovery. This should not be underrated, for it may make a vital difference to some depressed patients.

□ Such studies depend on the use of a control group. How might you construct a control to rule out placebo effects for ECT?

■ You could go through the procedure of applying the ECT, by placing the patients in the apparatus and giving the muscle relaxant, but leave the electrical current switched off, so that even those caring for the patients did not know whether they had received the treatment.

Such a procedure, if carried out deliberately, might be unethical without the informed consent of subjects, but there are reports of inadvertent 'experiments' of this sort when it was not known that the equipment was not working at the time it was being used. Even with the apparatus switched off, improvements in the individual's condition were said to occur.

The final form of direct intervention used in the treatment of mental distress is the removal of specific brain regions or the severing of connections between them, a treatment described as *psychosurgery*. You have already had reference to such treatments in the attempts to limit the spread of epilepsy, as in the case of the Montreal patient H.M. (Book 2, Section 11.4.2) and the split-brain procedures described in Section 6.3.1. Psychosurgery was introduced as a treatment for obsessive behaviour, schizophrenia and some forms of acute depression by Egon Moniz in Portugal in the 1930s. Moniz severed the tracts linking the prefrontal cortex with the rest of the brain (a procedure known as prefrontal leucotomy or lobotomy) and claimed it improved the condition of his patients. This claim was to win him a Nobel Prize. The approach was adopted in the UK and the USA in the 1940s. At its peak, some 1 200 people a year were being leucotomized in England and Wales. In the USA, Walter Freeman adopted a conveyer belt approach in which he performed the operation in bravura style, without recourse to anaesthetic. He used a modified icepick through the orbit of the eye in front of audiences of medical students. There was a slow decline in such treatments during the 1950s and 1960s, as follow-up studies suggested that there was little evidence of improvement. Furthermore, the negative consequences, which included emotional and intellectual retardation, could no longer be ignored. During the next decade, however, the use of more circumscribed types of psychosurgery began slowly to rise again; in particular, removal of portions of the limbic system, such as the amygdala (Book 2, Section 10.4.1 and Book 5, Section 4.4.1) have been employed on individuals regarded as being uncontrollably violent. Scarcely surprisingly, such operations have been subject to considerable criticism on ethical grounds, quite apart from the uncertainty of whether any improvement ensues. Improvement is of course a difficult concept. Thus one British psychosurgeon described the effects of psychosurgery on a woman who was a compulsive house-cleaner, spending all day in washing, tidying and retidying her house, and becoming very depressed about it. After the operation and rehabilitation the house-cleaning stopped for a while, but soon re-emerged, with the difference that the patient instead of being depressed was cheerful as she cleaned. This difference constituted, for the surgeon, the evidence of success.

So, perhaps it is not surprising that, just as biological psychiatrists are critical of psychotherapists, psychotherapists tend to be critical of biological psychiatry. They claim that it tends to medicalize a problem which is properly treated at the level of the whole person; that it subjects patients to powerful drugs or physiological regimes which may merely make them more docile without affecting the root cause of the problem. Above all, biological psychiatry is seen as offering a powerful means of social control, and there is often a difficulty in determining where the borderline between therapy and control should be placed, as you will see in Section 8.4 below.

8.3.4 Integrating therapies

The descriptions of differing approaches to therapy and treatment in the last three sections imply that there is little common ground between them. In fact, although it is true that strong advocates of either psychotherapy or biological psychiatry also tend to be strong critics of the opposing approach, many psychiatrists are more relaxed, and offer the possibility of mixed methods. In a hospital environment in particular, it is relatively common to find patients being treated by eclectic combinations of psychotropic drugs and psychotherapy. Such a mixed approach is best accommodated within an integrated identity theory of the type discussed in Section 8.2.4. If mind language and brain language are indeed both talking about the same unitary phenomenon, not only should changes in the brain (brought about for example by the use of psychotropic drugs) also imply changes in mind and behaviour, but changes in mind and behaviour (brought about by psychotherapy), should also be associated with changes in brain biochemistry and physiology. So, if depression is characterized by decreases in the activity of serotonin re-uptake sites, either in the brain or in platelets, and antidepressants work by interacting with these sites, then psychotherapy should also increase the activity of the sites in depressed people. Such a study has in fact been conducted at the Open University by Sarah Willis and Steven Rose, and Figure 7.3 in the previous chapter is derived from their findings. This figure shows the increase in the activity of the serotonin re-uptake sites (as measured by the binding of the drug imipramine) with time in mildly depressed people going through psychotherapy, but *not* using psychotropic drugs, at a London therapy centre. By contrast, the control group, derived from a 'normal' group of volunteers matched for age and sex with the clients, show no significant change in either biochemical or behavioural measures over this period.

□ Look back at Figure 7.3 (page 214). Does it prove that psychotherapy is effective?

■ No, for the reasons discussed in 8.3.2 above; the improvement might have been due to spontaneous remission.

Nonetheless, this study and its findings are suggestive of the possibility of bridging the theoretical gap between psychotherapy and biological psychiatry.

Summary of Section 8.3

Therapies available for mental distress fit within the framework of explanations offered in the previous chapter, being divisible between social, psychological life-history approaches, and those based on strictly biochemical and physiological assumptions. They also relate to the theories of mind and brain contrasted in Section 8.2. Although protagonists of psychotherapy and biological psychiatry are often strong critics of one another's positions, the integrative theory of Section 8.2 does offer a way of reconciling them, and some evidence pointing towards such a reconciliation is offered in Section 8.3.4.

8.4 Therapy and control

The technologies of biological psychiatry and of psychotherapy are powerful, and offer the prospect of hope for the alleviation of suffering from brain and mind distress which affects such a large proportion of the population. Like all powerful technologies, however, their use is two-edged. As has already been suggested, the borderline between therapy and control is a fine one. Indeed, for some schools of thought, some political groups and some religious sects, there is no borderline at all, and all psychiatric interventions are seen as an abuse of authority. One does not need to take such an absolutist position however, to recognize the problem. The Calvin Johnson case (Section 7.5.1), would seem a good example of the use of psychiatric classifications and treatments in the service of racist views about black people and Rastafarianism. It is by no means an isolated or artificial example, as data collected by the Institute of Race Relations in 1991 has shown.

□ How would the disproportionate diagnosis and treatment of depression among women be regarded within this framework?

■ As exemplifying masculine power to discount women by labelling them as mad.

8.4.1 Biological psychiatry and control

More extreme examples than these are unfortunately not rare. In the 1960s and 1970s it became common practice in the Soviet Union, for example, to use psychiatric measures against political dissidents. Criticism of the existing social order became regarded as a symptom of madness and the standard diagnosis was schizophrenia. Dissidents were placed in special forensic psychiatric units and treated with psychotropic drugs. A good inside account is given by the Russian dissident biochemist Zhores Medvedev in his book *A Question of Madness*, published in 1974. The abuses became so flagrant that for a period the Soviet Union was expelled from the World Psychiatric Association. However, it has not only been in the former Soviet Union that a confusion arose between madness and dissidence. The 1970s were also a period of major urban unrest in the USA. It was in the aftermath of some of the worst of the ghetto riots that a group of USA psychosurgeons argued that the only explanation for the anger and violence of the rioters must be that they were suffering from a brain disorder. The most likely disorder was considered to be some malfunction of the limbic system which made them abnormally aggressive. The proposed solution to inner city violence was to carry out amygdalectomies (removal of the amygdala) in selected 'ringleaders'. For a time at least this programme was put into practice within the prison system in California.

These are perhaps isolated cases. But the vast scale of the use of psychotropic drugs presents much broader problems. During the 1970s and 1980s the consumption of tranquillizers and antidepressants among the general population approached epidemic proportions. There were many cases of addiction and the brain damage that resulted from the long-term uses of these drugs had become apparent. (Prolonged use of chlorpromazine and related drugs results in destruction of some neural pathways in the brain and as a result a disorder called tardive dyskinesia, which is characterized by abnormal and uncontrollable movement and tics.)

At the same time, more and more categories of what might in the past have been regarded as normal behaviour have become diagnosed as representing a medical condition, a brain or mind disorder, in need of treatment. From the 1960s, USA clinicians began to refer to a new disorder, initially called hyperactivity but later given the grander name of *minimal brain dysfunction*, or MBD. There were no 'organic' symptoms related to this condition, which is supposed to be recognized purely in terms of behaviour. MBD is described as a disorder primarily affecting children (mainly boys) in the age group of 9–13 years. MBD children are poor learners at school, and are badly behaved and inattentive in class; easily distracted, they also seem to distract the teachers. In fact, by the 1970s MBD was even being diagnosed among schoolchildren in the USA on the basis of schoolteachers' reports alone. Its protagonists claimed that some one in ten children were suffering from the condition. The favoured treatment was the use of an amphetamine-like drug, Ritalin, which seemed to have the paradoxical effect of calming the hyperactive children. By the beginning of the 1980s, according to some reports, up to 600 000 schoolchildren in the USA were taking daily doses of Ritalin, at least in the school term, as MBD seemed to have the strange property of remitting during school holidays.

By the 1990s, when this course is being written, the enthusiasm for MBD as a disease category is less apparent than it had been a few years previously. However there are signs that a new disorder, this time affecting not schoolchildren but those in their forties and above, is coming to the fore. It is called Age-Associated Memory Loss (AAML), and its diagnostic signs are described as a self-observed increasing weaknesses in memory, for example for names and telephone numbers. Despite some resistance from psychologists and neurobiologists, AAML has now been accepted as a disease category by the USA regulatory body, the Food and Drug Authority, and the pharmaceutical companies, conscious of a potentially huge market, are developing new generations of drugs, or new uses for old agents, generically described as 'cognitive enhancers' or, more colloquially, as 'smart drugs'. Because of the evidence that Alzheimer's disease (Section 3.7.1), is associated with catastrophic loss of memory, and also involves the loss of neural pathways involving the cholinergic system, the agents being advocated as memory enhancers include a variety of cholinergic agonists. Others may affect attention and arousal, processes which are necessarily involved in learning. (It is because of their effect on such processes that, in the days when they were still readily obtainable on prescription, amphetamines were sometimes used by students revising for examinations.) How any such agents might be expected to help in retrieval of memories once established is, however, quite unclear. In any event, the illicit and semi-licit sales of drugs believed to act in this way is already very high, and if the pharmaceutical companies have judged their market correctly, the potential far exceeds that which MBD provided for Ritalin earlier. The MBD categorization never proved as popular in the UK as it did in the USA, and Ritalin was never prescribed in the same way here; what will happen with the new 'cognitive enhancers' remains to be seen. It should be emphasized however, that at the time of writing of this course there is no clear evidence for the genuine existence of such a catch-all disease category as AAML, and none at all that any of the drugs in current circulation with claimed properties as 'cognitive enhancers' have any significant effect on human memory processes.

8.4.2 Psychotherapy and control

Emphasis on the controlling power of the methods of biological psychiatry and the interests of the drug companies should not distract attention from the parallel powers of psychotherapy. While many such therapies put those undergoing them into a peculiarly vulnerable position in relation to their therapists, it is the use of psychotherapeutic techniques as a general mode of social control which is of concern here. Most attention has been directed towards behaviour therapy, because of its explicit uses of contingencies of reinforcement in its efforts to change a person's behaviour. Although the original intent of such methods may have seemed benign, in their general application the techniques of behaviour therapy readily transmute into procedures which are hard to distinguish from straightforward rewards and punishments. Thus behaviour modification is used in 'special schools' in the UK—that is schools for behaviourally disturbed children—where Ritalin might have been the treatment of choice in the USA. Such schools operate 'token economies'. 'Good behaviour' is rewarded by tokens which may be exchanged, say for chocolate, or permission to leave the school for a free afternoon. 'Bad behaviour' on the contrary is punished by denial of certain freedoms or privileges. Similar techniques, though with much more drastic extremes of punishment, such as solitary confinement, have been described as in use in a number of prisons. However, in 1991 there was a public scandal concerning the use of such confinement techniques, described as 'pin-down regimes', for youngsters who had been taken into care and placed in special institutions.

8.4.3 Ethical questions: mad or bad?

The motives of those developing and applying such therapies may be impeccable; they raise however, very sharp questions about the borderlines of the acceptable. Enthusiasts see them as cornerstones in the building of tranquil societies. José Delgado, who, first in the USA, later in Madrid, pioneered the use of implanted electrodes into specific brain regions of his patients as a way of modifying their behaviour, called his dream the creation of a 'psychocivilized society'. Skinner saw behaviourism as offering the possibility of a human society which went 'beyond freedom and dignity'. Yet one has to ask: whose civilisation; whose freedom; and whose dignity? Doubtless both Delgado and Skinner would have condemned the political psychiatry of the Soviet Union. However, Delgado would not have seen biological psychiatry as anything other than benign whereas Skinner would have seen the use of behaviour modification in Western societies as desirable. The problem is that in both cases, the question of who decides 'what is desirable', resides with those having the power and the technology at their disposal. In short, 'who controls the controllers'?

Of course such a question goes beyond issues of what has been called 'psychopolitics' and lies at the heart of democratic theory. Because the manipulation of mind and brain are so central to our very notions, as humans, of who we are and the freedom we have to choose and to act upon the world, the issue is perhaps more sharply posed here than anywhere else.

To take a familiar example, in murder trials a jury is often called upon to decide whether the accused was 'responsible for his actions'; that is, was sane and bad and so ought to be punished. If, on the basis of psychiatric evidence the accused person is seen not to be sane, but mad ('acting while the balance of the mind was disturbed'), this becomes a matter of medical treatment rather than punishment. One of the most notorious of such cases in recent years was that of the 'Yorkshire Ripper', Peter Sutcliffe, whose defence lawyers called on psychiatric evidence to prove that he was suffering from schizophrenia. It was not enough to argue that to commit such crimes must itself be a sign of madness; other diagnostic criteria were called for. In the event the jury was clear that the crimes deserved punishment rather than that their perpetrator required psychiatric treatment. But to make such distinctions requires that in everyday life one adopts a pragmatic Cartesian dualism, separating the mental world of 'intentionality' from the biological world of uncontrollable 'brute impulses' or disease, even though at the same time it is clear that rationally such a distinction is impossible.

Summary of Section 8.4

The new techniques of biological psychiatry (from drugs to electrodes to psychosurgery) and of psychotherapy (notably behaviour therapy and behaviour modification), offer the prospects of controlling the minds and brains of large sectors of the population. Such control may have therapeutic intent but it also can be used or abused for more explicitly social and political ends. Examples include the psychiatric incarceration of political dissidents in the Soviet Union, the medicalization of the protests of the inner city rioters in the USA, and the uses of behaviour modification as a means of reward and punishment in schools and prisons in the UK. The tendency to redefine behaviours as abnormal and therefore representing new disease categories is exemplified by the case of 'Minimal Brain Dysfunction' in USA schoolchildren and 'Age Associated Memory Loss' in the mature population. These developments offer far-reaching prospects for social control and raise major ethical and political questions.

8.5 New knowledge, new technology, new dilemmas

The issues raised so far are those which have arisen from actually existing technologies. Yet the rate of growth of the brain and behavioural sciences is so great that new problems are constantly emerging, and concerns which might have been seen a few years ago as science fiction speculations are now becoming imminent realities. This chapter will consider just two, both of which have already been introduced earlier in the present book. These are brain transplantation (Section 5.5) and gene therapy (which uses the techniques of genetic engineering; Section 4.4.1).

8.5.1 Prospects and implications of brain transplantation

Intracerebral transplantation in mammals dates back to the 1890s, when succesful brain grafts were achieved by Gilman Thomas. However, it was not until the 1970s that the issue was taken up seriously and only in the 1980s did the transplantation of cells and neural tissues into the developing and adult CNS become a widely used experimental tool. The research purposes of such studies were clear that they could be used to help understand the relationships between genes and environment during development, the roles of particular cell types, and the mechanisms of growth, plasticity and regeneration. In the 1970s and early 1980s, Anders Bjorklund and his colleagues in Lund, Sweden, were able to demonstrate that fetal brain grafts could establish functional afferent and efferent connections. Fred Gage (in the USA and at Lund), Stephen Dunnett (in the UK) and many others followed up this work by showing that, in animals with experimentally-induced lesions and hence behavioural deficits, some compensation could be achieved by fetal grafts of appropriate pituitary, hypothalamic, brainstem, dopaminergic or cholinergic neurons. The question of how, at the cellular level, the grafts might work still remains controversial. Nonetheless, it was obvious from the earliest days of this work that there could be major clinical implications if transplantation could succesfully be achieved.

□ Try to name some potential clinical applications for treatment or repair of damage or disease from your study of this course.

■ You might have thought of some of the following:

Repair of damage caused by stroke (Section 5.5)

Reversal of degenerative conditions such as Parkinson's disease (Sections 4.3.1 and 5.5)

Treatment of Alzheimer's disease (Section 3.7.1)

Treatment of the brain damage due to alcoholism—Korsakoff's syndrome (Section 5.2.6)

Although the clinical utility of grafting is still not fully established, it is clear that the technique raises major ethical issues. As the material used for grafting is derived from aborted or miscarried fetuses, who 'owns' such material? Does the woman who carried the fetus have the right to decide what should happen to its tissues? Is there a danger that women could be pressured into having abortions in order to provide a supply of fetal brains for transplantation into elderly or damaged people? These questions have been asked, and how they are answered generally reflects people's more general attitudes to abortion and to the experimental use of fetal material. Those opposed to abortion in principle, or who believe in 'fetal rights', have a different response from those who argue that the gift of tissue from an aborted fetus is rather like the gift of blood—that is, if the tissue can be used to benefit another human, then it should be. Other questions surround the extent to which it is possible for people who are candidates for experimental treatment by transplant technology to give informed consent to the operation. How can the risks and benefits of the treatment be assessed? What degree of funding priority should such work be given, granted the overstretched resources for medical and health care and for research? Transplant operations are always going to be expensive. Do they then give the best social value for money in a world of scarce resources, as opposed to investing in public and child health programmes. Or do they represent a misorientation of funds towards glamorous technology likely to benefit only a few—including the surgeons who conduct them? In the event that transplantation becomes a prospect in the treatment of Alzheimers' disease, granted the many tens of thousands of sufferers from the condition, and the small number of operations that are likely ever to be possible in any year, how are decisions as to who is a candidate for treatment to be made? Will they be cash-led, as in the case of a private health care system, or in some way determined by social need, as is the aim (if not always the practice) of a public health service?

8.5.2 The new genetics

As explained in Chapter 4, there are a number of disorders of brain and behaviour whose overriding causation is genetic. Although they only account for a small fraction of all known cases of brain/mind malfunction, with many others being primarily determined by chromosomal, developmental or environmental contingencies, the genetic disorders are dramatic and often devastating in their effects. Although the effects of some, like phenylketonuria, can be mitigated by environmental manipulation (such as diet) for others there is no known phenotypic treatment. Amongst the most severe, (discussed in Sections 4.3 and 4.4), are Huntington's and Lesch-Nyhan's diseases. Chapter 4 opened the discussion of the potential offered by modern genetics for the understanding and treatment of such conditions. This section will look at that potential in a slightly broader context and also consider its possible ethical implications.

As you will recall, the exact gene responsible for such conditions is not always known; more often what is known is the linkage of that gene. In the case of Lesch-Nyhan disease the gene itself is in fact known, whereas in Huntington's disease the gene is not known, but the linkage is.

□ Recall what linkage means.

■ It means that it is known on which specific chromosome the gene of interest is located, and that close to it there is a marker gene whose presence can be detected at the molecular or biochemical level. The two genes lie so close that they remain linked together during the recombination of alleles that goes on during meiosis (gamete production).

□ What are the implications of there being a known linked marker for detecting the mutant allele?

■ It means that a pregnant woman can be offered a screening process to detect the possible presence of the abnormality in the fetus. The same possibility is offered if the actual gene is known, as in the case of Lesch-Nyhan syndrome, provided it can be detected at the molecular or biochemical level.

The woman can thus be advised on the likelihood of her carrying a fetus with the genetic defect, and can be offered an abortion. She may accept this choice, or, of course, elect to carry the damaged fetus. Either decision brings with it a degree of moral and social responsibility. In the case of conditions like Lesch-Nyhan, where a child is born with little chance of surviving to adulthood, and with the dreadful behavioural traits of self-mutilation that at present seem beyond treatment, there may be little doubt as to what to do. But in the case of Huntington's disease, the offspring will be born and will live an essentially normal life until adulthood before suffering a decline into the symptoms of the disease described in Section 4.3.1. Here the choice for the parent may not seem so straightforward because there is always the *hope* that, by the time the child is adult, some form of treatment may have been devised. Yet the choice is not one for the parent alone, for the Huntington's sufferer is destined to become chronically incapacitated and therefore to be a substantial cost to the state in terms of care and medical resource. It is said to be hard for a person carrying the Huntington's genetic abnormality to obtain state employment in Germany, for example, as the government has decided that it does not want to carry the inevitable extra burden of prolonged invalidity pension as the disorder runs its course. It may be in the social interest, therefore, to encourage an abortion which the prospective parent may not want.

There are other subtler implications, too. Screening the fetus for Huntington's disease will also reveal whether one or other parent is also a carrier and therefore themselves doomed. This is something that they themselves may not have known and may not wish to know. Living in the shadow of a seemingly inevitable and terrible disease is not easy. What, for that matter, of a child born in the knowledge that its days are thus numbered? Would it not be preferable to be unaware, or even just not certain, than to live with that certainty? Such considerations are among those that have led some members of families at risk to refuse to be tested for the condition. But might the time come when employers or insurance companies will begin to demand that tests for possible genetic abnormalities be carried out before a person is employed or insured? There have been suggestions that this is already occurring for some conditions in the USA. The issues here are rather similar to those which have made insurance companies hostile to insuring people who have had tests for the AIDS virus (HIV), whether or not the tests have proved negative.

Conditions like Lesch-Nyhan or Huntington's are in some senses relatively easy to discuss because there seems to be an almost certain relationship between the existence of the genetic abnormality and the phenotypic condition. Other cases are more complex. Down's syndrome is a chromosomal abnormality (Book 4, Section 4.5.2) which is easy to detect and leads to characteristic phenotypic abnormalities including intellectual impairment. The degree of impairment is variable, so a woman will not be able to predict the extent of the deficit her Down's syndrome fetus might be carrying. In addition, Down's syndrome children, though impaired in some ways, are also remarkable for their sweet and placid disposition. The proper action to follow if you were told that you were pregnant with a Down's syndrome fetus, therefore, is by no means clear and is likely to be a source of agonized debate for any woman and her partner.

What is already true today about the complex consequences of screening for conditions like Down's syndrome, will apply even more strongly if reliable linkage to known markers should be found for any of the conditions, such as manic-depression or schizophrenia, discussed in Section 7.7. At best the advice that a parent could then be given is that the fetus has an $x\%$ probability of growing up to develop the condition. How should one act on that $x\%$? How far would the knowledge that your child might become schizophrenic affect your treatment of it as it grew up and therefore perhaps even help precipitate the condition you were trying to avoid? These are rhetorical questions, for the answers are unknown. But unless it is possible to provide some guidance as to how to act, the case for genetic screening for such conditions is a weak one.

Genetic counselling, aided by prenatal screening, offers the prospect of what has been described as *eugenics*—the elimination of 'undesirable' genotypes. But potentially the new genetics offers more than this. One prospect held out by the identification of the genetic abnormality associated with a particular condition is that the discovery will point to the gene product concerned and hence to the biochemical abnormality, which might then be corrected, as in the case of phenylketonuria. A more dramatic and much canvassed possibility however, is that the abnormal gene itself could be removed or replaced. The techniques for the insertion of new genes (that is, new DNA sequences prepared by genetic engineering) into mammalian embryos now exist. It would in principle be possible to ensure that all the cells of a growing embryo were corrected in this way. This is called *somatic gene therapy*, because its effect is to alter a particular gene in all the cells of the organism except the germ cells—those that produce the gametes—so there are no effects on the offspring.

After many initial reservations, licences are beginning to be issued for the experimental use of somatic gene therapy in humans for certain conditions. Two early candidates are cystic fibrosis and hypercholesterolemia; these are very common and serious genetic disorders, neither directly related to the brain or nervous system. Legislation controlling such work varies from country to country. The Germans have emphatically banned it, for instance, whereas it is now approved in the USA.

Somatic gene therapy is but the first step down the road of human genetic engineering, however. It is also in principle possible to insert new genes into the germ cells themselves. During the 1980s, the DNA responsible for coding for rat growth hormone was inserted into the genome of mice; the mice gave birth to offspring which produced the rat growth hormone, and hence grew considerably larger than their normal litter-mates. This direct manipulation of the genetic material, thus affecting future generations, would, if applied to humans, be described as *germ line gene therapy*.

Germ line gene therapy is regarded as unethical and is firmly banned everywhere. It is too early as yet to see whether the techniques of somatic gene therapy will prove of any practical use in humans; the uncertainties and the failure rates are bound to be high. One can imagine that in due course, once the responsible genes have been identified and localized, it may prove applicable to conditions like Huntington's and Lesch-Nyhan disease, though it is very difficult to see its wider application to any conditions other than such relatively straightforward and very damaging single gene defects. This is even more likely to be the case with germ line therapy; even the manipulated mice described above do not always subsequently 'breed true' and genes can be rejected or their effects cancelled out by the cell just as much as the immune system can ensure that tissue grafts are rejected. Genetics, molecular biology and development are more subtle processes than the reductionism of the genetic engineering enthusiasts often allows. It is, however, such new techniques that have led to long debates about the almost science fiction-like prospects of genetic manipulation not only to eliminate or repair damaged genotypes but to produce desired ones, for instance, for high intelligence.

☐ From your study of this course, do you think it might ever prove possible to use genetic engineering techniques to prevent an infant developing schizophrenia, or to ensure that it develops high intelligence?

■ Neither would seem possible. Insofar as either schizophrenia or intelligence may be under genetic influence, they are not the product of a single gene but of many genes, and it has already been said that the techniques of somatic gene therapy are likely to be confined, for the foreseeable future, to the modification of single genes. Further, the assumption behind the idea is that there is a simple relationship between genes and phenotypic characters, and you should have recalled the point, made several times in the course (e.g. Book 1, Section 3.2.3; Book 4, Chapter 4 and this book, Section 4.1) that there is no such direct one-for-one connection. The effects of any particular gene depend not only on the environment in which it is expressed but on the genetic background against which it is expressed.

One final point in this context; as with the transplant technologies, the potential of genetic engineering, though it has captured the imagination of researchers and lay public alike (and with the imagination, massive funding), has to be considered against many other aspects of the health budget. Lesch-Nyhan disease, for example, affects 1 in 20 000 live births each year in England and Wales. Though each case is tragic, the numbers are small. By far the commonest causes of neurological damage and hence later brain and mind disorder are complications in pregnancy, sometimes induced by environmental factors such as diet (for instance, lack of essential vitamins, such as folic acid, now known to be associated with spina bifida), or smoking, or poor living conditions. Perinatal mortality (deaths of young babies at or just after birth) is still quite high in the UK by comparison with Sweden for example and, furthermore, unlike the situation in Sweden, the rates are much higher among poorer people than the better-off. The death rate of babies in Bradford in Yorkshire in 1989 was four times that in Cambridge. Putting more funds into improving the health and welfare of pregnant women and of children may be a better bet in improving the brain and mind health of future generations than a continuing expansion of high technology medicine and genetic intervention for a few.

Summary of Section 8.5

This section reviews the future prospects of new technologies such as neural transplants and genetic engineering. It concludes that the technical obstacles to the succesful utilization of these techniques in the treatment of brain disorders are still formidable, and there remain serious questions both of an ethical nature, and about the social priorities involved in investing in these techniques rather than in measures more directly related to public health.

8.6 The future of the brain and behaviour sciences

As the end of the 20th century approaches, the brain and behavioural sciences are experiencing an incredible flowering. This is the 'Decade of the Brain' in the USA; it is the decade of the 'Human Frontier Science Program' for Japan. In Europe there is now a 'European Neuroscience Programme'. The numbers of active scientific researchers in the field has grown dramatically; the annual meetings of the American Society for Neuroscience attract between twelve and sixteen thousand participants. There are literally thousands of scientific journals regularly reporting the results of research in the fields of brain and behaviour. Those who work in this field feel alternately elated by the flow of new and exciting results and techniques, and desperate as they sense they are drowning beneath the flood. Seemingly intractable problems are resolved almost daily by the application of methods deriving from molecular genetics, positron imaging and computer simulation which would have seemed inconceivable only a few years ago. If unifying theories still lag behind the accumulation of data, nonetheless the excitement and sense of purpose amongst the research community is tangible. That the new sciences will add to the understanding of what it is to be human (and animal) and will clarify age-old mysteries of mind and consciousness is clear. Will they improve human welfare, help ensure the healthy development of creative children and the serene and wise old age of adults? That, sadly, is a different question, and requires answers that are beyond the scope of this course to provide.

Objectives for Chapter 8

When you have completed this chapter you should be able to:

8.1 Distinguish between dualist, reductionist and integrationist (identity theory) approaches to understanding the relationships between mind and brain. (*Question 8.1*)

8.2 Contrast the approaches of psychotherapy and biological psychiatry to the treatment of mental distress. (*Question 8.2*)

8.3 Describe the types of experimental design required to test the claims made for particular therapies, including the use of blind studies and placebos. (*Question 8.3*)

8.4 Discuss the therapeutic prospects and ethical dilemmas resulting from the introduction of potential new modes of intervention into human brain and behaviour, in particular behaviour modification, behaviour-altering drugs, brain transplants and genetic engineering. (*Question 8.4*)

Questions for Chapter 8

Question 8.1 (*Objective 8.1*)
'She is in love'. Give (a) a dualist, (b) a reductionist and (c) an integrationist interpretation of this sentence.

Question 8.2 (*Objective 8.2*)
Consider the case of Calvin Johnson, discussed in Section 7.5.1. How might (a) a psychotherapist, and (b) a biologically-inclined psychiatrist approach the problem of treating his condition.

Question 8.3 (*Objective 8.3*)
A claim is made that a new drug may be of particular value in alleviating the symptoms of depression. What type of experimental procedures would you need to go through in order to test this claim? Consider the role of double blind and placebos in such an experiment.

Question 8.4 (*Objective 8.4*)
As a member of the committee of a research funding charity, you receive a grant application to explore the effects of transplants of fetal brain tissue in alleviating the symptoms of obsessional-compulsive behaviour in adults. What types of consideration should you bring to bear in considering whether to fund such research?

Further reading

Churchland, P. S. (1986) *Neurophilosophy: towards a unified science of the mind-brain*, MIT Press, Cambridge, Mass.

Institute of Race Relations (1991) *Deadly Silence*, Institute of Race Relations, London.

Kitcher, P. (1985) *Vaulting Ambition*, MIT Press, Cambridge, Mass.

Medvedev, Z. and Medvedev, R. (1974) *A Question of Madness*, Macmillan.

Rose, S. P. R., Lewontin, R. C. and Kamin, L. (1984) *Not in our Genes*, Penguin, Harmondsworth.

Wilson, E. O. (1975) *Sociobiology: the new synthesis*, Harvard University Press, Cambridge, Mass.

GENERAL FURTHER READING

Thibodeau, G. A. and Patton, K. T. (1992) *The Human Body in Health and Disease*, Mosby–Year Book, Inc., St Louis, Missouri.

ANSWERS TO QUESTIONS

Chapter 1

Question 1.1

There are many problems with such a study. First, the control group is not hospitalized, so is likely to have a very different diet and life-style from the hospitalized group. So it would be necessary to match the patient group with a similar but non-schizophrenic hospitalized group. Second, there is no information as to whether the schizophrenic group is taking medication to treat the condition. If so, the metabolite could be a breakdown product of the drug rather than a factor in the disease. Third, the metabolite could be a symptom of, or consequence of, the schizophrenic condition rather than a cause (like a runny nose with a cold). You can probably think of other flaws in the study too.

Question 1.2

It is an inadequate model in that no human baby could really be raised in the conditions of extreme deprivation of Harlow's monkeys. Humans also have the capacity to communicate by speech, so their social interactions during development are very different. Human depression is of course a state of feeling, emotion and orientation that one cannot be sure exists in any readily analogous way in monkeys, where one can only infer what is felt from how they behave. Nonetheless, it might be one of the best models that can be devised.

Question 1.3

He might have done, but you cannot be sure. There might be some common feature in his, his father's and his grandfather's environment which is responsible. A good example is the condition known as 'Derbyshire neck'. This is a condition that used to be common in that county, characterized by a swelling of the thyroid gland and also by apparently slow and simple-minded behaviour. The condition was traced to a lack of iodine in the drinking water, leading to a loss of function of the thyroid gland (a gland in the neck producing an iodine-containing hormone, thyroxine, which controls aspects of body metabolism), and it disappeared in the affected individuals when iodine was added to their water.

Chapter 2

Question 2.1

Life expectancy is the average length of life in the species; lifespan is the maximum length of life.

Question 2.2

Below are three reasons. Even if elderly people suffering from diseases are excluded, changes in function, e.g. deterioration in hearing, accompany the ageing process and cannot be separated out. It is also difficult to separate out ageing from the effects of environmental factors such as life-style. Elderly individuals vary so much that it is difficult to know what is 'normal'.

Question 2.3

By using memory aids; by expending more time and effort; by opting out of very demanding activities; by adopting helpful strategies.

Question 2.4

The experimental method is objective and rigorous. Extraneous factors can be carefully controlled and precise measurements obtained, but it is rather artificial. The everyday approach is more 'natural', but less objective. Performance in everyday situations is affected by many variables which cannot be controlled.

Question 2.5

Playing chess, vocabulary, and wisdom are all relatively unaffected by ageing; memory of routes, judgements of colours and space, and ability to do two or more things at once are likely to be impaired by ageing.

Question 2.6

Memory tasks are easier if the older person has plenty of time to memorize the material and is given practice in the task. It also helps if the material is familiar. If memory is tested by asking subjects to recognize items, rather than to recall them, older people do better.

Question 2.7

Elderly people might under-perform in intelligence tests because of anxiety. The test materials and the tasks are unfamiliar and many of the tests are scored on speed of performance.

Question 2.8

Network models show how factors like memory capacity and speed of processing are not necessarily alternative explanations, but might all be operating within the network. Network models are also closer to the neurophysiology of the brain than the box-and-arrow type models in Figures 2.6 and 2.7.

Chapter 3

Question 3.1

Post-mortem material undergoes non-specific damage and thus may not be a true representation of the situation in life.

Question 3.2

(a) 1 and 2 both decrease in the normally ageing brain; 3 and 4 both increase.

(b) In the pathologically ageing brain 1 and 2 show a greater decrease, 3 a greater increase and 4 is not significantly different from that in the normally ageing brain.

Question 3.3

Buell and Coleman developed this model in the late 1970s. There is considerable evidence to show that some neurons die in normal elderly subjects, and that dendritic trees of surviving neurons are more extensive than in younger, adult subjects. Buell and Coleman suggested that, during ageing, compensatory mechanisms of synaptic growth and extension of surviving dendrites may act to maintain brain function.

Question 3.4

PET scanning on a routine basis has revealed that CBF declines slightly during normal ageing and that this reduction is more dramatic when accompanied by intellectual deterioration. It is thought that the reductions in CBF that occur with age and with dementia may be due to similar underlying structural changes but that these changes are much greater in demented people.

Question 3.5

Brain scanning techniques are non-invasive and relatively simple and can be performed quickly to produce immediate results.

Question 3.6

One hypothesis assumes that a population of 'anti-self' white cells may arise later in life as a result of mutation. There is evidence to suggest that mutations occur more often and are corrected less efficiently with increasing age. The second hypothesis assumes that the immune system always possesses an 'anti-self' element, but that in younger life its effect is inhibited whereas with increasing age the inhibition becomes less effective.

Question 3.7

Senile dementia is a neurodegenerative disease mainly involving the cerebral cortex, which occurs in individuals of 65 years or older.

Question 3.8

The main symptoms displayed by people with Alzheimer's disease are increasing memory impairment and perception disorder. They may also suffer from bouts of intense jealousy, show violent reactions to everyday events and they often become incontinent.

Question 3.9

Senile plaques are masses of abnormal, unmyelinated neuronal processes, thickened axons, degenerating dendrites and enlarged synaptic endings, which may involve many neurons. Lipofuscin granules are generally present and plaques often have dense, fibrillar cores and an increased number of mitochondria. Neurofibrillary tangles are composed of closely-packed neurofilaments that displace other cellular organelles within individual neurons. They too often contain lipofuscin deposits.

Chapter 4

Question 4.1

This observation may reflect differences in the nature and duration of the original exposure, but it also suggests that other causative factors are involved, which might be environmental, genetic or both.

Question 4.2
This is because lead continues to be stored in the body even after exposure has stopped so that the damage to the nervous system is permanent.

Question 4.3
The second child could be a phenocopy of Wilson's disease. This could arise from excess intake of copper.

Question 4.4
Your diagram should look like Figure 4.5.

Question 4.5
The person would almost certainly have suffered from hyperkinetic movement disorders. There would be reduced inhibitory activity in the globus pallidus with the ultimate effect of an increased activity from the thalamus to the cortex.

Question 4.6
All of the factors are implicated. Different individuals with different genetic backgrounds (a), have different sensitivities to the levels of potential neurotoxins (d). Some individuals are more efficient than others at metabolizing harmful substances (c) and/or storing them (b).

Question 4.7
Options (a) and (c) are correct.

Question 4.8
The marker gene, G, is not sufficiently reliable as a marker because it separates from the gene for muscular dystrophy in the pedigree.

Question 4.9
The correct option is (b). Both (a) and (c) are symptoms of people with Lesch-Nyhan disease but are also symptoms of other diseases.

Question 4.10
Although the primary function of the gene is known, that is, it determines the structure of the enzyme HPRT, it is not known how a deficiency in this enzyme causes neural disorder. Thus, there is a large gap in knowledge between the metabolic abnormality and the cause of the neurological disorder.

Chapter 5

Question 5.1
(a) is wrong. A stroke can result from either an occlusion or bleeding, but in each case an artery and not a vein is involved. All other statements are correct.

Question 5.2
(a) 3, (b) 1, (c) 2 and (d) 2. (See Figures 5.2 and 5.4.)

Question 5.3

(f) because neurons do not divide in adults. All the other mechanisms listed could take place to some extent.

Question 5.4

New connections involving the same two regions of the brain carry similar information to that carried by the damaged neurons. However, if neurons of damaged pathways were to make connections with axons from other brain regions, they may well receive anomalous information. This is particularly true of regions which are topographically organized. The same would follow if the repair involved the release of silent synapses from inihibition.

Question 5.5

The synapses in the striatum are not of the classical type, being large and diffuse with terminal varicosities. Other regions of the brain have precise connections, making it highly unlikely that grafting will be successful.

Chapter 6

Question 6.1

Statement 1 (a) and (b). Gall included language on the left (faculty 33) while the localizationists originally established their position with the localization of language in the left temporal and frontal lobes.

Statement 2 (a) and (c). Both the phrenologists (for example, Gall's passionate widow) and diffusionists (Lashley's law of mass action) agreed that it was the volume of tissue that was important to the performance of a behaviour pattern.

Statement 3 (a) and (b). Both the phrenologists and localizationists claimed that the integrity of specific regions is important for specific cognitive functions.

Statement 4 (a). Only the phrenologists made the claim for innate faculties localized in discrete regions of the cortex.

Statement 5 (c). The diffusionists originally thought that neurons were interconnected and that the cytoplasm was continuous throughout the nervous system, allowing flow of information bi-directionally along axons.

Statement 6 (a). Only the phrenologists proposed a reliable relationship between the regional development of the cortex and the curvature of the overlying cortex.

Question 6.2

The list in support of student 1 should contain 2 items:

(a) The phrenologists made the claim that psychological functions are localized in the human cortex; at first sight this is similar to the positions of both disconnectionists and cognitive neuropsychologists.

(b) The phrenologists made the claim that the mind is organized into discrete localized chunks; at first sight this is similar to the description of discrete cognitive modules proposed by the cognitive neuropsychologists.

The list of points supporting student 2 is potentially very long and includes the following major points:

(a) Disconnectionist analysis shows that aspects of human behaviour that might be described as a faculty such as language are localized, but in a quite different way from that proposed by the phrenologists, that is, as extended anatomical systems running through many cortical regions rather than being localized in a single region.

(b) The analyses of the cognitive neuropsychologists point to a quite different organization of mind that contrasts with that of the phrenologists. Although the cognitive modules proposed by the cognitive neuropsychologists do act independently of one another, they do not act as isolated faculties, nor have they been located anatomically. Cognitive modules are descriptions of the way that the resources needed to carry out cognitive tasks are organized, based on logical inferences of the way that behaviour fractures after brain damage. Thus, cognitive modules are not equivalent to the phrenologist's faculties.

Question 6.3

The significant feature revealed from such studies is that more than one consciousness may occur within an individual.

Question 6.4

Such models show that a description of cognitive processes requires a number of independently functioning modules. (They do not show where these are located within the brain.)

Chapter 7

Question 7.1

(a) Neurosis.

(b) Psychosis.

(c) You should be hesitant in your diagnosis. After all, it is just possible objectively that the individual in the second case has some prior knowledge that Martians have landed! The first individual might be suffering from vitamin deficiency or malnutrition or a drug overdose, or be anxious that if she goes outside the house she will be attacked by her husband. Without a more detailed investigation, one should not jump to conclusions. Even with a more detailed investigation, one should be careful!

Question 7.2

(a) There are of course genetic differences between men and women associated with the presence in men of a Y as opposed to a second X chromosome (Y and X chromosomes are described in Book 4, Section 4.5.5). However, while the admission rate for personality disorders in Northern Irish men is about twice that for Northern Irish women in England and Wales, that for English and Welsh men is roughly the same as that for English and Welsh women, and that for American men is just under half that for American women (see Table 7.5). No simple genetic differences between men and women could account for such variation which would seem to be more simply ascribed to culture.

(b) You cannot conclude that the Irish as a group have a genetic predisposition to anything on the basis of these data. As Chapter 1 points out, social and biological ascriptions of race do not match, so there is no clear 'Irish race' in the biological sense. Even if there were, you would need to know what the personality disorder admissions rates for the Irish in Ireland were, why there were substantial differences between the rate for people born in Northern Ireland and the Irish Republic, and what happened to the children of Irish parentage born in England and Wales. You would have to consider all these factors before you could begin to unpick some of the variables involved in interpreting these simple statistics. You would also need to know something of doctors' expectations of how they should diagnose people from a particular cultural background.

Question 7.3

(a) They are incompatible with the extreme form of relational theories espoused by Foucault (described in Section 1.3.1), who denies that the condition is located in the individual at all. They are, however, partially compatible with social explanations. Brown and Harris's 'vulnerability factors' might include those childhood experiences discussed by the psychoanalytic tradition, although they see these problems of interpersonal relationships within the family as being imposed by some external accident, such as death or economic insecurity. Within the psychoanalytic tradition, whether Freudian or Laingian, such external contingencies are supposed to be largely irrelevant — the economy, job security, adequate housing, all are almost taken for granted. In so far as personal life-history explanations are concerned with the remote, rather than the immediate past, and lead to the conclusion that therapeutic intervention demands the recovery of that past, they are incompatible. When Brown and Harris note that the future for many women is bleak, they clearly do not hold out much hope that if the depressed woman relives her birth experience, or comes to terms with her penis envy, then she will transform her future prospects.

(b) On the surface, they seem to bear no relation to biological explanations at all. Psychoanalysts and psychotherapists rarely discuss, or are concerned with, an individual's biology. Yet, in fact, the incompatibility is only on the surface. If childhood experience affects future behaviour, it could be because of the way in which that experience has shaped nerve pathways and patterns of activity in the brain.

Question 7.4

(a) is initial evidence that there may be a genetic factor in schizophrenia, but says nothing about the asymmetry hypothesis.

(b) is evidence that asymmetry may be associated with schizophrenia but to prove cause it would be necessary to show that the asymmetry existed prior to the behavioural pattern of schizophrenia rather than vice versa.

(c) is, on the face of it, evidence against the asymmetry hypothesis; as the text points out, that there are no gender differences in the incidence of schizophrenia.

(d) simply shows that asymmetry may be associated with behavioural disorders, but there is no evidence cited in the text to connect autism with schizophrenia.

Question 7.5

The symptoms of schizophrenia such as hallucinations, loss of affect, etc., seem to indicate a subtle alteration in nervous system functioning, and this type of alteration could in general be expected to be associated with changes in the efficacy of neural transmission in particular pathways such as those of the limbic system. Drugs which alleviate the symptoms of schizophrenia are also known to interact biochemically with neurotransmitters, functioning as antagonists, agonists or interfering with re-uptake mechanisms. Post-mortem studies of brains from schizophrenic people have been reported to show abnormal levels of some neurotransmitters or their receptors (notably dopamine and serotonin) in particular brain regions, especially limbic areas. Animals treated with agonists or antagonists of some of the suspected transmitters show behaviour patterns which might be interpreted as 'schizophrenia-like'. Against this evidence, there are however many contrary reports of failure to confirm these findings.

Question 7.6

For:

(a) Schizophrenia is clearly familial, and the closer related genetically two people are, the greater is the chance of concordance for the diagnosis. The greatest concordance occurs in identical twins.

(b) Adoption studies show a tendency for schizophrenia to be more common in the biological relatives of schizophrenic adoptive children than in the families in which the children are located.

(c) In at least one pedigree a linkage to a specific chromosome has been shown.

Against:

(a) A familial tendency to a condition does not prove a genetic factor, as families share common environments. The concordances, although clearly there, are not very strong. Pedigree studies do not reveal a clear-cut transmission of the disease, unlike, say, Huntington's, so, whatever the mode of genetic transmission, it must be rather complex and involve many genes. There is also a problem of being sure from medical records of past generations that the conditions described correspond to what is now called schizophrenia.

(b) The epidemiological data on the class and ethnic distribution of the diagnosis in the UK and USA are not compatible with any simple genetic model.

(c) It has not so far been possible to demonstrate genetic linkage except in the case of a small number of pedigrees.

Chapter 8

Question 8.1

(a) Dualist. Her mind (or soul) experiences an intense emotion of affection and desire for a particular person. There may be processes going on in her body as a result of being in love, but they are initiated or controlled by her mind.

(b) Reductionist. Information arriving at her sense organs—sight, sound, touch etc.—has resulted in set of neuronal processes in her limbic system and other brain regions, and changes in her body's hormonal state, which result in the sensation of sexual desire, which she calls love.

(c) Integrationist. Being in love is an experience resulting from seeing, hearing and otherwise sensing a particular person, which can be described in terms of information processing and cellular processes in her sense organs, changes in her body's hormonal state, etc. The complete cellular and biochemical description of these body and brain processes is identical with—another way of talking about—the mental and emotional state described as 'being in love'.

Question 8.2

(a) A psychotherapist would not necessarily make a specific diagnosis, such as schizophrenia, but would talk with Mr Johnson about his experiences and past, about his particular set of religious beliefs, about his experiences in childhood leading up to his religious conversion, and explore with him his own feelings about the circumstances in which he finds himself in conflict with the police. The psychotherapist would try to help Mr Johnson understand his own feelings and motivations better, perhaps so as to be able to control them better. The therapy might involve a variety of procedures. Mr Johnson might be seen by the therapist one or more times a week, alone or with a group, and the therapy might involve various forms of physical activity which would offer him ways to express or better understand his emotions. Mr Johnson might be seen as an out-patient, or if he were confined to hospital, then as an in-patient, possibly as part of a group therapy programme.

(b) A biologically-inclined psychiatrist would begin by making a diagnosis, to decide whether Mr Johnson should be regarded as schizophrenic or possibly manic-depressive. The diagnosis might be based on interviews and observation, but drugs would also be prescribed. If the diagnosis was schizophrenia, phenothiazine would probably be given. If the diagnosis was manic depression, then lithium would probably be prescribed. If one class of drugs was ineffective but the other produced some calming effect or alleviation of what the psychiatrist regarded as the symptoms, then this would influence the diagnosis of the condition.

Question 8.3

Assuming that the safety and possible efficacy of the proposed new drug had first been checked in animal experiments and that it was licensed for clinical trials, the normal procedure would be to take two groups of people who had given informed consent to take part in the trial (upwards of at least 20 per group) matched for age, sex and diagnosis. One group would be given the new drug and the other a placebo, or neutral substance. Alternatively, the control group would stay on their existing medication and the trial group would switch to the new one. Neither the people involved nor the health workers administering the trial should know which person is receiving which treatment This is called a double-blind trial. (If the health worker does know, it is called a single-blind trial, and has the weakness that the health worker's expectations may bias the result.) The two groups would stay on their respective treatments for an appropriate period (normally at least several weeks) and then their conditions would be compared using a standardized depression-rating scale as well as the clinical impressions of the health workers. Differences between the results of the two treatments would be evaluated statistically. In a more refined version of the trial, known as a cross-over procedure, the results for the two groups would be compared half way through the trial and then the trial and control groups would be switched around.

You might like to think about the ethics of this type of procedure, which although 'pure' scientifically has been criticized by some people for its use of placebos, especially in situations where patients may be desperate to find a cure or at least some alleviation of symptoms. The HIV–AIDS situation is the one in which this has caused most concern in recent years.

Question 8.4

You should consider the scientific, the social and the ethical issues involved in the proposal.

First, is the proposal well-formulated scientifically? What evidence is there that obsessive behaviour is the result of a brain disorder which might be treated in this way? Is the region of the brain or the neurotransmitter(s) involved in the condition known? What evidence is there from animal models which might have a bearing on the condition? What are the scientific arguments for using fetal material? Could any meaningful scientific conclusions be drawn from the study? For instance, if there was an improvement in the person's symptoms, how specific is this to the transplant—in other words, what sort of controls might be proposed? How would you evaluate the improvement? Clearly, given this experimental procedure you could not use the blind procedure described in the answer to Question 8.3.

Second, consider the social issues. Is obsessive behaviour a sufficiently serious or debilitating condition so as to warrant spending limited resources on such a novel research approach? Or is it sufficiently well-treated by current therapies as to make this new one unnecessary?

Third, consider the ethical issues. What would be the source of the fetal material? Would the donor of the tissue (the aborted or miscarried woman) need to consent to its use in this way? What sort of informed consent would be needed from the people who were going to receive the transplants? How would they be chosen? What might be the hazards of the procedures and what steps should be taken to minimize them or to compensate anyone who suffers from the procedure? Is it morally justifiable to attempt to try to change a person's behaviour and personality by using biological manipulation of this sort? What forms of counselling would people undergoing the procedures receive?

GLOSSARY FOR BOOK 6

affect Term used in psychology and psychiatry as a way of describing a person's responses of emotion and attention to events in the world around them. (Section 7.3.1)

cohort A generation of people (or animals) all born during the same period of time. (Section 2.3)

cross-sectional method An experimental method which involves making comparisons between groups of people from different sections of the population made at the same point in time. (Section 2.3)

dementias A group of mental disorders characterized by progressive loss of cognitive and other intellectual functions associated with pathological ageing. (Section 3.1)

denervation supersensitivity Increased sensitivity of the neurons of the CNS brought about by the proliferation of receptor molecules on the postsynaptic membrane following the loss of synaptic input.

epidemiology The study of how a condition is distributed within a population and how it changes over time, undertaken in order to understand the factors that contribute to the development of the condition. (Section 1.3.2)

incidence The number of new cases occurring per thousand of the population in any given period. (Section 1.4.2)

labelling Description of some aspect of an individual's condition such that it becomes a way of defining that individual's relationship with the rest of the world. (Section 1.3.1)

life expectancy The average number of years a person can expect to live from birth. (Section 2.1.1)

lifespan The length of life of the longest-lived member of the species. (Section 2.1.1)

linkage The situation in which alleles of different genes do not assort independently; they are held together, or linked. The degree of linkage is greater the closer together the genes are because, the closer they are, the less likely they are to be separated during the process of crossing over (described in Book 1, Section 3.2.4)

longitudinal method An experimental method which involves testing and retesting the same subjects at different ages over a long period of time. (Section 2.3)

neurological disorder A disorder of behaviour or personality brought about by damage to the brain. (Sometimes called an *organic disorder*.) (Section 1.2.3)

norm of reaction The range of phenotypes that may arise from the interplay between a given genotype and various environments. (Section 4.1)

prevalence The number of individuals per thousand of the population actually showing a disease or trait at any given point in time. (Section 4.3.3)

psychiatric disorder A disorder of behaviour or personality without obvious brain damage. (Sometimes called a *functional disorder*.) (Section 1.2.3)

psychological life-history explanation Explanation that seeks the origins of a disorder within an individual's own life history. (Section 1.3.3)

split-brain individual An individual in whom the corpus callosum has been cut. Under certain conditions, such individuals behave as if they had two brains with separate consciousnesses. (Section 6.3.1)

varicosity A nodule on an axonal branch that contributes to the postsynaptic influence of the neuron by releasing large quantities of transmitter. This type of synapse is characteristic of neurons projecting to the striatum from the substantia nigra. (Section 5.5.1)

ACKNOWLEDGEMENTS

The series *Biology: Brain and Behaviour* (for Open University course SD206) is based on and updates the material in course SD286. The present course team gratefully acknowledges the work of those involved in the previous course who are not also listed as authors in this book.

Grateful acknowledgement is made to the following sources for permission to reproduce material in this book:

FIGURES

Figure 1.2: courtesy of Professor L. Squire, Veterans Administration Hospital, San Diego, California; *Figures 2.2, 2.3, 2.8, 2.10, 2.12, 2.14:* adapted from Kausler, D. H. (1982) *Experimental Psychology and Human Aging*, John Wiley and Sons Inc, © Donald H. Kausler; *Figures 2.4, 2.13:* Salthouse, T. A. (1982) *Adult Cognition: An Experimental Psychology of Human Aging*, Springer-Verlag; *Figure 2.7:* adapted from Norman, D. A. (1980) *Cognitive Science*, **4**, pp. 1–32, Ablex Publishing Corporation; *Figure 2.11:* adapted from Light, L. L. and Burke, D. M. (1988) *Language, Memory and Aging*, Cambridge University Press; *Figure 2.15:* Cohen, G. (1983) *The Psychology of Cognition*, 2nd edn, Academic Press Inc; *Figure 3.2:* photograph printed courtesy of Amersham International plc. Glial fibrillary acidic protein detected in human brain astrocytes using a GFAP monoclonal antibody from Amersham International plc (product code RPN 1106). Detection was with Amersham International's streptavidin-biotinylated peroxidase complex (RPN 1051). The GFAP monoclonal antibody is specific to astrocytes and tumours of neurological origin, for example astrocytomas and ependymomas; *Figure 3.3:* Buell, S. J. and Coleman, P. D. (1979), *Science,* **206**, pp. 854–856, © 1979 by the American Association for the Advancement of Science; *Figures 3.4, 3.8:* Hoyer, S. (1982) *The Aging Brain*, Springer-Verlag; *Figures 3.6, 3.7:* Probst, A., Basler, B. B. and Ulrich, J. (1983) *Brain Research,* **268**, pp. 249–254, Elsevier Science Publishers BV; *Figure 4.1:* courtesy of the *New Scientist*/IPC Magazines; *Figure 4.7:* Pearlman, A. L. and Collins, R. C. (1990) *Neurobiology of Disease*, Oxford University Press; *Figures 4.8, 4.15:* adapted from McGeer, P. L., Itagaki, S., Akiyama, H. and McGeer, E. G. (1989) in *Parkinsonism and Aging* (ed. by Calne, D. B., Comi, G., Crippa, D., Horowiski, R. and Trabucchi, M.), Raven Press Ltd; *Figure 4.14:* adapted from Wolters, E. and Calne, D. B. (1989) in *Parkinsonism and Aging* (ed. by Calne, D. B., Comi, G., Crippa, D., Horowiski, R. and Trabucchi, M.), Raven Press Ltd; *Figures 5.1, 5.3, 5.6:* Kandel, E., Schwartz, J. and Jessel, T. (1991) *Principles of Neural Science,* 3rd edn, Elsevier Science Publishers BV; *Figures 5.2, 5.4, 5.8:* from Kolb, B. and Whishaw, I. Q (1990) *Fundamentals of Human Neuropsychology*, 3rd edn. Copyright © 1990 by W. H. Freeman and Company. Reprinted by permission; *Figure 5.7:* Teuber, H. L. (1975) in *Outcome of Severe Damage to the Nervous System, CIBA Foundation Symposium,* **34,** Elsevier Science Publishers BV; *Figure 5.10:* from Bradford, H. E. (1986) *Chemical Neurobiology*. Copyright © 1986 by Henry E. Bradford. Reprinted by permission of W. H. Freeman and Company; *Figure 6.1:* from Gregory, R. L. (1987) *The Oxford Companion to the Mind*, © Oxford University Press 1987. Reprinted by permission of Oxford University Press; *Figure 6.2:*

Colver Pictures; *Figure 6.3:* Corsi, P. (1991) *The Enchanted Loom,* Oxford University Press; *Figure 6.4:* Shallice, T. (1988) *From Neuropsychology to Mental Structure,* Cambridge University Press; *Figure 6.5b:* adapted from Sperry, R. W. (1964) The great cerebral commissure, *Scientific American*, January 1964; *Figure 6.6:* from Springer, S. and Deutsch, G. (1989) *Left Brain, Right Brain.* Copyright © 1989 by Sally P. Springer and Georg Deutsch. Reprinted by permission of W. H. Freeman and Company; *Figure 6.9:* adapted from Shallice, T. and Warrington, E. K. (1970) *Quarterly Journal of Experimental Psychology*, **27**, pp. 187–199, Lawrence Erlbaum Associates; *Figure 6.10:* Eysenck, M.W. and Keane, M. T. (1990) *Cognitive Psychology: A Student Handbook*, Lawrence Erlbaum Associates Ltd; *Figure 6.11:* Ellis, A. and Young, A. (1988) *Human Cognitive Neuropsychology*, Lawrence Erlbaum Associates Ltd; *Figure 7.1:* Curran, D. and Partridge, M. (1969) *Psychological Medicine*, Churchill Livingstone; *Figure 7.2:* Suddath, R. L. *et al.* (1990) *The New England Journal of Medicine*, **322**, pp. 789–794, Massachussetts Medical Society. Reprinted, by permission of *The New England Journal of Medicine*; *Figure 7.4:* McCarley, R. W., Faux, S. F., Shenton, M. E., Nestor, P. G. and Adams, J. (1991) *Schizophrenia Research*, **4,** pp. 209–231, Elsevier Science Publishers BV; *Figure 7.5:* Egeland, J. A. *et al.* (1987) *Nature*, **325**, pp. 783–787. Reprinted by permission from *Nature*, Copyright © 1987 Macmillan Magazines Ltd.

TABLE

Table 5.1: from Kolb, B. and Whishaw, I. Q. (1990) *Fundamentals of Human Neuropsychology,* 3rd edn. Copyright © 1990 by W. H. Freeman and Company. Reprinted by permission.

TEXT (Chapter 7)

Toates, F. (1990) *Obsessional Thoughts and Behaviour*, HarperCollins; Littlewood, R. and Lipsedge, M. (1982) *Aliens and Alienists*, Unwin Hyman; Pringle, J. (1972) Living with schizophrenia, *Mind and Mental Health Magazine*, Spring 1972, National Schizophrenia Fellowship.

INDEX

Note Entries in **bold** are key terms. Indexed information on pages indicated by *italics* is carried mainly or wholly in a figure or table.